Regulatory Non-Coding RNAs

Regulatory Non-Coding RNAs

Edited by Anna Robinson

SYRAWOOD
PUBLISHING HOUSE
New York

Published by Syrawood Publishing House,
750 Third Avenue, 9th Floor,
New York, NY 10017, USA
www.syrawoodpublishinghouse.com

Regulatory Non-Coding RNAs
Edited by Anna Robinson

International Standard Book Number: 978-1-64740-362-1 (Hardback)

Cataloging-in-publication Data

Regulatory non-coding RNAs / edited by Anna Robinson.
p. cm.
Includes bibliographical references and index.
ISBN 978-1-64740-362-1
1. Non-coding RNA. 2. Genetic regulation. 3. RNA. I. Robinson, Anna.
QP623 .R44 2023
572.88--dc23

TABLE OF CONTENTS

PREFACE

Every book is a source of knowledge and this one is no exception. The idea that led to the conceptualization of this book was the fact that the world is advancing rapidly; which makes it crucial to document the progress in every field. I am aware that a lot of data is already available, yet, there is a lot more to learn. Hence, I accepted the responsibility of editing this book and contributing my knowledge to the community.

Non coding RNAs (ncRNAs) consist of a large portion of poly-A tailed mature RNAs. They are single stranded molecules, which are involved in the regulation of gene expression at the post-transcriptional level in a cell. There are two types of ncRNAs, which are regulatory ncRNAs and infrastructural ncRNAs. Small interfering RNAs (siRNAs), microRNAs (miRNAs), long non-coding RNAs (lncRNAs), and piwi-interacting RNAs (piRNAs) are the four types of regulatory ncRNAs. The expression of thousands of genes is regulated through ncRNAs and this regulation can take place in either cis or trans. NcRNAs must be integrated into regulatory networks for the identification of molecular driver events. The two largest members of the ncRNA family, which includes long non coding RNA (lncRNA) and micro RNA (miRNA), play significant regulatory roles in gene expression. They are also involved in a variety of physiological processes including programmed cell death. This book unravels the recent studies on non coding RNA biology. It attempts to understand the regulation of non coding RNAs. For all those who are interested in ncRNAs, it can prove to be an essential guide.

While editing this book, I had multiple visions for it. Then I finally narrowed down to make every chapter a sole standing text explaining a particular topic, so that they can be used independently. However, the umbrella subject sinews them into a common theme. This makes the book a unique platform of knowledge.

I would like to give the major credit of this book to the experts from every corner of the world, who took the time to share their expertise with us. Also, I owe the completion of this book to the never-ending support of my family, who supported me throughout the project.

Editor

1

Regulatory Non-Coding RNAs in Pluripotent Stem Cells

Alessandro Rosa [1,2] and Ali H. Brivanlou [2,*]

1 Department of Biology and Biotechnology "Charles Darwin", Sapienza University of Rome, Rome 00185, Italy; E-Mail: alessandro.rosa@uniroma1.it

2 Laboratory of Molecular Vertebrate Embryology, The Rockefeller University, New York, NY 10065, USA

* Author to whom correspondence should be addressed; E-Mail: brvnlou@rockefeller.edu

---

Abstract: The most part of our genome encodes for RNA transcripts are never translated into proteins. These include families of RNA molecules with a regulatory function, which can be arbitrarily subdivided in short (less than 200 nucleotides) and long non-coding RNAs (ncRNAs). MicroRNAs, which act post-transcriptionally to repress the function of target mRNAs, belong to the first group. Included in the second group are multi-exonic and polyadenylated long ncRNAs (lncRNAs), localized either in the nucleus, where they can associate with chromatin remodeling complexes to regulate transcription, or in the cytoplasm, acting as post-transcriptional regulators. Pluripotent stem cells, such as embryonic stem cells (ESCs) or induced pluripotent stem cells (iPSCs), represent useful systems for modeling normal development and human diseases, as well as promising tools for regenerative medicine. To fully explore their potential, however, a deep understanding of the molecular basis of stemness is crucial. In recent years, increasing evidence of the importance of regulation by ncRNAs in pluripotent cells is accumulating. In this review, we will discuss recent findings pointing to multiple roles played by regulatory ncRNAs in ESC and iPSCs, where they act in concert with signaling pathways, transcriptional regulatory circuitries and epigenetic factors to modulate the balance between pluripotency and differentiation.

Keywords: embryonic stem cells; induced Pluripotent Stem Cells; microRNA; long non-coding RNA; pluripotency; reprogramming

---

1. Introduction

Pluripotency is defined, in a broad sense, as the ability of a cell to give rise to derivatives of the three germ layers, ectoderm, mesoderm and endoderm, and the germ line. In mammals, this property is restricted to embryonic cells until the stage of blastocyst, in which a small number of cells that constitute the Inner Cell Mass (ICM) are still pluripotent. Such pluripotent cells exist very transiently during embryonic development. Upon gastrulation and during subsequent embryonic development, they lose pluripotency and progressively acquire a specialized character. Embryonic stem cells (ESCs) can be derived from the ICM of human and mouse blastocysts; are pluripotent and can self-renew *in vitro* indefinitely.

ESC pluripotency is tightly regulated. Amongst several signaling pathways, the TGF-β pathway has been shown to play a central role [1]. Interestingly, the two branches of the pathway play different roles in human and in mouse. More precisely, in human ESCs (hESCs) the Nodal/Activin branch is both necessary and sufficient to sustain pluripotency [2,3], whereas in mouse ESC (mESCs) the bone morphogenetic protein (BMP) branch is required for maintaining self-renewal and prevent differentiation [4]. Downstream of signaling pathways, the maintenance of ESCs pluripotency is ensured by a regulatory circuitry including three main core transcription factors (TFs), Oct4, Sox2 and Nanog [5–7]. The three core TFs co-occupy a conspicuous set of target promoters and have a dual role. They can activate transcription of genes involved in the maintenance of pluripotency, including their own genes. At the same time, in association with Polycomb Repressive Complexes (PRC1 and PRC2), they silence a subset of lineage-specific genes that play a role in development [8,9]. In ESCs, the promoters of these genes present peculiar bivalent chromatin domains, in which epigenetic histone modifications normally associated with silent genes co-exist with marks of active transcription [10]. Such unique epigenetic profiles are required to keep key developmental genes "poised" in a repressed state that can be quickly turned on.

Pluripotent cells exist in two different states, defined as naïve and primed [11]. Mouse ESCs are considered to be in a naïve ground state of pluripotency that corresponds to the preimplantation epiblast. *In vitro*, they require Leukemia Inhibitory Factor (LIF) and BMP signaling for self-renewal and differentiate in presence of basic Fibroblast Growth Factor (bFGF). They express a typical set of markers and female cells have two active X chromosomes. Functionally, they are able to form chimeric embryos upon injection in a recipient blastocyst. Conversely, mouse Epiblast Stem Cells (EpiSCs) are in a primed state, which corresponds to the post-implantation epiblast. EpiSCs require bFGF and Nodal/Activin signaling for self-renew, present X chromosome inactivation in female cells, and fulfill some criteria for pluripotency, such as the ability to form teratomas, but cannot generate chimeric mice. There is some plasticity in pluripotent cells, as mESCs can be induced to differentiate into EpiSCs by culturing in presence of Activin and bFGF and EpiSCs can be converted to the ground state by transfection with Klf4 or Nanog and culturing in the presence of LIF and BMP [12,13]. Human ESCs share many features with EpiSCs but differ from mESCs in terms of morphology, culture requirements, expression markers and X chromosome inactivation. This evidence led to the hypothesis that hESCs are in a primed state of pluripotency [14].

The seminal work by Shinya Yamanaka showed that pluripotency could be induced in mouse somatic cells by ectopic expression of a defined set of reprogramming factors (RFs) [15]. Since then, reprogramming of somatic cells into induced Pluripotent Stem Cells (iPSCs) has been achieved in human as well [16,17]. Alternative sets of RFs that contain well-known pluripotency factors can be used. The most common are Oct4, Sox2, Klf4 and c-Myc (Yamanaka RFs) and Oct4, Sox2, Nanog and Lin28 (Thomson RFs). iPSCs share with ESCs the pluripotency, meaning that they could virtually be differentiated *in vitro* into all adult cell types. Moreover, they can be derived from human patients as patient-specific iPSCs (PS-iPSCs) that hold the same disease-causing genetic alteration [18]. The mechanisms underlying reprogramming have been deeply investigated and involve a profound change in cell identity. During reprogramming, the epigenetic landscape of the somatic cell of origin shifts to a state proper of the embryonic stem cell, including erasure of repressive marks on the chromatin of pluripotency genes and establishing of bivalent domains on lineage-specific genes [19].

The ability to differentiate into multiple tissues makes ESCs and iPSCs promising tools for regenerative medicine and cell-replacement therapy approaches [20]. However, to fully exploit their potential, the molecular basis of pluripotency must be deeply characterized. Non-coding RNA (ncRNA) molecules, previously regarded to exert only passive roles in the cell, are conversely primary players to define the cell identity. Rather than the coding portion of the genome, it is now clear that its non-coding counterpart is correlated with the greater complexity of higher eukaryotes [21]. Recently, ncRNAs are also emerging as new regulatory factors in pluripotent cells. Among small non-coding RNAs (<200 nucleotides), microRNAs (miRNAs) are now considered major regulators of development, metabolism, differentiation and homeostasis in all multicellular organisms [22–26]. miRNAs are also involved in several human diseases, including cancer [27]. Biogenesis of miRNAs requires a multistep process [23]. miRNAs are generally transcribed by RNA polymerase II as part of introns of mRNA genes, or from intergenic regions. The miRNA primary precursor (pri-miRNA) is then processed in the nucleus by the Microprocessor complex [28,29], comprised of the cleavage enzymes Drosha, DGCR8 and other factors, releasing a stem-loop precursor (pre-miRNA). The pre-miRNA is then exported in the cytoplasm and cleaved by the RNAse III enzyme Dicer, which is also involved in the maturation of short interfering RNAs (siRNAs) [30,31]. The mature miRNA is finally incorporated as a single strand in the RNA Induced Silencing Complex (RISC) [32]. Guided by the miRNA, RISC binds the 3'UTR and/or the coding sequences of target mRNAs resulting in inhibition of translation and/or degradation. A single miRNA can inhibit several targets and a single mRNA can be targeted by multiple miRNAs in a combinatorial way [33]. For target recognition by miRNAs, base pairing of a short "seed" sequence, located at the 5'end of the mature miRNA, is required [34]. Families of miRNAs comprise members with identical seed sequences and all miRNAs belonging to a family are thought to share the same targets. Multiple miRNAs can be excised from a single, multicistronic, pri-miRNA transcript. Such clusters can comprise multiple members of a miRNA family as well as unrelated miRNAs.

Despite lacking an open reading frame (ORF), long (>200 nucleotides) non-coding RNAs (lncRNAs) are transcribed by RNA polymerase II and have a structure that resemble protein-coding mRNAs. They are generally spliced from multi-exonic precursors, have a 5'cap and are polyadenylated [35,36]. Some lncRNAs are transcribed from loci which overlap or are very close to protein-coding genes, while others are intergenic (long intergenic non-coding RNAs, lincRNAs). Globally, lncRNA expression levels are only slightly lower than protein-coding transcripts [37,38]. Similar to protein-coding and miRNA genes, and differently from other structural non-coding RNAs such as ribosomal RNAs, the expression of lncRNAs is characterized by tissue specificity, and is dynamically regulated during differentiation [39–42]. Moreover, many lncRNA are conserved in terms of sequence and predicted secondary structure [37,38,43]. All together, these elements suggest that lncRNAs may represent a new, still largely unexplored class of functional molecules, potentially involved in multiple biological processes [44–46]. In the last years, the continuous development of powerful sequencing technologies and bioinformatics tools boosted the discovery of thousands novel lncRNAs. According to some estimates, the total number of human lincRNAs would be around 4500 [47]. So far, however, a function has been only assigned to a limited number of lncRNAs. Intracellular localization is often used as a predictive element to get insights into lncRNA molecular mechanisms [48]. Nuclear lncRNAs can regulate gene expression either in cis (on neighboring genes) or in trans (on distant genes). Some of them are able to modulate the activity of chromatin modifiers. A paradigmatic example for this class of lncRNAs is HOTAIR, which is required for Polycomb Repressive Complex 2 (PRC2) occupancy and histone H3 lysine-27 trimethylation of the HOXD locus [49]. Other nuclear lncRNAs act as antisense transcripts or as decoy for splicing factors [50,51]. Cytoplasmic lncRNAs may conversely function as endogenous "sponges" for miRNAs, thus releasing miRNA repression on target genes [52,53].

Here we will review recent findings that point to regulatory non-coding RNAs as important players in the maintenance of ESCs pluripotency and cell fate choices during differentiation. We will also discuss the roles played by ncRNAs during somatic cells reprogramming.

2. MicroRNAs in Embryonic Stem Cells

Among all miRNAs expressed in embryonic stem cells, a single family of miRNAs with the AAGUGC seed sequence is the most highly expressed and has the most functional impact. Members of this family are organized in two major clusters (Figure 1a). The conserved miR-302/367 cluster comprises four AAGUGC seed-containing miRNAs (miR-302a, miR-302b, miR-302c and miR-302d) and the unrelated miR-367. The second cluster is less conserved. In the mouse it is commonly referred to as the miR-290-295 cluster and includes six miRNAs with the AAGUGC seed (miR-290, miR-291a, miR-291b, miR-292, miR-294 and miR-295) and miR-293. The human orthologue comprises miR-371 and the AAGUGC seed-containing miR-372 and miR-373 (miR-371-373 cluster). Therefore, multiple members with the AAGUGC seed are present in each cluster, as well as other miRNAs with a different seed [25]. For simplicity, in this review we will refer to miRNAs containing the AAGUGC seed as miR-302 family.

The miR-302 family miRNAs are abundantly expressed in undifferentiated ESCs and decline upon differentiation [54–58]. According to the miRNA expression atlas [59], miR-302 family members are specifically expressed in embryonic cells in both mouse and human. In mouse, high-throughput sequencing indicated that miRNAs belonging to the miR-290-295 cluster represent almost one third of all miRNAs expressed in undifferentiated mESCs [60]. Similarly, undifferentiated human ESCs are dominated by the mir-302 cluster, which accounts for more than 60% of all expressed miRNAs [61]. miRNAs with the AAGUGC seed are also conserved in other vertebrates, such as Zebrafish, Xenopus and Chicken, where they seem to be specifically expressed during early embryonic development [25].

Figure 1. Clusters of miRNAs specifically expressed in embryonic stem cells (ESCs). (**a**) Schematic representations of the two ESC-specific miRNA clusters in mouse and human. miRNAs with the AAGUGC seed sequence within each cluster are underlined. The miR-302/367 cluster is highly conserved in mammals, whereas the other cluster is less conserved; (**b**) The two clusters are differentially expressed in ESCs during the conversion between the naïve and primed states of pluripotency.

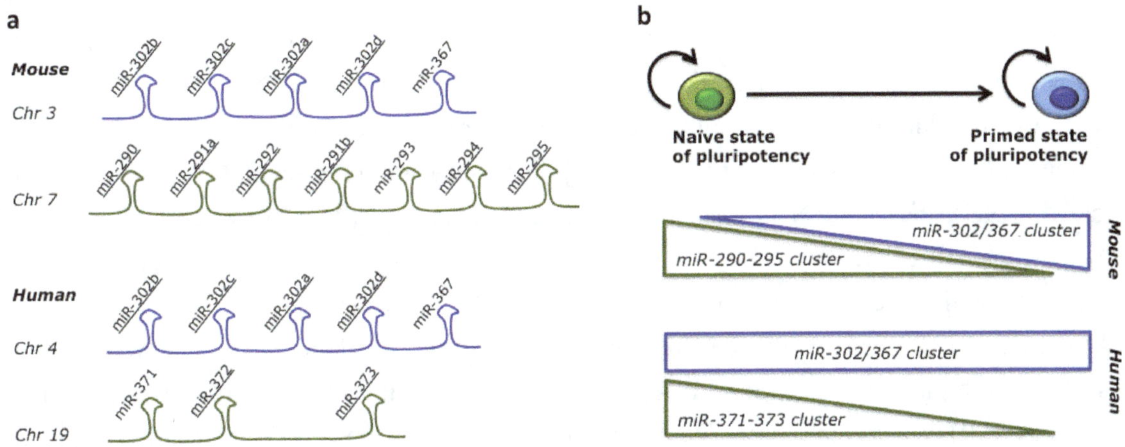

The specific expression of miR-302 family miRNAs is ensured by their regulation, at the transcriptional level, by the core ESC transcriptional regulatory circuitry [8,9,62,63]. The promoters of the miR-302 and the miR-290-295 clusters are bound by Oct4, Nanog, Sox2 and Tcf3, that also promote transcription of other unrelated miRNAs. The same transcription factors, in cooperation with Polycomb group proteins, repress transcription of lineage-specific miRNAs, such as the neural miRNAs miR-9 and miR-124 and the mesodermal miRNA miR-155 [9].

Interestingly, the conversion from a naïve to a primed state in mESCs correlates with a switch between the miR-290-295 and the miR-302/367 clusters in terms of miRNA abundance [64] (Figure 1b). mESCs express high levels of miR-290-295 that decline after conversion to EpiSCs and are replaced by an increase of miR-302/367. The total levels of miRNAs with the AAGUGC seed are maintained. Since miRNAs with the same seed should share the same targets, the significance of this switch remains obscure. In human ESCs, which correspond to a primed state of pluripotency, the levels of miR-302/367 are much higher than the levels of miR-371-373 and the switch to an earlier developmental state led to an increase of the levels of miR-371-373 [65–67].

2.1. miRNAs Regulate ESC Cell Cycle

Inactivation of genes involved in the processing of miRNAs represents a useful approach to study the global function of miRNAs. Dicer −/− pre-gastrulation embryos showed lack of Oct4-positive epiblast cells and could not undergo gastrulation [68], indicating that the miRNA pathway is essential for establishing the pool of pluripotent embryonic cells. As expected, mESCs could not be derived from Dicer mutant embryos. Surprisingly, Dicer −/− mESCs obtained by conditional gene targeting are viable, hold an appropriate morphology, and express normal levels of pluripotency markers [69]. However, the miRNA pathway is crucial for mESC differentiation both *in vitro* and *in vivo*, as assessed by defects in lineage markers expression and teratoma formation. Moreover, inactivation of

Dicer impairs the proliferation potential of mESCs [70]. As Dicer is involved in the processing of both miRNAs and other endogenous small RNAs, the observed phenotype could be due to the loss of both classes of Dicer targets. However, comparative analysis of Dicer +/+ and Dicer −/− mESCs by high-throughput sequencing indicated that very few short non-miRNA transcripts that are dependent on Dicer exist in these cells [60]. Moreover, a similar phenotype, albeit less severe, was observed in DGCR8 −/− mESCs [71], indicating that the defects detected in Dicer −/− mESCs are mainly due to the loss of miRNAs. A comparative transcriptome analysis revealed that in cells lacking all miRNAs there was a significant increase of transcripts harboring a GCACUU motif in the 3'UTR [72]. This sequence is complementary to the AAGUGC seed of miR-302 family miRNAs, thus confirming their prominent role in pluripotent stem cells. Accordingly, miR-302 family members could rescue the proliferation defects of DGCR8 mutant mESCs [73]. For their role in the regulation of the characteristic cell cycle of ESCs, these miRNAs have been called “ES-cell-specific cell-cycle-regulating” (ESCC) miRNAs [74]. Specifically, they target the Cyclin Dependent Kinase (CDK) inhibitors p21 (Cdkn1a), Rbl2 and Lats2, which are present at low levels in wt mESCs and overexpressed upon disruption of the miRNA pathway [72,73,75]. Other miRNAs may contribute to the silencing of cell cycle inhibitors in mESCs. For example, miR-320 and miR-702 target p21 and p57 [76]. Interestingly, these are non-canonical miRNAs, which require Dicer, but not DGCR8, for their biogenesis [77]. The activity of such non-canonical miRNAs might explain the more severe phenotype observed in mESCs upon loss of Dicer as compared to loss of DGCR8 [76].

Similar to mESCs, also hESCs are characterized by a shortened cell cycle, accumulating in the S phase at the expense of the G1 phase [78]. Regulation of embryonic stem cell cycle by miRNAs is evolutionary conserved, as knockdown of Dicer or Drosha impairs human ESCs proliferation [79]. This defect could be partially rescued by miR-302 family members and the unrelated miR-195. These miRNAs play distinct roles in regulating cell cycle progression. miR-195 facilitates the G2/M transition by targeting the WEE1 kinase, a negative regulator of the Cycline B/CDK complex. The miR-302 family targets p21 and promotes G1/S transition [79,80]. Another miRNA, miR-92b, participates in the regulation of the G1/S transition by targeting the CDK inhibitor p57 (Cdkn1c) [81]. Moreover, at least other two components of the cell cycle machinery, Cyclins D1 and D2, are under the control of miR-302 family members in hESCs [63,82]. Therefore, the miRNA pathway plays a major role in maintaining the peculiar cell cycle of pluripotent embryonic cells.

2.2. miRNAs Regulate ESC Pluripotency and Differentiation

A genome-wide approach has been recently performed in hESCs to systematically identify bona fide miRNA targets [61]. PAR-CLIP (photoactivatable ribonucleoside-enhanced cross-linking and immunoprecipitation) [83] for the RISC component AGO2, combined with perturbation of miRNA levels, led to the identification of a set of 146 high-confidence miR-302-367 direct targets. These included previously validated targets, as well as novel genes regulated by this cluster. This gene set is enriched in factors involved in the maintenance of pluripotency, such as regulators of cell cycle and proliferation, chromatin modification, metabolism and signaling [84].

Despite holding normal morphology and expression of pluripotency markers, both mESCs and hESCs depleted of miRNAs cannot properly differentiate [69,71,79]. The miR-302 family has been involved in the TGF-β/BMP signaling pathway, which regulates embryonic stem cells pluripotency and differentiation. During hESC differentiation, Nodal activity inhibits neuroectodermal specification and promotes mesendodermal formation, whereas overexpression of the Nodal inhibitor, Lefty, results in the opposite effect [85,86]. We have shown that, by directly inhibiting Lefty, miR-302 is necessary for proper mesoderm and endoderm specification [87]. The regulation of Nodal signaling by miR-302 seems evolutionary conserved, as the Xenopus and Zebrafish hortologues target Lefty during early embryogenesis [87,88]. In addition to the inhibition of Nodal antagonists, miR-302 could promote BMP signaling by targeting the inhibitors DAZAP2, SLAIN1, and TOB2 [61]. Double blockage of the two branches of the TGF-β pathway induces a neural fate in ESCs [89,90]. By inhibiting inhibitors of both branches, miR-302 plays a central role in negatively regulating neural induction in pluripotent stem cells (Figure 2). It has recently been shown that BMP signaling down-regulates the miR-302/367 cluster in human primary pulmonary artery smooth muscle cells (PASMCs), mouse mesenchymal cells and embryonic carcinoma p19 cells [91]. In a feedback loop, miR-302 targets the type II BMP receptor (BMPRII) in PASMCs, thus inhibiting BMP signaling. Further studies are necessary to address whether this reciprocal inhibition between BMP signaling and miR-302 also exist in ESCs.

Figure 2. Role of miR-302 during human ESCs (hESC) neural differentiation. Transcription from the miR-302 locus is activated by the ESC core transcriptional regulatory circuitry. miR-302 post-transcriptionally inhibits NR2F2 which in turn is an inhibitor of Oct4. In a negative feedback loop, Oct4 inhibits NR2F2 transcription. NR2F2 expression is necessary for proper activation of neuroectoderm genes. miR-302 also targets inhibitors of both branches of the TGFβ pathway: the Nodal inhibitors, Lefty1 and Lefty 2, and the bone morphogenetic protein (BMP) inhibitors, DAZAP2, SLAIN1 and TOB2. According to the neural default model, inhibition of both branches of the TGFβ pathway leads to neural induction. Thus, by targeting inhibitors of both branches miR-302 has a negative effect on neural induction.

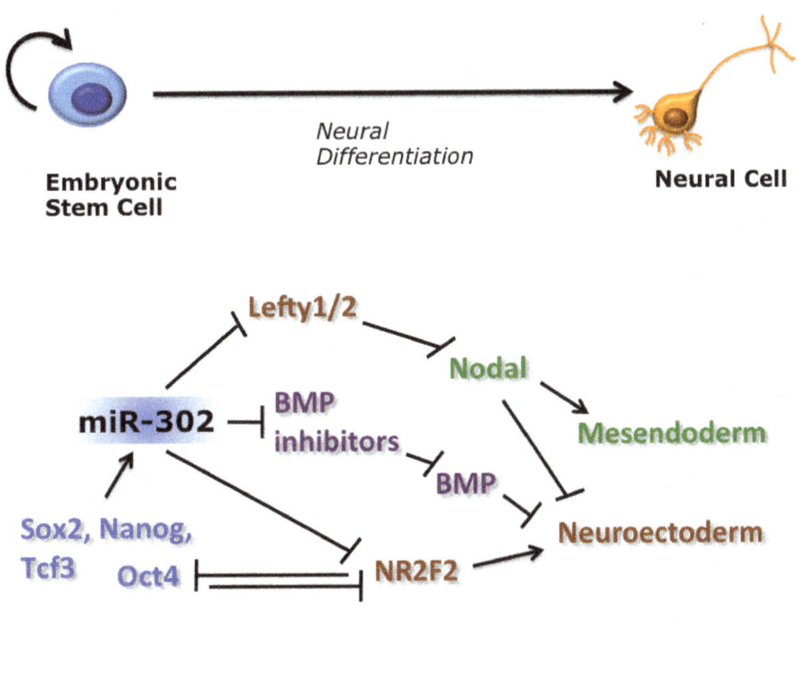

Interestingly, a regulatory loop comprising an unrelated miRNA, miR-125, and components of the BMP signaling has recently been described in mouse embryonic stem cells [92]. Such regulatory circuitry sets mESC sensitivity to BMP4. Therefore, the interplay between the TGF-β/BMP pathway and miRNAs may represent a more general regulatory mechanism modulating the response of ESC to extracellular stimuli. The interaction between miRNAs and signaling pathways underlying pluripotency might not be limited to TGF-β. In cancer cells, the Wnt/β-catenin pathway causes aberrant expression of the miR-371-373 cluster and these miRNAs, in turn, repress the activity of Wnt inhibitors, such as DKK1 [93]. It would be interesting to assess whether such interplay between the Wnt signaling and the miR-371-373 cluster has also a role in ESC pluripotency.

miR-302 is deeply integrated in the core transcriptional regulatory circuitry of ESCs. As mentioned above, the miR-302 host gene is under the control of Oct4, Nanog and Sox2, which ensure high miRNA levels in undifferentiated ESCs [9,62,63]. We have shown that both Oct4 (at the transcriptional level) and miR-302 (post-transcriptionally) repress a common target, NR2F2 (also known as COUP-TFII) [94]. NR2F2 in turn is an inhibitor of Oct4. It is activated during early neural ectoderm induction and is necessary for the proper expression of neural genes upon hESC differentiation [94]. Therefore, miR-302 and the two transcription factors, NR2F2 and Oct4, form a feedback regulatory circuitry that regulates hESC exit from pluripotency and neural fate specification (Figure 2). Individual pluripotent cell lines have different propensity to differentiate along specific lineages. Interestingly, miR-371-373 expression levels negatively correlate with the neurogenic differentiation propensity of hESC and hiPSC lines [67].

Epigenetic silencing of the Oct4 locus is necessary to ensure proper differentiation upon exit from pluripotency. This is achieved by de novo DNA methylation by DNMT3 factors. It has been shown that miR-290-295 play a role in this process by inhibiting Rbl2, that in turn is an inhibitor of DNMT3 [72]. In the absence of miR-290-295 there would be incomplete silencing of the Oct4 locus during differentiation. This would in part explain the differentiation defects observed in Dicer −/− cells, which maintain high levels of pluripotency factors and fail to activate lineage specific gene programs [69]. Other miRNAs may contribute to this phenotype, inhibiting the expression of pluripotency genes during mESC differentiation (Figure 3). For instance, miR-134, miR-296 and miR-470 target the core transcription factor trio, Oct4, Sox2 and Nanog [95]. Sox2 and Klf4 are also repressed by miR-200c, miR-183 and miR-203 [96]. It is interesting to notice that miR-203 has also a role in skin stem cell terminal differentiation by inhibiting p63 [97]. Pluripotency factors are under the control of miRNAs also in human ESC. An interesting regulatory loop has been shown for miR-145, which targets Oct4, Sox2 and Klf4 during differentiation, and in turn is repressed at the transcriptional level by Oct4 in undifferentiated cells [98]. Recently, it has been proposed that another role for miRNAs during ESC differentiation is to modulate the activity of chromatin modifiers. An interesting switch between different variants of the Polycomb Repressive Complex 1 (PRC1) has been shown during ESC differentiation. In particular, an ES-specific PRC1, containing the Cbx7 subunit, is replaced by a differentiation-specific PRC1, containing the subunits Cbx2/4/8 [99,100]. This switch is crucial, as Cbx proteins confer distinct target selectivity to the PRC1 complex. The different Cbx variants inhibit each other. In pluripotency conditions, Cbx7 negatively regulates the other Cbx that are conversely induced during differentiation, and repress Cbx7 transcription. miRNAs of the miR-125 and miR-181 families contribute to the downregulation of Cbx7 during mESC differentiation by directly targeting its 3'UTR [100]. Overexpression of these miRNAs in mESCs caused loss of pluripotency markers and increased expression of a subset of PRC1 target genes involved in lineage specification.

Figure 3. miRNAs target pluripotency genes. The activity of pluripotency genes, including core transcription factors, must be shut down as embryonic stem cells differentiate. Several miRNAs that are induced upon exit from pluripotency directly target these genes in a combinatorial way. Examples are shown in the table. M: mouse; H: human.

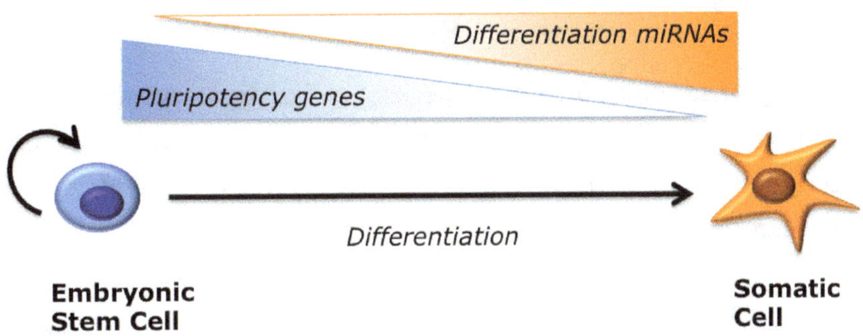

miRNAs	*Targets*	*Species*	*References*
miR-134, miR-296, miR-470	Oct4, Sox2, Nanog	M	Tay et al., 2008
miR-200c, miR-183, miR-203	Sox2, Klf4	M	Wellner et al., 2009
miR-145	Oct4, Sox2, Klf4	H	Xu et al., 2009
let-7	cMyc, Lin28, Sall4	M	Melton et al., 2010
miR-125, miR-181	Cbx7	M, H	O'Loghlen et al., 2012

Among other miRNAs with a role in ESC differentiation, the let-7 family plays a prominent role. This is a highly conserved family with orthologues in all metazoa. In mammals it comprises several let-7 species (let-7a to let-7i) and other miRNAs, such as miR-98 and miR-202 [101]. Members of the let-7 family are induced during development and differentiation, with a parallel reduction of their targets, and have a role in cancer [102]. In ESCs, levels of let-7 miRNAs are regulated at the post-transcriptional level. Whereas the primary transcript and the hairpin precursor accumulate in these cells, the production of the mature miRNA is blocked [103]. The underlying molecular mechanism relies on the recognition of the terminal loop of let-7 precursors by the RNA binding protein Lin28 [104–106]. Lin28 recruits the terminal uridylyl transferase TUT4, which in turn adds a poly-U tail, targeting the miRNA for degradation [107,108]. A feedback regulatory loop exists, in which let-7 negatively regulates Lin28 [109]. Lin28 is highly expressed in undifferentiated ESCs and declines during differentiation, when levels of mature let-7 increase. This switch is ensured by their mutual repression. As mentioned above, DGCR8 −/− ESCs maintain high levels of pluripotency genes expression. Upon transfection of let-7 family miRNAs, expression of Oct4, Sox2 and Nanog is inhibited, suggesting an anti-pluripotency activity for let-7. However, if miR-302 family miRNAs are co-transfected, they impair this activity and restore the levels of pluripotency markers [110].

Comparative transcriptome analysis showed that let-7 introduction in DGCR8 −/− mESCs led to the inhibition of genes that are transcriptionally activated by Myc factors. Both c-Myc and N-Myc are direct targets of let-7 [110]. Myc genes play a well-established role in ESC self-renewal [111]. Moreover, let-7 directly targets transcripts that are induced by other pluripotency factors, Oct4, Sox2, Nanog and Tcf3 [110]. Inhibition of the function of pluripotency factors, direct for Myc and indirect for other factors, explains the anti-pluripotency activity of let-7 (Figure 4). Conversely, Myc-activated genes and c-Myc itself are enriched among transcript that are upregulated in presence of miR-302 family members, suggesting that the pro-pluripotency activity of these miRNAs may be mediated, at least in part, by the indirect increase of Myc [110].

Figure 4. let-7 antagonizes pluripotency networks. Let-7 antagonizes indirectly the activity of the ESC core transcriptional regulatory circuitry by targeting multiple genes induced by the core transcription factors (TFs). Moreover, let-7 directly inhibits c-Myc, reducing transcription of its target genes.

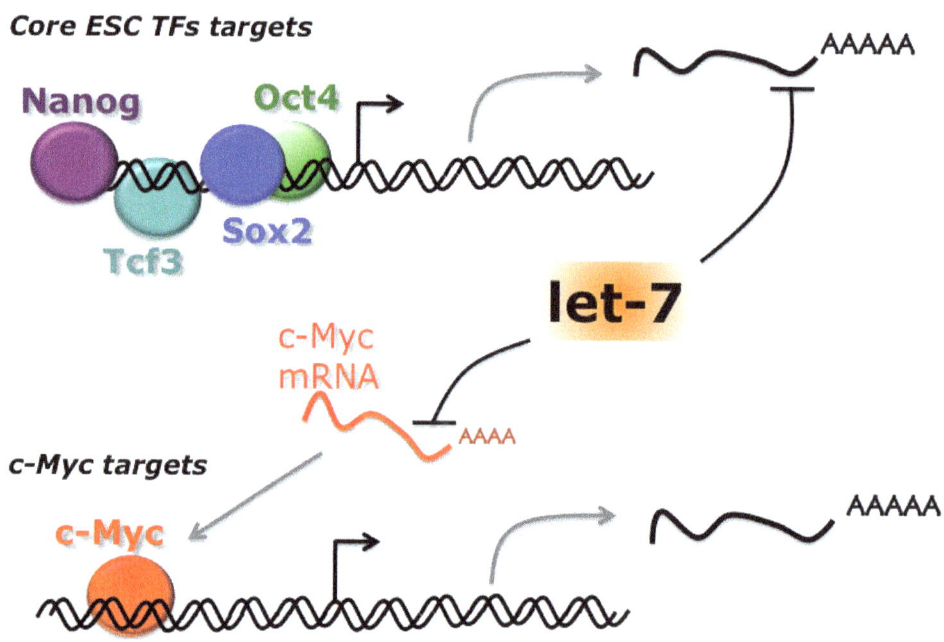

For both let-7 and the miR-302 families feedback loops with Myc are in play. The promoter of the miR-290-295 cluster is directly bound and activated by c-Myc and N-Myc [7], and the miR-302 cluster is also induced by Myc [112], establishing positive feedback loops. Conversely, in cancer cells c-Myc has been shown to bind the promoters and repress transcription of several let-7 genes [113]. Other pluripotency factors that are directly inhibited by let-7, and indirectly activated by miR-302 family, are Lin28 and Sall4, suggesting that these miRNAs exert their function via multiple pathways [110]. Therefore, according to the model depicted in Figure 5, the miR-302 and let-7 miRNA families play opposite, crucial roles in regulating ESC pluripotency and differentiation.

Figure 5. Opposing roles for the let-7 and miR-302 families. During ESC differentiation, the levels of pluripotency miRNAs belonging to the miR-302 family are decreased. In parallel, let-7 family members are induced. The miR-302/367 and miR-290-295 clusters are induced by core ESC TFs, and in a positive feedback loop indirectly promote their expression and transcription of their targets. Conversely, let-7 is post-transcriptionally inhibited by Lin-28. In cancer cells c-Myc inhibits let-7 at the transcriptional level. In a negative feedback loop, let-7 targets its inhibitors. Black arrows indicate direct activation; grey arrows indicate indirect activation; dashed lines indicate post-transcriptional inhibition.

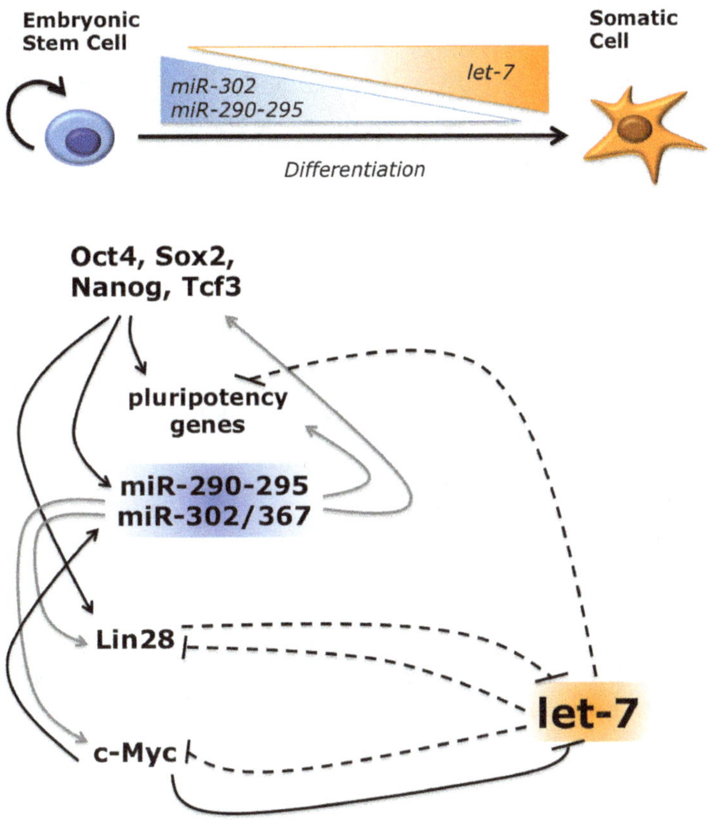

3. MicroRNAs and Reprogramming to iPSCs

ESCs and iPSCs express a similar signature group of miRNAs, including the miR-302 family, with small differences between the two cell types [114–116]. When genetically identical mouse ESCs and iPSCs were analyzed, these differences were circumscribed to miRNAs encoded in an imprinted locus on chromosome 12qF1 [117]. Imprinted loci are transcribed in a parental specific manner and contain clusters of protein-coding and noncoding genes [118]. The 12qF1 locus includes several maternally expressed genes, including two miRNA clusters, that are aberrantly silenced in many, but not all, iPSC lines [117]. Interestingly, iPSC lines in which this imprinted locus was silenced were unable to generate all-iPSC mice upon tetraploid blastocyst complementation, due to arrest of embryonic development around mid-gestation. Conversely, iPSC lines with normal expression of the imprinted locus generated viable all-iPSC mice. Therefore, the transcriptional status of imprinted genes in the 12qF1 locus, including miRNAs, marks the development potential of different iPSC clones

3.1. Several miRNAs Promote or Inhibit Reprogramming

Recently, several miRNAs have been directly involved in the reprogramming process (Figure 6). Knockdown of RISC components globally impairs miRNA activity and leads to a dramatic decrease of reprogramming activity [119]. Much work concerned the miRNAs belonging to the miR-302 family. When the minimal Oct4/Sox2/Klf4 (OSK) cocktail was provided to mouse embryonic fibroblasts (MEFs) in combination with these miRNAs, the efficiency of iPSCs generation was substantially increased [119,120]. Interestingly, in presence of the three factors plus c-Myc the reprogramming enhancement by the miR-302 family members was strongly reduced [120,121]. Other miRNAs may also promote reprogramming in the mouse. For instance, miR-93 and 106b (that belong to the same family and share 5/6 of the miR-302 seed) and the unrelated miR-138 enhance iPSC generation [118,122]. Similar to the mouse system, reprogramming of human fibroblasts was also enhanced when miR-302 family members are provided along with the reprogramming factors [123].

Figure 6. Role of miRNAs during reprogramming. The reprogramming process can be divided in three phases. Early events include inhibition of apoptosis and an increase in proliferation. A mesenchymal-to-epithelial transition (MET) occurs in the intermediate phase. Activation of the endogenous pluripotency program occurs as a late event. Several miRNAs have been shown to promote (blue) or inhibit (orange) reprogramming by facilitating or hampering the completion of these events. The miR-93 family includes miR-93 and miR-106b.

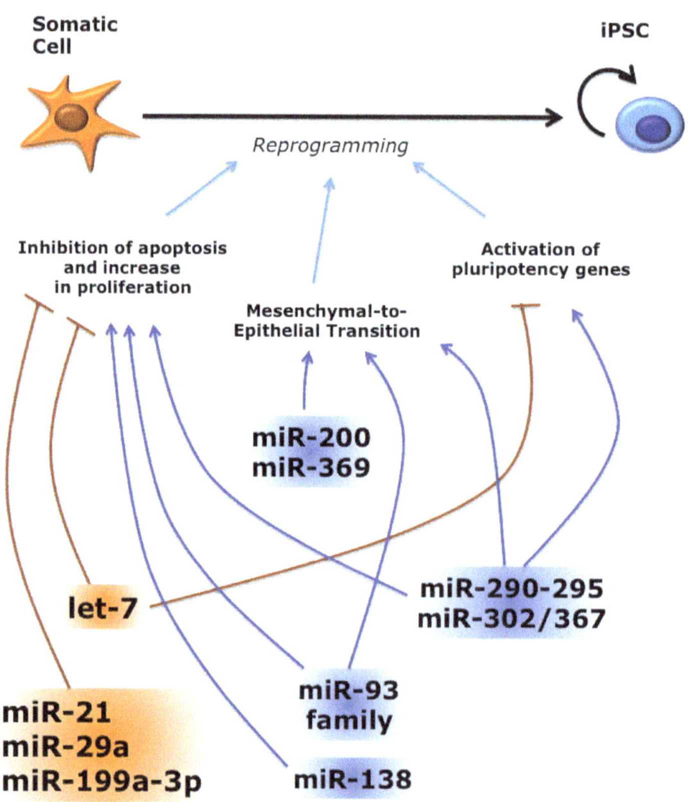

As expected, let-7 has an opposite role during reprogramming. MEFs express high levels of let-7 and this anti-pluripotency miRNA family must be shut down during conversion into iPSCs. In the presence of let-7 inhibitors, reprogramming by the OSK cocktail was enhanced [110]. When OSK plus c-Myc were used, this increase of efficiency by let-7 inhibition was much less pronounced. Again, let-7 inhibits c-Myc, so the inhibition of let-7 might just increase c-Myc activity. Lin28, which is a component of the Thomson reprogramming cocktail and is induced during reprogramming with the Yamanaka factors, is another target of let-7. As in the case of c-Myc, addition of Lin28 to the reprogramming cocktail accelerates proliferation [124]. However, there is not an increase in MEF proliferation upon let-7 inhibition [110], suggesting that other targets besides c-Myc and Lin28 contribute to the effects of let-7 on the reprogramming process. Besides let-7, other miRNAs negatively affect reprogramming. For instance, inhibition of miR-21, miR-29a and miR-199a-3p enhances reprogramming [125,126].

3.2. miRNAs Reprogram Somatic Cells in the Absence of Protein-Coding Reprogramming Factors

Given their multiple roles in ESCs, the increase of reprogramming efficiency by miRNAs in combination with reprogramming factors was not unexpected. More surprisingly, recent work has shown that cocktails of miRNAs, without canonical reprogramming factors, can be sufficient to reprogram both mouse and human somatic cells. The Morrisey lab demonstrated that the miR-302-367 cluster induced pluripotency with two orders of magnitude more efficiency than standard methods [127]. Importantly iPSCs generated with miR-302-367 fulfilled stringent criteria that define bona fide pluripotent stem cells, such as the ability to contribute to chimeras with germ line transmission (mouse iPSCs) and to form teratomas (human iPSCs). In these experiments, Valproic Acid (VPA)-mediated inhibition of Hdac2 was required for mouse, but not human, miRNA-mediated reprogramming. The miR-302-367 cluster contains four miR-302 members, with the AAGUGC seed, and the unrelated miR-367. Both were required for reprogramming, as miR-302 alone is not able to give rise to iPSCs in the absence of miR-367. Interestingly, when used in combination with reprogramming factors to enhance reprogramming, miR-302 alone was almost as effective as the intact miR-302-367 cluster, whereas miR-367 alone had no effect [121]. Moreover, miR-367, but not miR-302, can be substituted by other miRNAs in alternative reprogramming cocktails. For instance, it has been shown that the combination of miR-302, miR-200c and miR-369 can reprogram both human and mouse somatic cells [128]. In this case, instead of delivering miRNA genes via viral vectors, repeated transfections of mature synthetic miRNAs were used. This approach reduces the reprogramming efficiency but might be useful towards therapeutic applications of iPSCs, which would require cells devoid of exogenous reprogramming genes integration in the genome [18]. In contrast to the finding that combinations of multiple miRNA families are necessary for reprogramming [123,127,128], Lin *et al.* have reported that miR-302 alone could convert skin cancer and hair follicle cells into iPSCs in the absence of other miRNAs or RFs [129,130]. It remains unclear whether other somatic cell types can be reprogrammed under these conditions.

3.3. miRNAs Regulate Reprogramming by Multiple Pathways

The mechanisms underlying miRNA-mediated reprogramming have been only partially clarified. As mentioned before, introduction of miR-302 family members in DGCR8 −/− mESCs increases expression of endogenous c-Myc and N-Myc downstream genes [110]. However, it is unlikely that the pro-reprogramming activity of the miR-302 is merely mediated by downstream activation of c-Myc. In fact, the mechanisms by which these miRNAs and c-Myc enhance the reprogramming efficiency seem different, since the miRNAs, unlike c-Myc, do not accelerate doubling time of MEFs and produce a more homogenous population of fully reprogrammed cells [120]. Multiple pathways may be affected by overexpression of these miRNAs. For instance, it has been proposed that during reprogramming miR-302 regulates multiple genes involved in cell cycle regulation, epigenetic regulation, vesicular transport, cell signaling and mesenchymal-to-epithelial transition [123].

Other miRNAs that affect reprogramming are integrated in the p53 pathway. p53 activation elicits apoptosis and cell cycle arrest, representing a roadblock to reprogramming. Consistently, p53 inhibition promotes reprogramming [131]. The anti-reprogramming miR-21 and miR-29a sustain p53 activity. As in the case of let-7, c-Myc negatively regulates these miRNAs [125]. p53 positively regulates several anti-reprogramming miRNAs: it promotes transcription of miR-34a and miR-145, which target pluripotency genes [132], and upregulates at the post-transcriptional level miR-199a-3p that inhibits cell proliferation [126]. Conversely, miR-138 enhances reprogramming by directly targeting p53 [119,122].

A mesenchymal-to-epithelial transition (MET) is one of the early events occurring during fibroblasts reprogramming to iPSCs [133,134]. Since iPSCs have an epithelial character, MET is necessary for reprogramming of mesenchymal cells, such as fibroblasts, but not for keratinocytes and other somatic cells with an epithelial character. MET is under the control of the TGFβ signaling and several reprogramming factors and miRNAs are involved in this process [135]. In particular, TGFβ promotes expression of Snail, a mediator of epithelial-to-mesenchymal transition (EMT), whereas Sox2 and Oct4 inhibit Snail transcription. c-Myc downregulates TGFβ1 and TGFβ receptor 2 (TGFBR2), which is also a target of the miR-302 and miR-93 families [119,121,123,133]. Conversely, BMP signaling stimulates expression of the miR-200 family and miR-205, which repress other EMT factors, ZEB1 and SIP1 (ZEB2) [133,134,136]. In turn, ZEB1 can inhibit transcription of the miR-200 family in a negative feedback loop [137,138]. ZEB2 is also a predicted target of miR-369 that, together with miR-200c, is a component of a miRNA-only reprogramming cocktail [128]. Another gene involved in EMT, RHOC, is a direct target of miR-302 [123].

miR-302 overexpression sustains pluripotency markers in differentiating hESCs [87,139]. Similarly, during reprogramming, miR-302 may also indirectly promote the activation of endogenous core pluripotency genes by targeting their inhibitors. For instance, methyl-DNA binding domain protein 2 (MBD2), an epigenetic suppressor of Nanog, is a direct target of miR-302 [140]. Downregulation of MDB2 by miR-302 is necessary to achieve a fully reprogrammed iPSC state. We have previously demonstrated that miR-302 targets NR2F2, which in turn is a transcriptional inhibitor of Oct4 [94]. Recently, it has been shown that NR2F2 knockdown enhances reprogramming efficiency, thus mimicking miR-302 overexpression [141].

4. Long Non-Coding RNAs in Embryonic Stem Cells

A number of differentially expressed lncRNAs were detected in undifferentiated mESCs and upon induction of differentiation by microarray analysis [37]. They could be classified as pluripotency, early mesoderm, and hematopoietic lncRNAs. New intergenic transcripts can be predicted by taking advantage of chromatin immunoprecipitation followed by deep sequencing (ChIP-Seq). Genes actively transcribed by Pol II are characterized by a distinctive chromatin signature, consisting in trimethylation of lysine 4 of histone H3 in the promoter combined with trimethylation of lysine 36 of histone H3 in the transcribed region (K4-K36 domain). A K4-K36 domain located outside of known protein-coding loci would predict a putative novel long intergenic ncRNA. In mESCs and somatic cells, this approach led to the identification of over a thousand of novel lincRNAs that were further validated by microarray and northern blot [142]. Bioinformatics analysis showed that both the promoters and the transcribed sequences of these novel lincRNAs are conserved in mammals, suggesting that they may have biological functions. More recently, the number of mESC lincRNAs has been further expanded by taking advantage of a computational method that allows the reconstruction of the whole transcriptome (Scripture) from massive cDNA sequencing (RNA-Seq) data [38]. Overall, lincRNAs expressed in ESC show high evolutionary conservation and tissue specificity, and their expression levels are generally comparable to protein coding genes. Interestingly, Scripture identified hundreds of noncoding RNAs that are partially overlapped with protein-coding genes but transcribed in the opposite orientation. Such antisense transcripts are less conserved than long intergenic ncRNAs and likely represent a distinct class [38]. Analysis of intergenic K4-K36 domains indicated that lincRNAs are present also in human ESCs [47].

Many mESC lincRNA genes are regulated by core transcription factors that bind directly their promoters [142]. In particular, about 10% of Nanog and Oct4 binding sites are associated with lincRNA genes [143]. Knockdown of individual core TFs affects the expression of about 60% of lincRNAs [144]. As expected, mESC lincRNAs levels decrease upon differentiation, as the levels of mESC TFs decline. Also for some human ESC lincRNAs there is evidence for a direct regulation by core TFs [145,146].

lncRNAs Maintain Pluripotency in ESC

A biological role for mESC lincRNAs, initially postulated on the basis of their evolutionary conservation, has been recently experimentally demonstrated. RNAi against the lncRNA AK028326 (also known as Gomafu/Miat) resulted in morphological differentiation, reduced proliferation, decreased levels of Oct4 and other pluripotency markers and increase of trophoblast markers [143], suggesting that this lncRNA has a role in maintaining mESC pluripotency. AK028326 is activated by Oct4, establishing a positive feedback loop. RNAi against another mESC lncRNA, AK141205, also resulted in a decrease of Oct4, but in this case Nanog levels, morphology and proliferation were unaffected and differentiation genes were not induced [143]. Regulation of Oct4 by lncRNAs is conserved in human. Detailed analysis of transcripts generated by the Oct4 locus and Oct4 pseudogenes loci in human breast adenocarcinoma cells (MCF-7) has shown the existence of two transcripts, which were antisense to Oct4 (asOct4) and Oct4 pseudogene 5 (asOct4-pg5) [147]. It has been proposed that asOct4-pg5 is a functional lncRNA in MCF-7 cells, as it is able to recruit chromatin modifiers to the Oct4 promoter, leading to transcriptional silencing of the Oct4 gene by methylation of H3K27 and H3K9 by Ezh2 and G9a, respectively. It would be interesting to assess weather this regulatory mechanism has also a role in human pluripotent stem cells, which rely on Oct4 for the maintenance of pluripotency. Recent work has shown that three lncRNAs, lncRNA_ES1, lncRNA_ES2 and lncRNA_ES3, were specifically expressed in undifferentiated hESCs, and absent in somatic tissues [145]. For two of them, direct regulation by Oct4 and Nanog has been proposed. Functional analysis showed that the three ES lncRNAs are necessary to maintain pluripotency in hESCs, as their individual knockdown by RNAi resulted in downregulation of pluripotency markers and induction of lineage genes. Nuclear localization and association with PRC2 and Sox2 suggest that these lncRNAs may function by regulating gene expression at the transcriptional level [145].

Large-scale analysis suggests that the role of lncRNAs in pluripotent stem cells may be greater than expected. Loss-of-function by RNA interference was systematically achieved for 147 mESC lincRNAs [144]. Strikingly, for nearly all lincRNAs tested (93%), this produced a significant impact on global gene expression. On average, the number of genes affected by lincRNA knockdown was comparable to that obtained by the knockdown of previously characterized key ESC regulatory factors, suggesting that lincRNAs play major roles in mESCs. This was confirmed by further analysis, showing that knockdown of a subset of lincRNAs affected expression of pluripotency markers, including Oct4, Sox2 and Nanog, and led to a change in mESC morphology. Moreover, loss of function of many lincRNAs triggered the activation of specific neuroectoderm, endoderm, mesoderm, or trophectoderm genes [144]. These results indicate that lincRNAs are necessary for two crucial aspects of mESC pluripotency: the maintenance of the specific genetic program associated with the undifferentiated state, and the repression of genes involved in lineage differentiation.

Mechanistically, mESC lncRNAs may exert their function by modulating the activity of chromatin-modifying factors. Both "readers", "writers" and "erasers" histone modifiers and the chromatin-associated DNA binding protein Yy1 have been found in association with mESCs lncRNAs [37,144]. For example, the lncRNAs Evx1as and Hoxb5/6as, which are transcribed antisense to the homeotic genes Evx1 and Hoxb5-Hoxb6, are associated with the Trithorax protein MLL1. Interestingly, many lincRNAs are bound by multiple chromatin regulatory complexes, which form consistent combinations of readers, writers and erasers. This evidence suggests that in mESCs lincRNAs may function as a bridge to tether distinct functionally related complexes, as proposed in other systems [148]. Such molecular scaffolding function may play a crucial role in maintaining the epigenetic state of pluripotent stem cells.

5. Long Non-Coding RNAs in Induced Pluripotent Stem Cells

A subset of human lincRNAs demonstrated expression at higher levels in iPSCs with respect to both somatic cells of origin and hESCs [146]. Among them, there is at least one with a role in reprogramming, named lincRNA-RoR (or linc-RoR) for "Regulator of Reprogramming" (Figure 7). Knockdown of linc-RoR resulted in a significant decrease of iPSC colonies formation, whereas overexpression enhanced reprogramming efficiency. This pro-reprogramming effect can be explained by recent work showing that linc-RoR is a negative regulator of p53 [149]. Mechanistically, linc-RoR inhibits translation of p53 mRNA in the cytoplasm and this inhibitory activity is dependent on the interaction with heterogeneous nuclear ribonucleoprotein I (hnRNP I). Interestingly, linc-RoR transcription is induced by p53, establishing an autoregulatory feedback loop [149]. Another function recently proposed for linc-RoR may underlie its pro-reprogramming activity [150]. By sequestering a set of miRNAs that target Oct4, Nanog and Sox2, Linc-RoR may protect pluripotency factors from miRNA inhibition by acting as a competing endogenous RNA (ceRNA) [52].

Figure 7. Role of lincRNA-RoR during reprogramming. LincRNA-RoR may promote reprogramming by two different mechanisms. It inhibits p53, which in turn negatively affects the initial steps of reprogramming by inducing apoptosis. It also acts as a competing endogenous RNA and releases from repression the endogenous pluripotency genes Oct4, Sox2 and Klf4, which are targeted by miR-145. MiR-145 is also inhibited by Oct4 in a negative feedback loop. Other miRNA may be "sponged" by lincRNA-RoR (not shown).

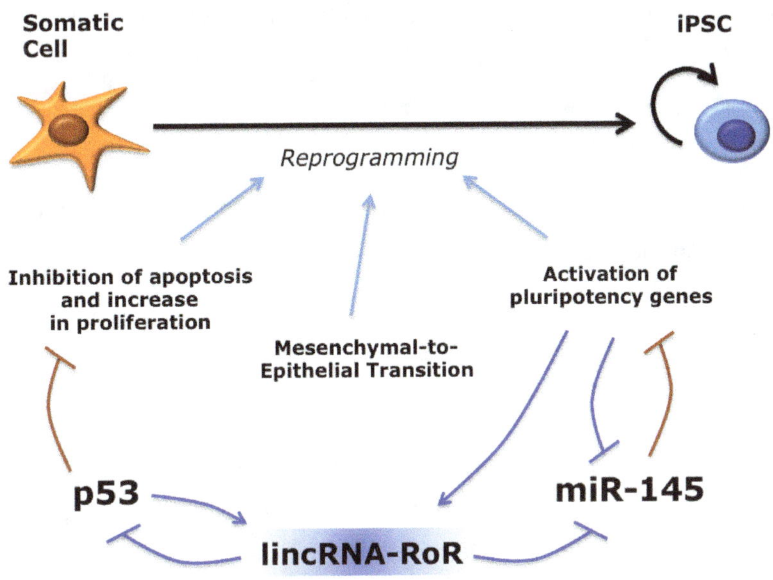

6. Conclusions and Future Perspectives

In ESCs and iPSCs, pluripotency is the result of: (i) the interplay between multiple signaling pathways; (ii) a core transcriptional regulatory circuitry; (iii) epigenetic control of gene expression by specific chromatin modification patterns; (iv) the maintenance of a peculiar cell cycle. Recent work suggests that regulatory ncRNAs have a role in modulating all these biological pathways. Similar to the core set of transcription factors, ESCs are characterized by a core set of miRNAs, belonging to the miR-302 family. These miRNAs are hierarchically on top of multiple targets, which participate in the maintenance of self-renewal, and regulate exit from pluripotency and cell fate choice. Moreover, miR-302 family members are required for efficient reprogramming of somatic cells and, in combination with other miRNAs, are sufficient for iPSC generation in the absence of canonical reprogramming factors. A similar role has been shown for other miRNAs in different systems. For instance, neural-specific miRNAs can induce trans-differentiation of fibroblasts into induced neurons without other factors [151]. Since miRNAs can be provided as mature RNA molecules, this approach could lead to the generation of therapeutic relevant cells devoid of exogenous genes integrations and genomic lesions.

The functions of lncRNAs in pluripotent cells are less characterized. So far, large datasets generated by RNA-seq projects have been followed only by a limited number of functional studies. Such are hampered by the fact that unlike protein-coding genes and miRNA the function of a lncRNA cannot be normally predicted by the analysis of its sequence [152]. However, the few examples in which the function of lncRNAs has been investigated suggest that they might be primary regulators of pluripotency, differentiation and reprogramming. In particular, since lncRNAs have been show to act as molecular scaffolds of chromatin regulators, they may have a huge impact in the context of stem cells where the maintenance of a characteristic epigenetic state of the chromatin is central for pluripotency and differentiation. So far, ESC lncRNAs have been associated with maintenance of pluripotency factors, repression of differentiation, and inhibition of apoptosis. Their possible role in other crucial pathways underlying pluripotency has not yet been investigated. For instance, it is unknown whether lncRNAs could be involved in tuning signaling pathways, whereas evidence exists for miRNAs. Finally, as cytoplasmic lncRNAs may work as inhibitors of miRNAs, they could add another layer of complexity to the regulatory circuitries underlying pluripotency.

Acknowledgments

We are grateful to A. Fatica, V. de Turris, J. Lenzi, R. De Santis and M. Heke for critical reading of the manuscript. A.H.B. is supported by NIH grant #5R01GM097615.

References

1. Warmflash, A.; Arduini, B.L.; Brivanlou, A.H. The molecular circuitry underlying pluripotency in embryonic stem cells. *Wiley Interdiscip. Rev. Syst. Biol. Med.* **2012**, *4*, 443–456.
2. James, D.; Levine, A.J.; Besser, D.; Hemmati-Brivanlou, A. TGFbeta/activin/nodal signaling is necessary for the maintenance of pluripotency in human embryonic stem cells. *Development* **2005**, *132*, 1273–1282.
3. Vallier, L.; Alexander, M.; Pedersen, R.A. Activin/Nodal and FGF pathways cooperate to maintain pluripotency of human embryonic stem cells. *J. Cell Sci.* **2005**, *118*, 4495–4509.
4. Ying, Q.L.; Nichols, J.; Chambers, I.; Smith, A. BMP induction of Id proteins suppresses differentiation and sustains embryonic stem cell self-renewal in collaboration with STAT3. *Cell* **2003**, *115*, 281–292.
5. Boyer, L.A.; Lee, T.I.; Cole, M.F.; Johnstone, S.E.; Levine, S.S.; Zucker, J.P.; Guenther, M.G.; Kumar, R.M.; Murray, H.L.; Jenner, R.G.; *et al.* Core transcriptional regulatory circuitry in human embryonic stem cells. *Cell* **2005**, *122*, 947–956.
6. Kim, J.; Chu, J.; Shen, X.; Wang, J.; Orkin, S.H. An extended transcriptional network for pluripotency of embryonic stem cells. *Cell* **2008**, *132*, 1049–1061.
7. Chen, X.; Xu, H.; Yuan, P.; Fang, F.; Huss, M.; Vega, V.B.; Wong, E.; Orlov, Y.L.; Zhang, W.; Jiang, J.; *et al.* Integration of external signaling pathways with the core transcriptional network in embryonic stem cells. *Cell* **2008**, *133*, 1106–1117.
8. Boyer, L.A.; Plath, K.; Zeitlinger, J.; Brambrink, T.; Medeiros, L.A.; Lee, T.I.; Levine, S.S.; Wernig, M.; Tajonar, A.; Ray, M.K.; *et al.* Polycomb complexes repress developmental regulators in murine embryonic stem cells. *Nature* **2006**, *441*, 349–353.

9. Marson, A.; Levine, S.S.; Cole, M.F.; Frampton, G.M.; Brambrink, T.; Johnstone, S.; Guenther, M.G.; Johnston, W.K.; Wernig, M.; Newman, J.; *et al.* Connecting microRNA genes to the core transcriptional regulatory circuitry of embryonic stem cells. *Cell* **2008**, *134*, 521–533.
10. Spivakov, M.; Fisher, A.G. Epigenetic signatures of stem-cell identity. *Nat. Rev. Genet.* **2007**, *8*, 263–271.
11. Nichols, J.; Smith, A. Naive and primed pluripotent states. *Cell Stem Cell* **2009**, *4*, 487–492.
12. Guo, G.; Yang, J.; Nichols, J.; Hall, J.S.; Eyres, I.; Mansfield, W.; Smith, A. Klf4 reverts developmentally programmed restriction of ground state pluripotency. *Development* **2009**, *136*, 1063–1069.
13. Silva, J.; Nichols, J.; Theunissen, T.W.; Guo, G.; van Oosten, A.L.; Barrandon, O.; Wray, J.; Yamanaka, S.; Chambers, I.; Smith, A. Nanog is the gateway to the pluripotent ground state. *Cell* **2009**, *138*, 722–737.
14. Pera, M.F.; Tam, P.P.L. Extrinsic regulation of pluripotent stem cells. *Nature* **2010**, *465*, 713–720.
15. Takahashi, K.; Yamanaka, S. Induction of pluripotent stem cells from mouse embryonic and adult fibroblast cultures by defined factors. *Cell* **2006**, *126*, 663–676.
16. Takahashi, K.; Tanabe, K.; Ohnuki, M.; Narita, M.; Ichisaka, T.; Tomoda, K.; Yamanaka, S. Induction of pluripotent stem cells from adult human fibroblasts by defined factors. *Cell* **2007**, *131*, 861–872.
17. Yu, J.; Vodyanik, M.A.; Smuga-Otto, K.; Antosiewicz-Bourget, J.; Frane, J.L.; Tian, S.; Nie, J.; Jonsdottir, G.A.; Ruotti, V.; Stewart, R.; *et al.* Induced pluripotent stem cell lines derived from human somatic cells. *Science* **2007**, *318*, 1917–1920.
18. Stadtfeld, M.; Hochedlinger, K. Induced pluripotency: History, mechanisms, and applications. *Genes Dev.* **2010**, *24*, 2239–2263.
19. Papp, B.; Plath, K. Reprogramming to pluripotency: Stepwise resetting of the epigenetic landscape. *Cell Res.* **2011**, *21*, 486–501.
20. Cohen, D.E.; Melton, D. Turning straw into gold: Directing cell fate for regenerative medicine. *Nat. Rev. Genet.* **2011**, *12*, 243–252.
21. Mattick, J.S. The central role of RNA in human development and cognition. *FEBS Lett.* **2011**, *585*, 1600–1616.
22. Bartel, D.P. MicroRNAs: Target recognition and regulatory functions. *Cell* **2009**, *136*, 215–233.
23. Bushati, N.; Cohen, S.M. microRNA functions. *Annu. Rev. Cell Dev. Biol.* **2007**, *23*, 175–205.
24. Fatica, A.; Rosa, A.; Ballarino, M.; de Marchis, M.L.; Rasmussen, K.D.; Bozzoni, I. Role of microRNAs in myeloid differentiation. *Biochem. Soc. Trans.* **2008**, *36*, 1201–1205.
25. Rosa, A.; Brivanlou, A.H. microRNAs in early vertebrate development. *Cell Cycle* **2009**, *8*, 3531–3520.
26. Rottiers, V.; Näär, A.M. MicroRNAs in metabolism and metabolic disorders. *Nat. Rev. Mol. Cell Biol.* **2012**, *13*, 239–250.
27. Iorio, M.V.; Croce, C.M. MicroRNA dysregulation in cancer: Diagnostics, monitoring and therapeutics. A comprehensive review. *EMBO Mol. Med.* **2012**, *4*, 143–159.
28. Denli, A.M.; Tops, B.B.J.; Plasterk, R.H.A.; Ketting, R.E.F.; Hannon, G.J. Processing of primary microRNAs by the Microprocessor complex. *Nature* **2004**, *432*, 231–235.

29. Gregory, R.I.; Yan, K.-P.; Amuthan, G.; Chendrimada, T.; Doratotaj, B.; Cooch, N.; Shiekhattar, R. The Microprocessor complex mediates the genesis of microRNAs. *Nature* **2004**, *432*, 235–240.
30. Hutvágner, G.; Zamore, P.D. A microRNA in a multiple-turnover RNAi enzyme complex. *Science* **2002**, *297*, 2056–2060.
31. Hammond, S.M. Dicing and slicing: The core machinery of the RNA interference pathway. *FEBS Lett.* **2005**, *579*, 5822–5829.
32. Gregory, R.I.; Chendrimada, T.P.; Cooch, N.; Shiekhattar, R. Human RISC couples microRNA biogenesis and posttranscriptional gene silencing. *Cell* **2005**, *123*, 631–640.
33. Rajewsky, N. microRNA target predictions in animals. *Nat. Genet.* **2006**, *38*, S8–S13.
34. Lewis, B.P.; Burge, C.B.; Bartel, D.P. Conserved seed pairing, often flanked by adenosines, indicates that thousands of human genes are microRNA targets. *Cell* **2005**, *120*, 15–20.
35. Djebali, S.; Davis, C.A.; Merkel, A.; Dobin, A.; Lassmann, T.; Mortazavi, A.; Tanzer, A.; Lagarde, J.; Lin, W.; Schlesinger, F.; *et al.* Landscape of transcription in human cells. *Nature* **2012**, *489*, 101–108.
36. Sone, M.; Hayashi, T.; Tarui, H.; Agata, K.; Takeichi, M.; Nakagawa, S. The mRNA-like noncoding RNA Gomafu constitutes a novel nuclear domain in a subset of neurons. *J. Cell Sci.* **2007**, *120*, 2498–2506.
37. Dinger, M.E.; Amaral, P.P.; Mercer, T.R.; Pang, K.C.; Bruce, S.J.; Gardiner, B.B.; Askarian-Amiri, M.E.; Ru, K.; Soldà, G.; Simons, C.; *et al.* Long noncoding RNAs in mouse embryonic stem cell pluripotency and differentiation. *Genome Res.* **2008**, *18*, 1433–1445.
38. Guttman, M.; Garber, M.; Levin, J.Z.; Donaghey, J.; Robinson, J.; Adiconis, X.; Fan, L.; Koziol, M.J.; Gnirke, A.; Nusbaum, C.; *et al.* Ab initio reconstruction of cell type-specific transcriptomes in mouse reveals the conserved multi-exonic structure of lincRNAs. *Nat. Biotechnol.* **2010**, *28*, 503–510.
39. Kapranov, P.; Cheng, J.; Dike, S.; Nix, D.A.; Duttagupta, R.; Willingham, A.T.; Stadler, P.F.; Hertel, J.; Hackermüller, J.; Hofacker, I.L.; *et al.* RNA maps reveal new RNA classes and a possible function for pervasive transcription. *Science* **2007**, *316*, 1484–1488.
40. Nakaya, H.I.; Amaral, P.P.; Louro, R.; Lopes, A.; Fachel, A.A.; Moreira, Y.B.; El-Jundi, T.A.; da Silva, A.M.; Reis, E.M.; Verjovski-Almeida, S. Genome mapping and expression analyses of human intronic noncoding RNAs reveal tissue-specific patterns and enrichment in genes related to regulation of transcription. *Genome Biol.* **2007**, *8*, R43.
41. Mercer, T.R.; Dinger, M.E.; Sunkin, S.M.; Mehler, M.F.; Mattick, J.S. Specific expression of long noncoding RNAs in the mouse brain. *Proc. Natl. Acad. Sci. USA* **2008**, *105*, 716–721.
42. Ravasi, T.; Suzuki, H.; Pang, K.C.; Katayama, S.; Furuno, M.; Okunishi, R.; Fukuda, S.; Ru, K.; Frith, M.C.; Gongora, M.M.; *et al.* Experimental validation of the regulated expression of large numbers of non-coding RNAs from the mouse genome. *Genome Res.* **2006**, *16*, 11–19.
43. Ponjavic, J.; Ponting, C.P.; Lunter, G. Functionality or transcriptional noise? Evidence for selection within long noncoding RNAs. *Genome Res.* **2007**, *17*, 556–565.
44. Carninci, P.; Kasukawa, T.; Katayama, S.; Gough, J.; Frith, M.C.; Maeda, N.; Oyama, R.; Ravasi, T.; Lenhard, B.; Wells, C.; *et al.* The transcriptional landscape of the mammalian genome. *Science* **2005**, *309*, 1559–1563.
45. Mattick, J.S. A new paradigm for developmental biology. *J. Exp. Biol.* **2007**, *210*, 1526–1547.

46. Carninci, P.; Hayashizaki, Y. Noncoding RNA transcription beyond annotated genes. *Curr. Opin. Genet. Dev.* **2007**, *17*, 139–144.
47. Khalil, A.M.; Guttman, M.; Huarte, M.; Garber, M.; Raj, A.; Rivea Morales, D.; Thomas, K.; Presser, A.; Bernstein, B.E.; van Oudenaarden, A.; *et al.* Many human large intergenic noncoding RNAs associate with chromatin-modifying complexes and affect gene expression. *Proc. Natl. Acad. Sci. USA* **2009**, *106*, 11667–11672.
48. Wang, K.C.; Chang, H.Y. Molecular mechanisms of long noncoding RNAs. *Mol. Cell* **2011**, *43*, 904–914.
49. Rinn, J.L.; Kertesz, M.; Wang, J.K.; Squazzo, S.L.; Xu, X.; Brugmann, S.A.; Goodnough, L.H.; Helms, J.A.; Farnham, P.J.; Segal, E.; *et al.* Functional demarcation of active and silent chromatin domains in human HOX loci by noncoding RNAs. *Cell* **2007**, *129*, 1311–1323.
50. Beltran, M.; Puig, I.; Peña, C.; García, J.M.; Alvarez, A.B.; Peña, R.; Bonilla, F.; de Herreros, A.G. A natural antisense transcript regulates Zeb2/Sip1 gene expression during Snail1-induced epithelial-mesenchymal transition. *Genes Dev.* **2008**, *22*, 756–769.
51. Tripathi, V.; Ellis, J.D.; Shen, Z.; Song, D.Y.; Pan, Q.; Watt, A.T.; Freier, S.M.; Bennett, C.F.; Sharma, A.; Bubulya, P.A.; *et al.* The nuclear-retained noncoding RNA MALAT1 regulates alternative splicing by modulating SR splicing factor phosphorylation. *Mol. Cell* **2010**, *39*, 925–938.
52. Salmena, L.; Poliseno, L.; Tay, Y.; Kats, L.; Pandolfi, P.P. A ceRNA Hypothesis: The rosetta stone of a hidden RNA language? *Cell* **2011**, *146*, 353–358.
53. Cesana, M.; Cacchiarelli, D.; Legnini, I.; Santini, T.; Sthandier, O.; Chinappi, M.; Tramontano, A.; Bozzoni, I. A long noncoding RNA controls muscle differentiation by functioning as a competing endogenous RNA. *Cell* **2011**, *147*, 358–369.
54. Houbaviy, H.B.; Murray, M.F.; Sharp, P.A. Embryonic stem cell-specific microRNAs. *Dev. Cell* **2003**, *5*, 351–358.
55. Laurent, L.C.; Chen, J.; Ulitsky, I.; Mueller, F.-J.; Lu, C.; Shamir, R.; Fan, J.-B.; Loring, J.F. Comprehensive microRNA profiling reveals a unique human embryonic stem cell signature dominated by a single seed sequence. *Stem Cells* **2008**, *26*, 1506–1516.
56. Bar, M.; Wyman, S.K.; Fritz, B.R.; Qi, J.; Garg, K.S.; Parkin, R.K.; Kroh, E.M.; Bendoraite, A.; Mitchell, P.S.; Nelson, A.M.; *et al.* MicroRNA discovery and profiling in human embryonic stem cells by deep sequencing of small RNA libraries. *Stem Cells* **2008**, *26*, 2496–2505.
57. Morin, R.D.; O'Connor, M.D.; Griffith, M.; Kuchenbauer, F.; Delaney, A.; Prabhu, A.L.; Zhao, Y.; McDonald, H.; Zeng, T.; Hirst, M.; *et al.* Application of massively parallel sequencing to microRNA profiling and discovery in human embryonic stem cells. *Genome Res.* **2008**, *18*, 610–621.
58. Ciaudo, C.; Servant, N.; Cognat, V.; Sarazin, A.; Kieffer, E.; Viville, S.; Colot, V.; Barillot, E.; Heard, E.; Voinnet, O. Highly dynamic and sex-specific expression of microRNAs during early ES cell differentiation. *PLoS Genet.* **2009**, *5*, e1000620.
59. Landgraf, P.; Rusu, M.; Sheridan, R.; Sewer, A.; Iovino, N.; Aravin, A.; Pfeffer, S.; Rice, A.; Kamphorst, A.O.; Landthaler, M.; *et al.* A mammalian microRNA expression atlas based on small RNA library sequencing. *Cell* **2007**, *129*, 1401–1414.
60. Calabrese, J.M.; Seila, A.C.; Yeo, G.W.; Sharp, P.A. RNA sequence analysis defines Dicer's role in mouse embryonic stem cells. *Proc. Natl. Acad. Sci. USA* **2007**, *104*, 18097–18102.

61. Lipchina, I.; Elkabetz, Y.; Hafner, M.; Sheridan, R.; Mihailovic, A.; Tuschl, T.; Sander, C.; Studer, L.; Betel, D. Genome-wide identification of microRNA targets in human ES cells reveals a role for miR-302 in modulating BMP response. *Genes Dev.* **2011**, *25*, 2173–2186.
62. Barroso-delJesus, A.; Romero-López, C.; Lucena-Aguilar, G.; Melen, G.J.; Sanchez, L.; Ligero, G.; Berzal-Herranz, A.; Menendez, P. Embryonic stem cell-specific miR302-367 cluster: Human gene structure and functional characterization of its core promoter. *Mol. Cell Biol.* **2008**, *28*, 6609–6619.
63. Card, D.A.G.; Hebbar, P.B.; Li, L.; Trotter, K.W.; Komatsu, Y.; Mishina, Y.; Archer, T.K. Oct4/Sox2-regulated miR-302 targets cyclin D1 in human embryonic stem cells. *Mol. Cell. Biol.* **2008**, *28*, 6426–6438.
64. Jouneau, A.; Ciaudo, C.; Sismeiro, O.; Brochard, V.; Jouneau, L.; Vandormael-Pournin, S.; Coppée, J.-Y.; Zhou, Q.; Heard, E.; Antoniewski, C.; *et al.* Naive and primed murine pluripotent stem cells have distinct miRNA expression profiles. *RNA* **2012**, *18*, 253–264.
65. Stadler, B.; Ivanovska, I.; Mehta, K.; Song, S.; Nelson, A.; Tan, Y.; Mathieu, J.; Darby, C.; Blau, C.A.; Ware, C.; *et al.* Characterization of microRNAs involved in embryonic stem cell states. *Stem Cells Dev.* **2010**, *19*, 935–950.
66. Hanna, J.; Cheng, A.W.; Saha, K.; Kim, J.; Lengner, C.J.; Soldner, F.; Cassady, J.P.; Muffat, J.; Carey, B.W.; Jaenisch, R. Human embryonic stem cells with biological and epigenetic characteristics similar to those of mouse ESCs. *Proc. Natl. Acad. Sci. USA* **2010**, *107*, 9222–9227.
67. Kim, H.; Lee, G.; Ganat, Y.; Papapetrou, E.P.; Lipchina, I.; Socci, N.D.; Sadelain, M.; Studer, L. miR-371-3 expression predicts neural differentiation propensity in human pluripotent stem cells. *Cell Stem Cell* **2011**, *8*, 695–706.
68. Bernstein, E.; Kim, S.Y.; Carmell, M.A.; Murchison, E.P.; Alcorn, H.; Li, M.Z.; Mills, A.A.; Elledge, S.J.; Anderson, K.V.; Hannon, G.J. Dicer is essential for mouse development. *Nat. Genet.* **2003**, *35*, 215–217.
69. Kanellopoulou, C.; Muljo, S.A.; Kung, A.L.; Ganesan, S.; Drapkin, R.; Jenuwein, T.; Livingston, D.M.; Rajewsky, K. Dicer-deficient mouse embryonic stem cells are defective in differentiation and centromeric silencing. *Genes Dev.* **2005**, *19*, 489–501.
70. Murchison, E.P.; Partridge, J.F.; Tam, O.H.; Cheloufi, S.; Hannon, G.J. Characterization of Dicer-deficient murine embryonic stem cells. *Proc. Natl. Acad. Sci. USA* **2005**, *102*, 12135–12140.
71. Wang, Y.; Medvid, R.; Melton, C.; Jaenisch, R.; Blelloch, R. DGCR8 is essential for microRNA biogenesis and silencing of embryonic stem cell self-renewal. *Nat. Genet.* **2007**, *39*, 380–385.
72. Sinkkonen, L.; Hugenschmidt, T.; Berninger, P.; Gaidatzis, D.; Mohn, F.; Artus-Revel, C.G.; Zavolan, M.; Svoboda, P.; Filipowicz, W. microRNAs control de novo DNA methylation through regulation of transcriptional repressors in mouse embryonic stem cells. *Nat. Struct. Mol. Biol.* **2008**, *15*, 259–267.
73. Wang, Y.; Baskerville, S.; Shenoy, A.; Babiarz, J.E.; Baehner, L.; Blelloch, R. Embryonic stem cell-specific microRNAs regulate the G1-S transition and promote rapid proliferation. *Nat. Genet.* **2008**, *40*, 1478–1483.
74. Wang, Y.; Blelloch, R. Cell cycle regulation by microRNAs in embryonic stem cells. *Cancer Res.* **2009**, *69*, 4093–4096.

75. Benetti, R.; Gonzalo, S.; Jaco, I.; Muñoz, P.; Gonzalez, S.; Schoeftner, S.; Murchison, E.; Andl, T.; Chen, T.; Klatt, P.; *et al.* A mammalian microRNA cluster controls DNA methylation and telomere recombination via Rbl2-dependent regulation of DNA methyltransferases. *Nat. Struct. Mol. Biol.* **2008**, *15*, 268–279.
76. Kim, B.-M.; Choi, M.Y. Non-canonical microRNAs miR-320 and miR-702 promote proliferation in Dgcr8-deficient embryonic stem cells. *Biochem. Biophys. Res. Commun.* **2012**, *426*, 183–189.
77. Babiarz, J.E.; Ruby, J.G.; Wang, Y.; Bartel, D.P.; Blelloch, R. Mouse ES cells express endogenous shRNAs, siRNAs, and other Microprocessor-independent, Dicer-dependent small RNAs. *Genes Dev.* **2008**, *22*, 2773–2785.
78. White, J.; Dalton, S. Cell cycle control of embryonic stem cells. *Stem Cell Rev.* **2005**, *1*, 131–138.
79. Qi, J.; Yu, J.-Y.; Shcherbata, H.R.; Mathieu, J.; Wang, A.J.; Seal, S.; Zhou, W.; Stadler, B.M.; Bourgin, D.; Wang, L.; *et al.* microRNAs regulate human embryonic stem cell division. *Cell Cycle* **2009**, *8*, 3729–3741.
80. Dolezalova, D.; Mraz, M.; Barta, T.; Plevova, K.; Vinarsky, V.; Holubcova, Z.; Jaros, J.; Dvorak, P.; Pospisilova, S.; Hampl, A. microRNAs regulate p21(Waf1/Cip1) protein expression and the DNA damage response in human embryonic stem cells. *Stem Cells* **2012**, *30*, 1362–1372.
81. Sengupta, S.; Nie, J.; Wagner, R.J.; Yang, C.; Stewart, R.; Thomson, J.A. microRNA 92b controls the G1/S checkpoint gene p57 in human embryonic stem cells. *Stem Cells* **2009**, *27*, 1524–1528.
82. Lee, N.S.; Kim, J.S.; Cho, W.J.; Lee, M.R.; Steiner, R.; Gompers, A.; Ling, D.; Zhang, J.; Strom, P.; Behlke, M.; *et al.* miR-302b maintains “stemness” of human embryonal carcinoma cells by post-transcriptional regulation of Cyclin D2 expression. *Biochem. Biophys. Res. Commun.* **2008**, *377*, 434–440.
83. Hafner, M.; Landthaler, M.; Burger, L.; Khorshid, M.; Hausser, J.; Berninger, P.; Rothballer, A.; Ascano, M.; Jungkamp, A.-C.; Munschauer, M.; *et al.* Transcriptome-wide identification of RNA-binding protein and microRNA target sites by PAR-CLIP. *Cell* **2010**, *141*, 129–141.
84. Lipchina, I.; Studer, L.; Betel, D. The expanding role of miR-302-367 in pluripotency and reprogramming. *Cell Cycle* **2012**, *11*, 1517–1523.
85. Smith, J.R.; Vallier, L.; Lupo, G.; Alexander, M.; Harris, W.A.; Pedersen, R.A. Inhibition of Activin/Nodal signaling promotes specification of human embryonic stem cells into neuroectoderm. *Dev. Biol.* **2008**, *313*, 107–117.
86. Vallier, L.; Reynolds, D.; Pedersen, R.A. Nodal inhibits differentiation of human embryonic stem cells along the neuroectodermal default pathway. *Dev. Biol.* **2004**, *275*, 403–421.
87. Rosa, A.; Spagnoli, F.M.; Brivanlou, A.H. The miR-430/427/302 family controls mesendodermal fate specification via species-specific target selection. *Dev. Cell* **2009**, *16*, 517–527.
88. Choi, W.-Y.; Giraldez, A.J.; Schier, A.F. Target protectors reveal dampening and balancing of Nodal agonist and antagonist by miR-430. *Science* **2007**, *318*, 271–274.
89. Chambers, S.M.; Fasano, C.A.; Papapetrou, E.P.; Tomishima, M.; Sadelain, M.; Studer, L. Highly efficient neural conversion of human ES and iPS cells by dual inhibition of SMAD signaling. *Nat. Biotechnol.* **2009**, *27*, 275–280.

90. Ozair, M.Z.; Noggle, S.; Warmflash, A.; Krzyspiak, J.E.; Brivanlou, A.H. SMAD7 Directly converts human embryonic stem cells to telencephalic fate by a default mechanism. *Stem Cells* **2013**, *31*, 35–47.
91. Kang, H.; Louie, J.; Weisman, A.; Sheu-Gruttadauria, J.; Davis-Dusenbery, B.N.; Lagna, G.; Hata, A. Inhibition of microRNA-302 (miR-302) by bone morphogenetic protein 4 (BMP4) facilitates the BMP signaling pathway. *J. Biol. Chem.* **2012**, *287*, 38656–38664.
92. Parisi, S.; Battista, M.; Musto, A.; Navarra, A.; Tarantino, C.; Russo, T. A regulatory loop involving Dies1 and miR-125a controls BMP4 signaling in mouse embryonic stem cells. *FASEB J.* **2012**, *26*, 3957–3968.
93. Zhou, A.-D.; Diao, L.-T.; Xu, H.; Xiao, Z.-D.; Li, J.-H.; Zhou, H.; Qu, L.-H. β-Catenin/LEF1 transactivates the microRNA-371-373 cluster that modulates the Wnt/β-catenin-signaling pathway. *Oncogene* **2012**, *31*, 2968–2978.
94. Rosa, A.; Brivanlou, A.H. A regulatory circuitry comprised of miR-302 and the transcription factors OCT4 and NR2F2 regulates human embryonic stem cell differentiation. *EMBO J.* **2010**, *30*, 237–248.
95. Tay, Y.; Zhang, J.; Thomson, A.M.; Lim, B.; Rigoutsos, I. microRNAs to Nanog, Oct4 and Sox2 coding regions modulate embryonic stem cell differentiation. *Nature* **2008**, *455*, 1124–1128.
96. Wellner, U.; Schubert, J.; Burk, U.C.; Schmalhofer, O.; Zhu, F.; Sonntag, A.; Waldvogel, B.; Vannier, C.; Darling, D.; zur Hausen, A.; *et al.* The EMT-activator ZEB1 promotes tumorigenicity by repressing stemness-inhibiting microRNAs. *Nat. Cell Biol.* **2009**, *11*, 1487–1495.
97. Yi, R.; Poy, M.N.; Stoffel, M.; Fuchs, E. A skin microRNA promotes differentiation by repressing "stemness". *Nature* **2008**, *452*, 225–229.
98. Xu, N.; Papagiannakopoulos, T.; Pan, G.; Thomson, J.; Kosik, K. microRNA-145 regulates OCT4, SOX2, and KLF4 and represses pluripotency in human embryonic stem cells. *Cell* **2009**, *137*, 647–658.
99. Morey, L.; Pascual, G.; Cozzuto, L.; Roma, G.; Wutz, A.; Benitah, S.A.; Di Croce, L. Nonoverlapping functions of the polycomb group cbx family of proteins in embryonic stem cells. *Cell Stem Cell* **2012**, *10*, 47–62.
100. O'Loghlen, A.; Muñoz-Cabello, A.M.; Gaspar-Maia, A.; Wu, H.-A.; Banito, A.; Kunowska, N.; Racek, T.; Pemberton, H.N.; Beolchi, P.; Lavial, F.; *et al.* MicroRNA regulation of Cbx7 mediates a switch of polycomb orthologs during ESC differentiation. *Cell Stem Cell* **2012**, *10*, 33–46.
101. Ambros, V. The evolution of our thinking about microRNAs. *Nat. Med.* **2008**, *14*, 1036–1040.
102. Büssing, I.; Slack, F.J.; Großhans, H. let-7 microRNAs in development, stem cells and cancer. *Trends Mol. Med.* **2008**, *14*, 400–409.
103. Thomson, J.M.; Newman, M.; Parker, J.S.; Morin-Kensicki, E.M.; Wright, T.; Hammond, S.M. Extensive post-transcriptional regulation of microRNAs and its implications for cancer. *Genes Dev.* **2006**, *20*, 2202–2207.
104. Viswanathan, S.R.; Daley, G.Q.; Gregory, R.I. Selective blockade of microRNA processing by Lin28. *Science* **2008**, *320*, 97–100.
105. Viswanathan, S.R.; Daley, G.Q. Lin28: A microRNA regulator with a macro role. *Cell* **2010**, *140*, 445–449.

106. Nam, Y.; Chen, C.; Gregory, R.I.; Chou, J.J.; Sliz, P. Molecular basis for interaction of let-7 microRNAs with Lin28. *Cell* **2011**, *147*, 1080–1091.
107. Heo, I.; Joo, C.; Kim, Y.-K.; Ha, M.; Yoon, M.-J.; Cho, J.; Yeom, K.-H.; Han, J.; Kim, V.N. TUT4 in concert with Lin28 suppresses microRNA biogenesis through pre-microRNA uridylation. *Cell* **2009**, *138*, 696–708.
108. Hagan, J.P.; Piskounova, E.; Gregory, R.I. Lin28 recruits the TUTase Zcchc11 to inhibit let-7 maturation in mouse embryonic stem cells. *Nat. Struct. Mol. Biol.* **2009**, *16*, 1021–1025.
109. Rybak, A.; Fuchs, H.; Smirnova, L.; Brandt, C.; Pohl, E.E.; Nitsch, R.; Wulczyn, F.G. A feedback loop comprising lin-28 and let-7 controls pre-let-7 maturation during neural stem-cell commitment. *Nat. Cell Biol.* **2008**, *10*, 987–993.
110. Melton, C.; Judson, R.L.; Blelloch, R. Opposing microRNA families regulate self-renewal in mouse embryonic stem cells. *Nature* **2010**, *463*, 621–626.
111. Cartwright, P.; McLean, C.; Sheppard, A.; Rivett, D.; Jones, K.; Dalton, S. LIF/STAT3 controls ES cell self-renewal and pluripotency by a Myc-dependent mechanism. *Development* **2005**, *132*, 885–896.
112. Lin, C.-H.; Jackson, A.L.; Guo, J.; Linsley, P.S.; Eisenman, R.N. Myc-regulated microRNAs attenuate embryonic stem cell differentiation. *EMBO J.* **2009**, *28*, 3157–3170.
113. Chang, T.-C.; Yu, D.; Lee, Y.-S.; Wentzel, E.A.; Arking, D.E.; West, K.M.; Dang, C.V.; Thomas-Tikhonenko, A.; Mendell, J.T. Widespread microRNA repression by Myc contributes to tumorigenesis. *Nat. Genet.* **2008**, *40*, 43–50.
114. Chin, M.H.; Mason, M.J.; Xie, W.; Volinia, S.; Singer, M.; Peterson, C.; Ambartsumyan, G.; Aimiuwu, O.; Richter, L.; Zhang, J.; *et al.* Induced pluripotent stem cells and embryonic stem cells are distinguished by gene expression signatures. *Cell Stem Cell* **2009**, *5*, 111–123.
115. Wilson, K.D.; Venkatasubrahmanyam, S.; Jia, F.; Sun, N.; Butte, A.J.; Wu, J.C. microRNA profiling of human-induced pluripotent stem cells. *Stem Cells Dev.* **2009**, *18*, 749–757.
116. Neveu, P.; Kye, M.J.; Qi, S.; Buchholz, D.E.; Clegg, D.O.; Sahin, M.; Park, I.-H.; Kim, K.-S.; Daley, G.Q.; Kornblum, H.I.; *et al.* MicroRNA profiling reveals two distinct p53-related human pluripotent stem cell states. *Cell Stem Cell* **2010**, *7*, 671–681.
117. Stadtfeld, M.; Apostolou, E.; Akutsu, H.; Fukuda, A.; Follett, P.; Natesan, S.; Kono, T.; Shioda, T.; Hochedlinger, K. Aberrant silencing of imprinted genes on chromosome 12qF1 in mouse induced pluripotent stem cells. *Nature* **2010**, *465*, 175–181.
118. Peters, J.; Robson, J.E. Imprinted noncoding RNAs. *Mamm. Genome* **2008**, *19*, 493–502.
119. Li, Z.; Yang, C.-S.; Nakashima, K.; Rana, T.M. Small RNA-mediated regulation of iPS cell generation. *EMBO J.* **2011**, *30*, 823–834.
120. Judson, R.; Babiarz, J.; Venere, M.; Blelloch, R. Embryonic stem cell-specific microRNAs promote induced pluripotency. *Nat. Biotechnol.* **2009**, *27*, 459–461.
121. Liao, B.; Bao, X.; Liu, L.; Feng, S.; Zovoilis, A.; Liu, W.; Xue, Y.; Cai, J.; Guo, X.; Qin, B.; *et al.* MicroRNA cluster 302-367 enhances somatic cell reprogramming by accelerating a mesenchymal-to-epithelial transition. *J. Biol. Chem.* **2011**, *286*, 17359–17364.
122. Ye, D.; Wang, G.; Liu, Y.; Huang, W.; Wu, M.; Zhu, S.; Jia, W.; Deng, A.-M.; Liu, H.; Kang, J. MiR-138 promotes induced pluripotent stem cell generation through the regulation of the p53 signaling. *Stem Cells* **2012**, *30*, 1645–1654.

123. Subramanyam, D.; Lamouille, S.; Judson, R.L.; Liu, J.Y.; Bucay, N.; Derynck, R.; Blelloch, R. Multiple targets of miR-302 and miR-372 promote reprogramming of human fibroblasts to induced pluripotent stem cells. *Nat. Biotechnol.* **2011**, *29*, 443–448.
124. Hanna, J.; Saha, K.; Pando, B.; van Zon, J.; Lengner, C.J.; Creyghton, M.P.; van Oudenaarden, A.; Jaenisch, R. Direct cell reprogramming is a stochastic process amenable to acceleration. *Nature* **2009**, *462*, 595–601.
125. Yang, C.-S.; Li, Z.; Rana, T.M. microRNAs modulate iPS cell generation. *RNA* **2011**, *17*, 1451–1460.
126. Wang, J.; He, Q.; Han, C.; Gu, H.; Jin, L.; Li, Q.; Mei, Y.; Wu, M. p53-facilitated miR-199a-3p regulates somatic cell reprogramming. *Stem Cells* **2012**, *30*, 1405–1413.
127. Anokye-Danso, F.; Trivedi, C.M.; Juhr, D.; Gupta, M.; Cui, Z.; Tian, Y.; Zhang, Y.; Yang, W.; Gruber, P.J.; Epstein, J.A.; Morrisey, E.E. Highly efficient miRNA-mediated reprogramming of mouse and human somatic cells to pluripotency. *Cell Stem Cell* **2011**, *8*, 376–388.
128. Miyoshi, N.; Ishii, H.; Nagano, H.; Haraguchi, N.; Dewi, D.L.; Kano, Y.; Nishikawa, S.; Tanemura, M.; Mimori, K.; Tanaka, F.; *et al.* Reprogramming of mouse and human cells to pluripotency using mature microRNAs. *Cell Stem Cell* **2011**, *8*, 633–638.
129. Lin, S.-L.; Chang, D.C.; Chang-Lin, S.; Lin, C.-H.; Wu, D.T.S.; Chen, D.T.; Ying, S.-Y. Mir-302 reprograms human skin cancer cells into a pluripotent ES-cell-like state. *RNA* **2008**, *14*, 2115–2124.
130. Lin, S.-L.; Chang, D.C.; Lin, C.-H.; Ying, S.-Y.; Leu, D.; Wu, D.T.S. Regulation of somatic cell reprogramming through inducible mir-302 expression. *Nucleic Acids Res.* **2011**, *39*, 1054–1065.
131. Menendez, S.; Camus, S.; Izpisúa Belmonte, J.C. p53: Guardian of reprogramming. *Cell Cycle* **2010**, *9*, 3887–3891.
132. Jain, A.K.; Allton, K.; Iacovino, M.; Mahen, E.; Milczarek, R.J.; Zwaka, T.P.; Kyba, M.; Barton, M.C. p53 regulates cell cycle and microRNAs to promote differentiation of human embryonic stem cells. *PLoS Biol.* **2012**, *10*, e1001268.
133. Li, R.; Liang, J.; Ni, S.; Zhou, T.; Qing, X.; Li, H.; He, W.; Chen, J.; Li, F.; Zhuang, Q.; *et al.* A mesenchymal-to-epithelial transition initiates and is required for the nuclear reprogramming of mouse fibroblasts. *Cell Stem Cell* **2010**, *7*, 51–63.
134. Samavarchi-Tehrani, P.; Golipour, A.; David, L.; Sung, H.-K.; Beyer, T.A.; Datti, A.; Woltjen, K.; Nagy, A.; Wrana, J.L. Functional genomics reveals a BMP-driven mesenchymal-to-epithelial transition in the initiation of somatic cell reprogramming. *Cell Stem Cell* **2010**, *7*, 64–77.
135. Massagué, J. TGFβ signalling in context. *Nat. Rev. Mol. Cell Biol.* **2012**, *13*, 616–630.
136. Gregory, P.A.; Bert, A.G.; Paterson, E.L.; Barry, S.C.; Tsykin, A.; Farshid, G.; Vadas, M.A.; Khew-Goodall, Y.; Goodall, G.J. The miR-200 family and miR-205 regulate epithelial to mesenchymal transition by targeting ZEB1 and SIP1. *Nat. Cell Biol.* **2008**, *10*, 593–601.
137. Burk, U.; Schubert, J.; Wellner, U.; Schmalhofer, O.; Vincan, E.; Spaderna, S.; Brabletz, T. A reciprocal repression between ZEB1 and members of the miR-200 family promotes EMT and invasion in cancer cells. *EMBO Rep.* **2008**, *9*, 582–589.
138. Bracken, C.P.; Gregory, P.A.; Kolesnikoff, N.; Bert, A.G.; Wang, J.; Shannon, M.F.; Goodall, G.J. A double-negative feedback loop between ZEB1-SIP1 and the microRNA-200 family regulates epithelial-mesenchymal transition. *Cancer Res.* **2008**, *68*, 7846–7854.

139. Barroso-delJesus, A.; Lucena-Aguilar, G.; Sanchez, L.; Ligero, G.; Gutierrez-Aranda, I.; Menendez, P. The Nodal inhibitor Lefty is negatively modulated by the microRNA miR-302 in human embryonic stem cells. *FASEB J.* **2011**, *25*, 1497–1508.
140. Lee, M.R.; Prasain, N.; Chae, H.-D.; Kim, Y.-J.; Mantel, C.; Yoder, M.C.; Broxmeyer, H.E. Epigenetic regulation of nanog by MiR-302 cluster-MBD2 completes induced pluripotent stem cell reprogramming. *Stem Cells* **2012**, *31*, 666–681.
141. Hu, S.; Wilson, K.D.; Ghosh, Z.; Han, L.; Wang, Y.; Lan, F.; Ransohoff, K.J.; Burridge, P.; Wu, J.C. microRNA-302 increases reprogramming efficiency via repression of NR2F2. *Stem Cells* **2013**, *31*, 259–268.
142. Guttman, M.; Amit, I.; Garber, M.; French, C.; Lin, M.F.; Feldser, D.; Huarte, M.; Zuk, O.; Carey, B.W.; Cassady, J.P.; *et al.* Chromatin signature reveals over a thousand highly conserved large non-coding RNAs in mammals. *Nature* **2009**, *458*, 223–227.
143. Sheik Mohamed, J.; Gaughwin, P.M.; Lim, B.; Robson, P.; Lipovich, L. Conserved long noncoding RNAs transcriptionally regulated by Oct4 and Nanog modulate pluripotency in mouse embryonic stem cells. *RNA* **2010**, *16*, 324–337.
144. Guttman, M.; Donaghey, J.; Carey, B.W.; Garber, M.; Grenier, J.K.; Munson, G.; Young, G.; Lucas, A.B.; Ach, R.; Bruhn, L.; *et al.* lincRNAs act in the circuitry controlling pluripotency and differentiation. *Nature* **2011**, *477*, 295–300.
145. Ng, S.-Y.; Johnson, R.; Stanton, L.W. Human long non-coding RNAs promote pluripotency and neuronal differentiation by association with chromatin modifiers and transcription factors. *EMBO J.* **2011**, *31*, 522–533.
146. Loewer, S.; Cabili, M.N.; Guttman, M.; Loh, Y.-H.; Thomas, K.; Park, I.H.; Garber, M.; Curran, M.; Onder, T.; Agarwal, S.; *et al.* Large intergenic non-coding RNA-RoR modulates reprogramming of human induced pluripotent stem cells. *Nat. Genet.* **2010**, *42*, 1113–1117.
147. Hawkins, P.G.; Morris, K.V. Transcriptional regulation of Oct4 by a long non-coding RNA antisense to Oct4-pseudogene 5. *Transcription* **2010**, *1*, 165–175.
148. Tsai, M.-C.; Manor, O.; Wan, Y.; Mosammaparast, N.; Wang, J.K.; Lan, F.; Shi, Y.; Segal, E.; Chang, H.Y. Long noncoding RNA as modular scaffold of histone modification complexes. *Science* **2010**, *329*, 689–693.
149. Zhang, A.; Zhou, N.; Huang, J.; Liu, Q.; Fukuda, K.; Ma, D.; Lu, Z.; Bai, C.; Watabe, K.; Mo, Y.-Y. The human long non-coding RNA-RoR is a p53 repressor in response to DNA damage. *Cell Res.* **2012**, *23*, 340–350.
150. Wang, Y.; Xu, Z.; Jiang, J.; Xu, C.; Kang, J.; Xiao, L.; Wu, M.; Xiong, J.; Guo, X.; Liu, H. Endogenous miRNA sponge lincRNA-RoR regulates Oct4, nanog, and Sox2 in human embryonic stem cell self-renewal. *Dev. Cell* **2013**, *25*, 69–80.
151. Yoo, A.S.; Sun, A.X.; Li, L.; Shcheglovitov, A.; Portmann, T.; Li, Y.; Lee-Messer, C.; Dolmetsch, R.E.; Tsien, R.W.; Crabtree, G.R. microRNA-mediated conversion of human fibroblasts to neurons. *Nature* **2011**, *476*, 228–231.
152. Guttman, M.; Rinn, J.L. Modular regulatory principles of large non-coding RNAs. *Nature* **2012**, *482*, 339–346.

2

Regulation of *MIR* Genes in Response to Abiotic Stress in *Hevea brasiliensis*

Virginie Gébelin [1,†], Julie Leclercq [1,†], Songnian Hu [2], Chaorong Tang [3] and Pascal Montoro [1,*]

[1] Centre de Coopération Internationale en Recherche Agronomique Pour le Développement, Unité Mixte de Recherche Amélioration Génétique et Adaptation des Plantes Méditerranéennes et Tropicales, Avenue Agropolis, Montpellier F-34398, France; E-Mails: gebelin.v@gmail.com (V.G.); julie.leclercq@cirad.fr (J.L.)

[2] Beijing Institute of Genomics, Chinese Academy of Sciences, Beijing 100029, China; E-Mail: husn@big.ac.cn

[3] Rubber Research Institute, Chinese Academy of Tropical Agricultural Sciences, Danzhou 531737, China; E-Mail: chaorongtang@126.com

† These authors contributed equally to this work.

* Author to whom correspondence should be addressed; E-Mail: pascal.montoro@cirad.fr

---

Abstract: Increasing demand for natural rubber (NR) calls for an increase in latex yield and also an extension of rubber plantations in marginal zones. Both harvesting and abiotic stresses lead to tapping panel dryness through the production of reactive oxygen species. Many microRNAs regulated during abiotic stress modulate growth and development. The objective of this paper was to study the regulation of microRNAs in response to different types of abiotic stress and hormone treatments in *Hevea*. Regulation of *MIR* genes differs depending on the tissue and abiotic stress applied. A negative co-regulation between *HbMIR398b* with its *chloroplastic HbCuZnSOD* target messenger is observed in response to salinity. The involvement of *MIR* gene regulation during latex harvesting and tapping panel dryness (TPD) occurrence is further discussed.

Keywords: gene expression; miRNA; *MIR* gene; abiotic stress; rubber tree; tapping panel dryness

---

1. Introduction

Hevea brasiliensis is the only source of natural rubber (NR), which is produced in latex cells in the phloem. Latex harvesting consists in tapping the soft bark of rubber trees. Ethephon is applied to stimulate production. Increasing demand for NR calls for an increase in latex yield, but also an extension of crops to marginal zones. Such marginal zones are subject to stronger and more detrimental abiotic constraints for latex production (positive cold, frost-prone, drought, *etc.*). Exposure to abiotic stress in addition to latex harvesting stress affects latex production and tree productivity [1]. Laticifer cells are the site of numerous types of stress that lead to oxidative stress. The plant's first response to wounding is to produce jasmonate [2]. The wound and associated jasmonate production also cause oxidative stress in the plant and activate antioxidant systems to overcome it [3]. Oxidative stress, whether produced in response to tapping and/or environmental constraints, may lie behind problems linked to latex flow [4]. It is reflected in an imbalance between ROS (reactive oxygen species) and the detoxification systems existing in laticifers. ROS then attack the membranes of cell organelles, causing the release of coagulant factors existing in the lutoids, which gives rise to *in situ* coagulation of rubber particles. This complex physiological disorder is called TPD (Tapping Panel Dryness). This phenomenon remains reversible up to a certain limit. It amounts to laticifer fatigue, called dry cut. In extreme cases, halted flow is followed by the degenerescence and death of the laticifers and, consequently, severe cortical necrosis (Brown Bast) [5]. Annual losses due to TPD are estimated at 10% to 40% [1].

MicroRNAs (miRNAs) form a small class of non-coding RNAs of 21 to 24 nucleotides involved in numerous physiological processes in plants including response to biotic and abiotic stresses through post-transcriptional regulation. miRNA biogenesis requires several stages and several enzymes from *MIR* gene transcription to the generation of mature miRNA [6–10]. The mature miRNA enters the complex called RISC (RNA-induced silencing complex) thereby enabling regulation of the target gene. In the RISC complex, the mature miRNA associates with proteins of the Argonaute family (AGO) to direct the regulation of the target gene [11,12]. In the RISC complex, the mature miRNA sequence links up with the sequence of the messenger RNA through perfect or imperfect sequence homology. This action causes a silencing phenomenon in plants, called PTGS (post-transcriptional gene silencing) [13]. In plants, most target messenger RNAs contain a single site complementary to the miRNA, corresponding to the cleavage site [9], suggesting that plant microRNAs use RNA cleavage rather than preventing translation. Long microRNAs are taken in charge by Dicer-like 3 (DCL3), associate with AGO4 and then guide methylation of the DNA of target genes in rice and mosses [14,15].

Many microRNAs regulated by abiotic stress have been identified in the model species (for review, see [16]). The first involvements of microRNAs in response to stress were described by Jones Rhoades and Bartel who, in *Arabidopsis thaliana*, predicted genes targeted by microRNAs such as superoxide dismutase, the laccases and ATP sulfurylases (APS) [17]. MicroRNA expression was then studied in different contexts of abiotic stress in numerous species (for review, see [16]). Most microRNAs conserved between species target transcription factors, such as miR/target gene pairs *miR156/SBP*, *miR159/319/MYB-TCP*, *miR160/ARF*, *miR166/HDZIPIII*, *miR169/NFY* subunit.

MicroRNAs are also involved in the regulation of antioxidant activities and particularly in the regulation of reactive oxygen species-scavenging enzymes. The production of ROS is linked to abiotic and biotic stress status in plants [18–20] and the excess ROS can lead to several types of cell damage, such as peroxidation of membrane lipid compounds, degradation of polysaccharides, denaturing of enzymes and lesions in DNA [21]. The miR398 targets the *CSD1* and *CSD2* genes encoding cytosolic and chloroplastic CuZnSOD in *Arabidopsis thaliana* [22]. The level of expression of these microRNAs and of their target is regulated in response to stress, involving a modulation of growth and development during stress. For example, miR398 is under-expressed in response to stress which leads to an accumulation of CuZnSOD enabling detoxification of an excess of reactive oxygen species in *Arabidopsis thaliana* [22].

To date, 28 sequences of *MIR* are accessible in miRBase [23–25] for the rubber tree. They represent 10 conserved families and 14 newly identified families potentially *Hevea*-specific miR [26,27], added to which four additional precursors have been identified recently [28]. Our previous studies on mature trees demonstrated the functionality of miRNA biogenesis in the latex cells. Moreover, we highlighted the reduction of small RNA size from 24 to 21 nucleotides in TPD-affected trees, which could not be explained by a general RNA degradation [28]. From this previous work, we hypothesized that the change in size could also be effective for mature miRNAs and consequently involved to independent miRNA pathway, one leading to targeted transcript cleavage and the other to methylation of targeted genes (for review [29]). Among all pre-miRNAs tested, only the relative *Hbpre-MIR159b* abundance was up-regulated between TPD-affected and healthy trees. Targets of miR159 were predicted and might be involved in various activities such as rubber biosynthesis, antioxidant activity and transcription regulation activity [26,28,30].

The objective of this paper was to study the regulation of microRNAs in response to different types of abiotic stress (cold, saline stress, wounding) or hormone treatments (ethylene and methyl jasmonate). Given some mature miRNAs could be produced by several *MIR* genes, this study focused on the expression analysis pattern using real-time RT-PCR with specific primers for each *MIR* genes. The relative accumulation of 19 premiRs was analyzed by real-time RT-PCR on various tissues from juvenile plant material in order to facilitate the application of treatments. Negative co-regulation of *Hbpre-miR398b* and its target *chloroplastic CuZnSOD* is revealed in response to saline stress.

2. Results and Discussion

2.1. Relative Accumulation of premiR in Response to Abiotic Stress in in Vitro *Plantlets*

Relative accumulation of premiR of 20 selected *MIR* genes was analysed in different tissues (leaf, bark and root; Table 1). In control plants, the conserved *MIR* genes were generally very weakly expressed such as *HbMIR156*, *HbMIR319* and *HbMIR398a* and *HbMIR398b* (from 6.78×10^{-4} in roots for *HbMIR398a* to 9.22×10^{-2} for *HbMIR319* in bark). By contrast, the *HbMIR408a* and *HbMIR408b* genes were strongly expressed (from 8.47×10^{-1} to 3.54). Among the *MIR* genes newly identified in *Hevea brasiliensis*, *HbMIR6482*, *HbMIR6483* and *HbMIR6485* were also strongly expressed (from 2.20 for *HbMIR6482* to 7.44×10 for *HbMIR6485* in leaves).

In the same *in vitro* plantlets, statistical analyses revealed a difference in gene expression at tissue level (Table 1). Indeed, three genes *HbMIR166b*, *HbMIR319* and *HbMIR6482* had a relative differential expression level in leaves, bark and roots under untreated conditions. *HbMIR166b* and

HbMIR319 were more strongly expressed in bark while *HbMIR6482* was more strongly expressed in roots. For these three genes, the lowest expression level was found in leaves (Table 1). In addition, seven genes significantly displayed a relative differential expression level, namely *HbMIR159a*, *HbMIR398a*, *HbMIR398b*, *HbMIR408b*, *HbMIR6483*, *HbMIR6484* and *HbMIR6485*. The *MIR398c* transcript was only present in trace form under the majority of conditions.

Table 1. Relative premiR abundance of 13 conserved *MIR* genes and 7 *MIR* genes newly identified in *Hevea brasiliensis* in the leaves, bark and roots of one-year-old *in vitro* plantlets in response to cold and saline (NaCl) treatments compared with non-treated plants (C: control), nc: not calculated. Data with the same letter are not significantly different at the level of 5%.

Gene	Tissue	Expression value for control	Cold/C	*p*-value	NaCl/C	*p*-value
HbpremiR156	leaf	$1.72 \times 10^{-3\ a}$	1.79	0.52	0.17	0.08
	bark	$1.80 \times 10^{-3\ a}$	0.91	0.72	0.20	0.08
	root	$2.63 \times 10^{-3\ a}$	0.70	0.39	0.39	0.13
HbpremiR159a	leaf	$4.68 \times 10^{-2\ b}$	3.45	0.008	0.04	0.002
	bark	$7.45 \times 10^{-2\ a}$	1.25	0.22	0.17	<0.001
	root	$5.94 \times 10^{-2\ a,b}$	0.32	0.02	0.10	0.01
HbpremiR159b	leaf	$1.56 \times 10^{-1\ a}$	4.06	0.004	0.05	0.003
	bark	$2.21 \times 10^{-1\ a}$	1.59	0.03	0.23	<0.0001
	root	$1.94 \times 10^{-1\ a}$	0.44	0.23	0.11	0.02
HbpremiR166a	leaf	$1.19 \times 10^{-2\ a}$	1.38	0.38	2.08	0.08
	bark	$1.78 \times 10^{-2\ a}$	2.00	0.18	0.70	0.16
	root	$4.16 \times 10^{-2\ a}$	0.60	0.62	2.76	0.13
HbpremiR166b	leaf	$2.21 \times 10^{-1\ c}$	1.34	0.31	0.20	0.01
	bark	$5.20 \times 10^{0\ a}$	0.64	0.01	0.14	0.003
	root	$9.67 \times 10^{-1\ b}$	0.48	0.39	1.01	0.80
HbpremiR319	leaf	$5.73 \times 10^{-3\ c}$	0.29	0.08	0.23	0.02
	bark	$9.22 \times 10^{-2\ a}$	1.30	0.90	0.07	<0.0001
	root	$1.76 \times 10^{-2\ b}$	nc	–	nc	–
HbpremiR396	leaf	$4.73 \times 10^{-1\ a}$	1.34	0.28	0.10	0.002
	bark	$1.41 \times 10^{0\ a}$	0.93	0.61	0.17	<0.0001
	root	$2.33 \times 10^{0\ a}$	0.28	0.19	0.12	0.05
HbpremiR398a	leaf	$6.93 \times 10^{-3\ a}$	0.34	0.12	0.21	0.02
	bark	$2.27 \times 10^{-3\ a,b}$	0.23	0.31	0.49	0.64
	root	$6.78 \times 10^{-4\ b}$	0.47	0.39	0.10	0.20
HbpremiR398b	leaf	$3.27 \times 10^{-2\ b}$	0.95	0.92	0.87	0.61
	bark	$8.68 \times 10^{-2\ a}$	0.52	0.15	0.07	0.01
	root	$3.19 \times 10^{-2\ b}$	0.62	0.34	0.15	0.03
HbpremiR398c	leaf	trace	nc	–	nc	–
	bark	trace	nc	–	nc	–
	root	trace	nc	–	nc	–
HbpremiR408a	leaf	$8.47 \times 10^{-1\ a}$	0.47	0.20	0.10	<0.0001
	bark	$1.50 \times 10^{0\ a}$	0.23	0.11	0.07	<0.001
	root	$1.48 \times 10^{0\ a}$	0.22	0.02	0.06	0.01

Table 1. *Cont.*

Gene	Tissue	Expression value for control	Cold/C	*p*-value	NaCl/C	*p*-value
HbpremiR408b	leaf	1.60×10^{0} [b]	0.48	0.21	0.12	<0.0001
	bark	3.54×10^{0} [a]	0.22	0.11	0.08	0.01
	root	2.75×10^{0} [a,b]	0.22	0.01	0.08	0.001
HbpremiR476	leaf	2.71×10^{-2} [a]	0.66	0.39	0.16	0.01
	bark	2.84×10^{-2} [a]	0.36	0.01	0.37	0.002
	root	8.49×10^{-2} [a]	0.20	0.11	0.41	0.23
HbpremiR6482	leaf	2.20×10^{0} [c]	0.09	0.001	0.30	0.06
	bark	9.86×10^{0} [b]	0.21	0.002	0.45	0.01
	root	7.06×10 [a]	0.05	0.004	0.05	0.003
HbpremiR6483	leaf	5.47×10 [a]	0.59	0.50	0.03	0.002
	bark	1.56×10 [a]	0.70	0.43	0.18	0.01
	root	5.54×10^{-1} [b]	1.05	0.72	0.51	0.59
HbpremiR6484	leaf	7.17×10^{-1} [a]	0.70	0.26	0.22	0.003
	bark	7.47×10^{-1} [a]	0.65	0.06	0.55	0.02
	root	3.47×10^{-1} [b]	0.25	0.02	0.35	0.05
HbpremiR6485	leaf	7.44×10 [a]	0.73	0.49	0.03	0.000
	bark	3.14×10 [a]	0.82	0.77	0.22	0.02
	root	6.06×10^{0} [b]	0.73	0.14	0.27	0.01
HbpremiRn11	leaf	trace	nc	–	nc	–
	bark	trace	nc	–	nc	–
	root	trace	nc	–	nc	–
HbpremiRn12	leaf	7.26×10^{-2} [a]	1.25	0.50	6.45	<0.0001
	bark	8.80×10^{-2} [a]	0.75	0.29	5.93	<0.001
	root	1.47×10^{-1} [a]	0.48	0.10	4.17	0.01
HbpremiRn13	leaf	3.77×10^{2} [a]	1.17	0.60	0.34	0.11
	bark	3.41×10^{2} [a]	1.17	0.74	0.46	0.001
	root	2.65×10^{2} [a]	0.81	0.51	1.90	0.06

The regulation of *MIR* genes in response to cold and salinity stresses was shown as the ratio between the mean of three treated *in vitro* plantlets and three untreated *in vitro* plantlets (Table 1). An analysis of the relative expression of the *MIR* genes showed that 8 *MIR* genes were significantly regulated in response to cold (Table 1). In leaves, both *HbMIR159* genes displayed a significant increase in the relative number of transcripts in response to cold while the relative expression of *HbMIR6482* was greatly inhibited. In bark, the relative expression of 3 *MIR* genes (*HbMIR166b*, *HbMIR476* and *HbMIR6482*) was repressed while that of *HbMIR159b* was slightly stimulated. In roots, 5 *MIR* genes (*HbMIR159a*, *HbMIR408a*, *HbMIR408b*, *HbMIR6482* and *HbMIR6484*) displayed a significant reduction in transcripts in response to cold. The drop in temperature led to a drop in transcripts for *HbmiR6482* in the 3 tissues. *HbMIR159a* displayed an expression profile that was antagonistic between leaves and roots. With the exception of *HbMIR159a* and *HbMIR159b*, the significantly regulated *MIR* genes were repressed in response to cold stress.

The *in vitro* plantlets were also subjected to saline stress by watering with a 300 mM NaCl solution rather than water twice a day for 24 h. In response to NaCl, 16 of the 20 genes displayed a significant repression in their level of relative transcript abundance. Only *HbmiRn12* was highly induced (4 to 6 times) in most tissues.

2.2. Relative Accumulation of preMIR in Response to Ethylene (ET), Methyl Jasmonate (MeJA) and Wounding in Three-Month-Old Epicormic Shoots

Based on previous studies on budded plant material [31], the effect of ET, MeJA and wounding was studied on three-month-old epicormic shoots. In control plants, the relative level of premiR was highly accumulated for *HbMIR6485* and *HbMIRn13* (1.60×10 and 3.92×10^2 respectively) and at low level for *HbMIR156*, *HbMIR319* and *HbMIR398a* (4.42×10^{-4}, 1.15×10^{-2} and 3.61×10^{-4}, respectively; Table 2). This result was similar to the basal expression profile found in the *in vitro* plantlets. 7 *MIR* genes had a relative expression level that differed significantly between bark and leaves (Table 2). They were *HbMIR159a*, *HbMIR159b*, *HbMIR166b*, *HbMIR319*, *HbMIR396*, *HbMIR398a* and *HbMIR6485*. In comparison, the aforementioned genes were also regulated differently depending on the tissues in the *in vitro* plantlets, apart from *HbMIR396*. Again, the *MIR398c* transcript was only present in trace form under the majority of conditions.

Table 2. Relative premiR abundance of 13 conserved *MIR* genes and 7 *MIR* genes newly identified in *Hevea brasiliensis* in the leaves and bark collected on 3-month-old epicormic shoot of budded plants in response to ethylene (ET), methyl jasmonate (MeJA) and wounding (W) compared with non-treated plants (C: control), nc: not calculated. Data with the same letter are not significantly different at the level of 5%.

Gene	Tissue	Expression value for control	ET/C	*p*-value	MeJA/C	*p*-value	W/C	*p*-value
HbpremiR156	leaves	$4.42 \times 10^{-4\ a}$	1.02	0.94	6.23	0.06	2.92	0.68
	bark	$6.80 \times 10^{-4\ a}$	1.63	0.26	0.66	0.62	1.68	0.35
HbpremiR159a	leaves	$2.34 \times 10^{-1\ a}$	1.22	0.91	1.40	0.04	1.04	0.88
	bark	$4.41 \times 10^{-2\ b}$	0.40	0.30	0.75	0.99	0.33	0.20
HbpremiR159b	leaves	$6.28 \times 10^{-1\ a}$	0.81	0.38	1.64	0.02	0.76	0.14
	bark	$1.96 \times 10^{-1\ b}$	0.50	0.35	0.86	0.92	0.41	0.22
HbpremiR166a	leaves	$4.02 \times 10^{-2\ a}$	0.49	0.39	2.84	0.05	0.98	0.84
	bark	$1.62 \times 10^{-2\ a}$	0.35	0.17	1.00	0.99	1.06	0.72
HbpremiR166b	leaves	$5.98 \times 10^{0\ a}$	1.50	0.24	2.45	0.22	1.78	0.07
	bark	$1.00 \times 10^{-1\ b}$	1.08	0.67	1.81	0.15	0.32	0.03
HbpremiR319	leaves	$2.80 \times 10^{-1\ a}$	0.73	0.63	1.46	0.33	0.87	0.83
	bark	$1.15 \times 10^{-2\ b}$	0.31	0.03	1.28	0.41	0.18	0.02
HbpremiR396	leaves	$2.87 \times 10^{0\ a}$	0.62	0.20	3.19	0.09	0.57	0.22
	bark	$4.80 \times 10^{-1\ b}$	0.27	0.02	0.50	0.13	0.43	0.07
HbpremiR398a	leaves	$3.61 \times 10^{-4\ b}$	0.22	0.05	0.69	0.53	1.85	0.34
	bark	$1.66 \times 10^{-2\ a}$	3.59	0.08	3.38	0.21	2.99	0.19
HbpremiR398b	leaves	$1.89 \times 10^{-3\ b}$	0.52	0.67	0.09	0.41	0.05	0.27
	bark	$5.01 \times 10^{-2\ a,b}$	15.19	0.09	0.61	0.46	4.66	0.60
HbpremiR398c	leaves	trace	0.91	0.82	nc	–	0.04	0.16
	bark	trace	nc	–	nc	–	nc	–

Table 2. *Cont.*

Gene	Tissue	Expression value for control	ET/C	*p*-value	MeJA/C	*p*-value	W/C	*p*-value
HbpremiR408a	leaves	3.49×10^{0} [a]	0.40	0.61	0.05	0.21	0.04	0.002
	bark	1.29×10^{0} [a]	0.50	0.64	0.002	0.52	0.05	0.62
HbpremiR408b	leaves	5.40×10^{0} [a]	0.42	0.55	0.02	0.01	0.04	0.001
	bark	2.52×10^{0} [a]	0.73	0.30	0.003	0.30	0.05	0.91
HbpremiR476	leaves	1.49×10^{-1} [a]	0.84	0.52	17.93	0.001	0.98	0.73
	bark	1.73×10^{-2} [a]	0.26	0.10	6.18	0.06	0.32	0.30
HbpremiR6482	leaves	4.41×10^{0} [a]	1.14	0.60	1.60	0.26	7.19	0.05
	bark	1.63×10^{0} [a]	0.62	0.48	1.32	0.53	2.68	0.14
HbpremiR6483	leaves	5.56×10^{0} [a]	2.97	0.18	4.14	0.13	0.81	0.85
	bark	8.63×10^{0} [a]	1.13	0.67	1.34	0.69	0.68	0.53
HbpremiR6484	leaves	6.10×10^{-1} [a]	0.92	0.65	1.67	0.14	0.48	0.12
	bark	8.92×10^{-1} [a]	0.99	1.00	1.15	0.72	0.73	0.41
HbpremiR6485	leaves	1.60×10 [b]	1.19	0.58	2.62	0.004	0.61	0.11
	bark	3.00×10 [a]	1.00	0.94	0.82	0.44	0.52	0.06
HbpremiRn11	leaves	trace	nc	–	3.52	0.08	1.35	0.98
	bark	trace	0.97	0.93	0.56	0.24	0.60	0.39
HbpremiRn12	leaves	6.19×10^{-2} [a]	1.23	0.65	1.51	0.12	7.12	0.13
	bark	6.13×10^{-2} [a]	1.20	0.58	0.84	0.96	1.42	0.44
HbpremiRn13	leaves	3.16×10^{2} [a]	1.23	0.39	1.24	0.55	1.41	0.96
	bark	3.92×10^{2} [a]	1.85	0.16	0.91	0.55	1.85	0.63

The regulation of *MIR* genes in response to ethylene, MeJA and wounding was shown as the ratio between the mean of three treated *in vitro* plantlets and three untreated *in vitro* plantlets. 3 *MIR* genes were significantly regulated by the ethylene treatment (Table 2). In leaves, only the relative expression of the *HbMIR398a* gene was repressed. In bark, 2 *MIR* genes were negatively regulated by ethylene, and the expression of *HbMIR319* and *HbMIR396* was inhibited (Table 2). Application of methyl jasmonate induced regulation of the relative expression of 6 *MIR* genes in leaves and not in bark (Table 2). The *HbMIR159a*, *HbMIR159b*, *HbMIR166a*, *HbMIR476* and *HbMIR6485* genes displayed a relative increase in transcripts, while *HbMIR408b* displayed a significant large drop in transcripts in response to external methyl jasmonate (Table 2). In response to wounding, 5 genes were significantly regulated (Table 2). In leaves, *HbMIR408a* and *HbMIR408b* displayed a large reduction in their relative expression levels while *HbMIR6482* significantly showed an increase in relative abundance for these transcripts (Table 2). In bark, the *HbMIR166b* and *HbMIR319* genes displayed a significant drop in their relative transcript abundance (Table 2). It should be noted that the *HbMIR319* gene was regulated in the same way in response to both ethylene and wounding.

The response to jasmonate in leaves seemed to differ from the other abiotic stresses. In fact, the expression of 5 *MIR* genes (*MIR159a*, *MIR159b*, *MIR166*, *MIR476* and *MIR6485*) was stimulated during the treatment. MiR159 and miR476 are known to target type MYB transcription factors [17,32] and PPRs (pentatricopeptide repeat proteins) [33], respectively. These targets were not identified during the prediction of targets of miR159 and miR476 in *Hevea*. In contrast, miR166 is known to target proteins of the HD-ZIPIII family in *Arabidopsis* and also in *Hevea* [17,26].

The expression of *MIR* genes was differentially regulated depending on the stress and the tissue (bark and leaf), suggesting specific response pathways for each type of stress (Figure 1a,b).

Figure 1. Venn diagram representing the regulation of the *MIR* genes common to all the treatments tested. (**a**): in leaves; and (**b**): in bark.

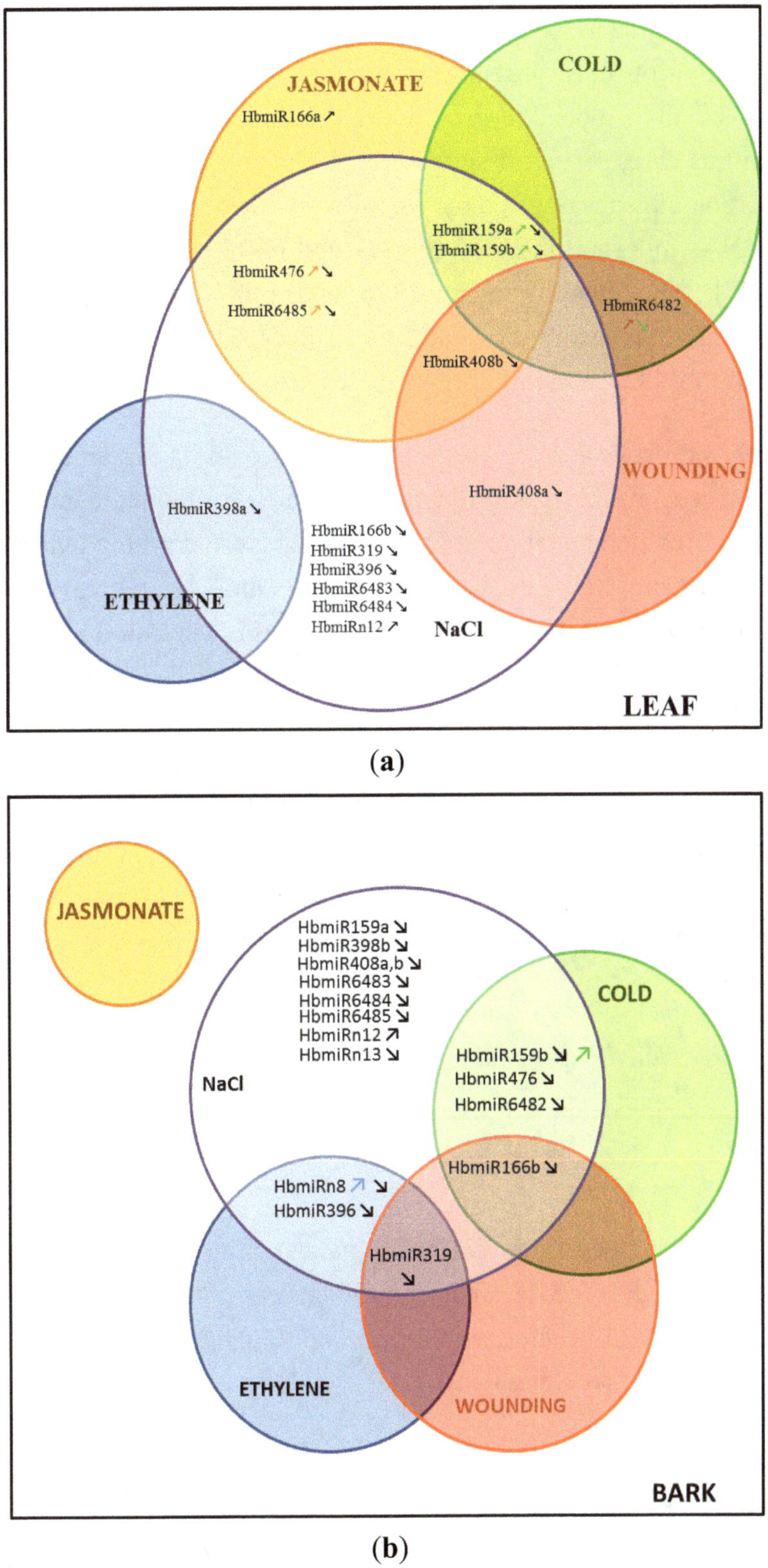

In maize, conversely, the comparison of expression pattern of mature miRNAs in plant subjected to saline stress and a viral infection lead the authors to suggest some cross-talk between abiotic and biotic stress [34]. In addition to differential expression depending on the tissue and stress applied, some differences between species are worth noting. In *Hevea brasiliensis*, the response to cold stress differs from that found in *Arabidopsis thaliana* [35]. In *Arabidopsis*, miR166b is accumulated in response to cold stress whereas in *Hevea*, repression of the *HbMIR166b* gene is found. Inhibition of the relative expression of the *HbMIR476* gene in *Hevea* bark shows an expression profile similar to that in poplar [36]. For saline stress, regulation of the expression of microRNAs appears to be specific to each species. In *Hevea*, saline stress led to global repression of the expression of all the *MIR* genes studied, except *HbMIRn12*. In this case, the result suggests that response mechanisms were brought into play by blocking the repression of genes targeted by microRNAs. By contrast, in *Arabidopsis*, accumulation of the mature microRNAs miR156, miR159, miR319 and miR396 and a reduction in the quantity of miR398 are seen [37]. The quantity of miR396 and miR156 is reduced in rice and maize respectively [38]. In poplar, the quantity of miR530a, miR1445, miR1446a-e, miR1447 and miR1711 is reduced while that of miR482.2 and miR1450 is increased [36].

The difference observed between *Hevea* and other plants could be explained by post-transcriptional regulations acting between pri-*MIR* transcript and the release of mature miRNA [39]. Moreover, most of studies reported on total mature miRNAs using Northern-blot hybridization highlighting the final products of several *MIR* genes. In our study, the use of real-time RT-PCR allowed identifying the specific expression of each *MIR* gene isoform. The microRNAs newly identified in *Hevea brasiliensis* have a high level of expression compared to that of conserved microRNAs under normal conditions. Different elements tend to show that the new microRNAs (young miRNAs) have little or no function [40]. Although conserved microRNAs are generally more abundant than the new microRNAs [41], an abundance of the latter does not mean greater regulation of the targets. Regulation of the targets of new microRNAs is not affected in mutant transgenic plants in the miRNA biogenesis pathway. The new microRNAs would seem not to be integrated yet in the regulation networks or may only function in a precise spatio-temporal context, or during response to a specific type of biotic or abiotic stress [40].

2.3. Analysis of the Co-Regulated Expression of the HbMIR398a, HbMIR398b *and* HbMIR398c *Genes and Their Putative Target Gene Chloroplastic HbCuZnSOD in Response to Abiotic Stress and to Hormone Treatments*

Three miR/target pairs have been experimentally validated in the rubber tree (miR156/Squamosa promoter binding protein, miR160/ARF and miR398/chloroplastic CuZnSOD) [26]. We chose to monitor expression of the *MIR398/chloroplastic CuZnSOD* pair. HbmiR398 is generated by three *HbMIR398* genes (3 genes, a, b and c; Figure 2).

In order to monitor expression of the *chloroplastic HbCuZnSOD* target, two primer pairs were designed, one flanking the cleavage site and the other at 3' UnTranslated Region (3'UTR) (Table 3).

Figure 2. Stem-loop structure of three precursors of *MIR398* in *Hevea brasiliensis*. The mature microRNA sequence is shown in dark.

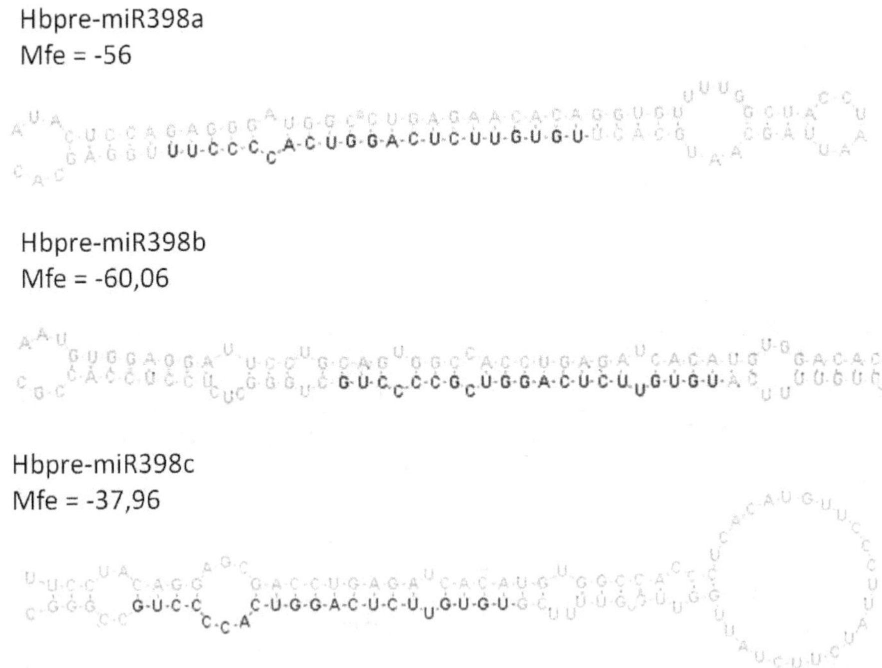

Table 3. List of primers and their sequences used for *MIR* gene expression analysis by RT-PCR. All primers are presented from 5' to 3' end.

Pre-microRNA	Forward primer	Reverse primer	PCR efficiency
Hbpre-miR156	TGGTGATGTTGTTGACAGAAGATAGAGAGC	GCACAAAGGAGTGAGATGCAGAGTCC	1.79
Hbpre-miR159a	GGTTAAGAAGTGGAGCTCCTTGAAGTC	GCTCCCTTCAATCCAAACAAGGATC	1.958
Hbpre-miR159b	GTGGAGCTCCTTGAAGTCCAATAGAGG	AGAGCTCCCTTCAATCCAAACAAGG	1.881
Hbpre-miR166a	TTCTTTTTGAGGGGAATGTTGTCTGG	GGAATGAAGCCTGGTCCGAGGAG	1.820
Hbpre-miR166b	GGGGAATGTTGTCTGGTTCGATG	TCAAATCAAACCCTGTTGGGGG	1.738
Hbpre-miR319	CCAGTCACGGTGGGCAATGGG	GGAGCTCCCTTCAGTCCAAGTACAGG	1.847
Hbpre-miR396	TGACCCTCTTCGTATTCTTCCACAGC	CCCACAGCTTTATTGAACCGCAAC	1.782
Hbpre-mir398a	TGAGAACACAGGTGTTTTGGCTACC	GTGCTCCAAAGGGGTGACCTGAG	1.879
Hbpre-mir398b	ACCTGAGATCACATGTGGACACCC	GCGGTGGAGGAGAGCCCAG	1.939
Hbpre-mir398c	TGGCCACCCTCACATGTTCCC	CCGGCAGGGGTGACCTGAG	1.965
Hbpre-miR408a	ACTGGGAACAGGCAGAGCATGG	GCCACAAGCCAGGGAAGAGGC	1.723

Table 3. *Cont.*

Pre-microRNA	Forward primer	Reverse primer	PCR efficiency
Hbpre-miR408b	GACATACAAAGACTGGGAAC AGGCAG	GCCACAAGCCAGGGAAGAG GC	1.792
Hbpre-miR476	GCCTTGTATGTTTCATTTAGTA ATCCTTCT	GATAATCCTTCTATGCAAAGT CTTTTATGC	1.732
Hbpre-miR6482	ACCAGGAACTGGTATCAACCC AGC	TGCTACCAATGAATCGGACCC ACC	1.837
Hbpre-miR6483	CAGTAAATAGCAGTATCGTGG ATAGGG	GTCCAATCATTGATCCTGAAA ATTTCTAC	1.828
Hbpre-miR6484	TGGATTGGAGCCCAATACTGT GAC	CTGCTCCATTGATTTTACCATC TATGC	1.873
Hbpre-miR6485	ACCTAGGATGTAGAAGAGCA TAAC	ACTACATGAGTGGATATATAG GAATCC	1.787
Hbpre-miRn11	GTATCAACGCAGATGTGCCGC C	CCCCAGCCAAACTCCCCACC	1.828
Hbpre-miRn12	AGCTTTCACCCAATAACCTTT GCAGT	GCTCTTCCAATTCCTATCCAA AGTGGT	1.78
Hbpre-miRn13	TGTGTTGGCCTTCGGGATCGG	CGAATGCCCCCGACTGTCCC	1.889
Hb-RH2b	GAGGTGGATTGGCTAACTGAG AAG	GTTGAACATCAAGTCCCCGAG C	1.68
HbCuZnSOD (flanking miRNA site)	GCTCTATCTCTCGCCGCCGCC TCC	CCGCAATTGTTGCTTCTGCC	1.785
HbCuZnSOD (3'UTR)	TGGCAGAAGCAACAATTGCGG	GCAGGGAACAATGGCTGCC	2

In response to abiotic stress, the transcripts of the *chloroplastic HbCuZnSOD* were accumulated significantly in roots in response to cold, in bark and roots in response to saline stress, and particularly after 24 h of treatment revealing that these transcripts are not cleaved by miR398 (Table 4, Tables S1 and S2). When the relative abundance of *chloroplastic HbCuZnSOD* transcripts was visualized (cleaved and non-cleaved with the primers at 3'UTR), the relative expression level for *chloroplastic HbCuZnSOD* transcripts decreased in leaves after 24 h of saline stress. However, that relative expression level rose significantly in response to saline stress in roots after 24 h of saline stress, and in wounded leaves (Table 4).

The relative accumulation of premiRs was analyzed for the three *MIR398* genes. The *MIR398c* transcript was only present in trace form under the majority of conditions. The relative expression of the *MIR398a* and *MIR398b* genes was significantly repressed in response to stress. In fact, in leaves, the relative expression of the *MIR398a* gene was repressed in response to 24 h of saline and ethylene treatment. Likewise, the relative expression of the *MIR398b* gene was repressed in bark and roots after 24 h of saline stress treatment.

When the relative expression level for the *HbMIR398* gene was compared with the expression data for the target, the significant reduction in the expression of the *HbMIR398b* gene in bark and roots in response to 24 h of saline stress was accompanied by a significant increase in the relative expression level for the non-cleaved *chloroplastic HbCuZnSOD* transcripts (Figure 3).

Table 4. Expression profiles of the *HbMIR398a*, *HbMIR398b* and *HbMIR398c* genes and their *chloroplastic HbCuZnSOD* target by real time-PCR in leaves, bark and roots of one-year-old *Hevea brasiliensis in vitro* plantlets in response to a positive cold, NaCl, ethylene, methyl jasmonate (MeJA) and wounding treatments. The statistically significant ratios (p-value < 0.05; Tables S1 and S2) with a value over 1 (**in red**) and under 1 (**in green**), represent over-expression and under-expression, respectively, in response to the treatments. The non-significantly regulated genes are shown **in yellow** (nc: not calculated).

Treatment	Tissue	Ratio of relative transcript accumulation (Treated/Control)				
		chloro CuZnSOD (flanking miRNA site)	chloro CuZnSOD (3'UTR)	*preMIR398a*	*preMIR398b*	*preMIR398c*
Cold	leaf	2.71	3.76	0.34	0.95	nc
	bark	8.98	8.90	0.23	0.52	nc
	root	3.41	1.46	0.47	0.62	nc
NaCl	leaf	0.84	0.29	0.21	0.87	nc
	bark	13.99	5.21	0.49	0.07	nc
	root	37.45	17.83	0.1	0.15	nc
Ethylene	leaf	nc	0.42	0.22	0.52	0.91
	bark	0.49	0.32	3.59	15.19	nc
MeJA	leaf	1.47	nc	0.69	0.09	nc
	bark	0.79	1.00	3.38	0.61	nc
Wounding	leaf	2.61	7.07	1.85	0.05	0.04
	bark	0.85	0.88	2.99	4.66	nc

Figure 3. Graph showing the expression values for the *chloroplastic HbCuZnSOD* and *HbMIR398b* genes in response to saline stress in bark and roots after 24 h of treatment. The expression values represent the mean and standard deviation of three treated or untreated plants, for each tissue. Data with the same letter are not significantly different at the level of 5%.

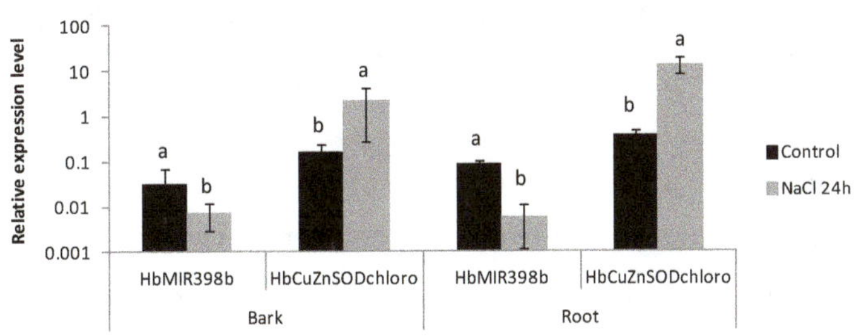

Co-expression of *MIR398* and of the gene encoding chloroplastic CuZnSOD was analyzed in response to several treatments. This enzyme takes part in ROS detoxification through the transformation of two superoxide ions H_2O_2 and H_2O. In *Arabidopsis*, miR398 targets cytosolic and chloroplastic superoxide dismutase (CSD1 and CSD2, [22]). Under oxidative stress conditions (heavy metal, strong sunlight and methyl viologen treatments), negative regulation of CSD1 and CSD2 by miR398 is halted in leaves, leading to an accumulation of enzymes and a drop in toxic free radicals for the cell. Tolerance of oxidative stress is consequently increased [22]. Interestingly, the *chloroplastic CuZnSOD* is the only one to have been validated in *Hevea* for miR398, the cytosolic form was not predicted to be targeted by miR398 and the absence of cleavage was experimentally validated [26]. Some other targets have been predicted for this microRNA but have yet to be experimentally validated [26]. Co-regulation was only visible in response to saline stress. A reduction in the expression of the *HbMIR398b* gene was associated with an accumulation of *chloroplastic HbCuZnSOD* transcripts in bark and roots but not in leaves. It is noteworthy to specify that the bark of the young *Hevea in vitro* plantlets was also chlorophyllous. For root tissues, other plastid organelles might be present. In addition, the level of expression for the *MIR398* genes was not always negatively correlated with the level of *chloroplastic HbCuZnSOD* transcripts, which has already been found in other species and suggests the existence of two regulation mechanisms, dependent on and independent of miR398 [42].

3. Experimental Section

3.1. Plant Material and Treatments

In vitro plantlets were regenerated by indirect secondary somatic embryogenesis from maternal inner integument of immature fruits from rubber clone PB 260 [43]. These *in vitro* plantlets were acclimatized and reared in the greenhouse for one year at a temperature of 28 °C. At that stage, the treatments were applied on three biological replicates. For cold stress, the plants were placed at 4 °C for 12 h. Saline stress was applied to the *in vitro* plantlets by watering with a 300 mM solution of NaCl rather than just water twice a day for the duration of the treatment (24 h).

Three-month-old epicormic shoot from grafted plants were used for ethylene (ET), methyl-jasmonate (MeJA) and wounding treatments. Plants from the *Hevea* clone PB 260 were grafted onto rootstock GT1. This plant material included a mature growth unit corresponding to a three-month-old epicormic shoot after cutting back. The treatment was applied for 4 h on the whole plant at 8:00 a.m. For application of the ethylene and methyl jasmonate treatments, each plant was placed in a hermetically sealed plexiglas box with a volume of 300 L leaving the doors open for 24 h prior to treatment to limit the effects of transport or any wounds. For application of the ethylene treatment, the gas was injected to obtain an ethylene concentration of 1 ppm (300 μL of pure ethylene) in the box. For the methyl jasmonate treatment, 20 μL of methyl jasmonate solution (>95%) (Sigma, St. Louis, MO, USA) was diluted in 500 μL of absolute ethanol and placed on Whatman paper inside the box, in order to release the methyl jasmonate in gas form. For these two stress conditions, samples were taken 4 h after treatment. The plants used as controls for these treatments were also placed in a hermetically sealed box without gas injection. In that way, it was possible to eliminate the impact of the box when comparing the two treatments (control and treated). For the wounding treatment, the plants were wounded on their leaves by applying pressure every 2 cm with serrated tweezers and on the bark by making cuts with a scalpel 1 cm apart on the stem of the scion.

Table 1 summarizes the different treatments applied to the rubber *in vitro* and grafted plants. Some leaf, bark and root samples were taken after 12 h for the *in vitro* plantlets, and only leaves and bark for the budded plants whose root system was not clonal. All the samples were immersed directly in liquid nitrogen after being taken and were stored at −80 °C pending RNA extraction.

3.2. Extraction and Purification of Total RNAs

The samples (leaf, bark and root) were ground in liquid nitrogen. The resulting fine powder was placed on ice in an extraction buffer (25 mM guanidium isothiocyanate 5M NaAc, 0.88% of sarcosine, 0.9% of polyvinylpyrrolidone (PVP) and 1% of β-mercapto-ethanol). The samples were mixed for 30 s to optimize membrane lysis. An initial centrifugation (30 min at 4 °C at 10,000 *g*) was applied to precipitate cell debris. The supernatant was deposited on a caesium cushion (24 mM NaAc, 45.6 mM CsCl). The tubes were then balanced and centrifuged for 20 h at 20 °C at 25,000 rpm in a Beckmann Coulter ultracentrifuge (L7, Beckmann Coulter, Brea, CA, USA) using the SW41 rotor (Beckmann Coulter). Only dense RNAs were able to cross the caesium chloride cushion. After 20 h, the supernatant was removed by pipetting, leaving 1 cm height of solution, which was eliminated by tipping up the tube. This precaution was taken to avoid contamination by genomic DNA. The pellet was dried for 5 min then rinsed in 70% ethanol then taken up in 200 μL of sterile water. The RNAs obtained were quantified then checked by electrophoresis (Mini-Sub Cell, Biorad, Richmond, CA, USA) on agarose gel (1% *w*/*v*, 1× Tris Acetate Ethylenediaminetetraacetic acid) and stained with ethidium bromide for visualization under ultraviolet. The RNAs were then aliquoted and stored at −80 °C.

3.3. Checking for the Presence of Genomic DNA and DNAse Treatment

A polymerase chain reaction (PCR) was carried out to check for the presence of genomic DNA in the samples of extracted total RNA. Amplification was carried out in a final volume of 50 μL. This was done with a volume of RNA corresponding to 50–100 ng to which was added a reaction mixture comprising 10× reaction buffer (10 mM Tris, pH = 8.3), $MgCl_2$ (1.5 mM), dNTP (0.04 mM), gene primers encoding actin (Forward: (5'-TCCATAATGAAGTGTGATGT-3', Reverse: (5'-GGACCTGACTCGTCATACTC-3', chosen astride an intron) (0.04 μM) and Taq polymerase (1 U/μL). A sample of *Hevea* genomic DNA was used as the positive control. The samples were incubated for 1 min and 30 s at 94 °C followed by 40 cycles, each composed of 30 s of denaturing at 94 °C, 30 s of hybridization at 58 °C and 90 s of elongation at 72 °C, with the reaction ending on a final elongation of 10 min at 72 °C. The amplification products were deposited on an agarose gel (1% *w*/*v*, 1× Tris Acetate Ethylenediaminetetraacetic acid) then stained with ethidium bromide. The RNA samples that proved to be contaminated by genomic DNA were treated by digestion with DNase (Turbo DNA-FREE kit from Ambion, Life technologies, Carlsbad, CA, USA). The reaction took place from 8 ng of total RNAs in a final volume of 40 μL composed of DNase buffer (10×, 4 μL, Ambion), DTT (0.1 M, 2 μL, Invitrogen, Life technologies, Carlsbad, CA, USA), RNase inhibitor (RiboLock, Fermentas, Burlington, ON, Canada, 40 U/μL) and TURBO DNase (2 U/μL). The samples were then incubated for 15 min at 37 °C. Next, a DNase inactivator (TURBO DNase Inactivation Reagent, Ambion) was added to the RNA samples. Enzyme inactivation took place for 2 min at ambient temperature by mixing the tubes. The supernatant containing the purified RNAs was recovered by centrifugation (12,000 *g*, 2 min, 4 °C).

3.4. cDNA Synthesis and Real-Time PCR

A microgram of total RNAs was used to synthesize some cDNAs in a reaction volume of 20 μL using RevertAid™ M-MuLV reverse transcriptase (MBI, Fermentas, Burlington, Canada) following the supplier's recommendations. The primers used to amplify the stem-loop structures are listed in Table 2. The PCR conditions comprised a denaturing cycle at 95 °C for 2 min, then 45 cycles comprising denaturing at 95 °C for 20 s and a hybridization and elongation stage at 60 °C for 20 s. The PCR reaction took place in a volume of 6 μL containing 2 μL cDNA diluted 25 times, 1 μL of primers at 5 μM, and 3 μL of 2× SYBR green mixture (LightCycler® 480 SYBR Green I Master, Roche Applied Sciences, Bâle, Switzerland). All the primers were validated by generating a standard curve from the dilution in series of 8 points, in triplicate, to calculate the efficiency of each primer pair (Table 3). In addition, amplification specificity was checked by generating the melting curves that needed to display a unique peak, and all PCR products were checked by sequencing. Relative quantification was carried out with Light Cycler 480 software (version 1.5.0, Roche Applied Sciences, Bâle, Switzerland), using *RH2b* [44] as the reference.

3.5. Statistical Analysis

The statistical analyses (analysis of variance followed by a Fisher comparison of means test) were done with XLSTAT (version 2011.4.02, Addinsoft, Paris, France) using expression data normalized by the LOG10(X) function. It was considered as up-regulation when the ratio was >1.0 and down-regulation when the ratio was <1.0. The p value corresponds to the Fisher test of the ANOVA.

4. Conclusions

Studying the regulation of *MIR* genes under harvesting stress (tapping, ethephon stimulation) and ROS-induced TPD in mature rubber trees remains difficult. However, some recent studies revealed that only one *MIR* genes is differentially regulated upon TPD. The *MIR159b* gene was shown up-regulated upon TPD occurrence [24]. The expression of this gene was increased in response to cold in leaves and bark, and in response to the jasmonic acid treatment in leaves of juvenile plantlets. Conversely, it was inhibited in all the tissues in response to salinity.

In order to get a full understanding of mechanisms involved in latex production and TPD syndrome, a complete validation of miRNA/target messenger pairs is first needed using by high throughput "degradome" analysis [45]. Combination of analyses on juvenile and mature plant materials will help developing model of *MIR* gene regulations under abiotic stress and further characterization of the TPD-regulated miRNAs and their targets.

Acknowledgments

This work was supported by the Institut Français du Caoutchouc. The authors thank Peter Biggins for his revision of the English, and Maryannick Rio, Florence Dessailly, Florence Martin and Eve Lorenzini for technical help.

References

1. Chen, S.; Peng, S.; Huang, G.; Wu, K.; Fu, X.; Chen, Z. Association of decreased expression of a Myb transcription factor with the TPD (tapping panel dryness) syndrome in *Hevea brasiliensis*. *Plant Mol. Biol.* **2003**, *51*, 51–58.
2. Strassner, J.; Schaller, F.; Frick, U.B.; Howe, G.A.; Weiler, E.W.; Amrhein, N.; Macheroux, P.; Schaller, A. Characterization and cDNA-microarray expression analysis of 12-oxophytodienoate reductases reveals differential roles for octadecanoid biosynthesis in the local *versus* the systemic wound response. *Plant J.* **2002**, *32*, 585–601.
3. Orozco-Cardenas, M.L.; Narvaez-Vasquez, J.; Ryan, C.A. Hydrogen peroxide acts as a second messenger for the induction of defense genes in tomato plants in response to wounding, systemin, and methyl jasmonate. *Plant Cell* **2001**, *13*, 179–191.
4. Chrestin, H. La stimulation à l'éthrel de l'hévéa jusqu'où ne pas aller trop loin. *Caoutchouc et Plastiques* janvier- fevrier **1985**, *1985*, 75–78.
5. Faÿ, D. Histo- and cytopathology of trunk phloem necrosis, a form of rubber tree (*Hevea brasiliensis* Müll. Arg.) tapping panel dryness. *Aust. J. Bot.* **2011**, *59*, 563–574.
6. Bartel, D.P. MicroRNAs: Genomics, biogenesis, mechanism, and function. *Cell* **2004**, *116*, 281–297.
7. Chen, X. MicroRNA biogenesis and function in plants. *FEBS Lett* **2005**, *579*, 5923–5931.
8. Bonnet, E.; van de Peer, Y.; Rouze, P. The small RNA world of plants. *New Phytol.* **2006**, *171*, 451–468.
9. Zhang, B.; Pan, X.; Cobb, G.P.; Anderson, T.A. Plant microRNA: A small regulatory molecule with big impact. *Dev. Biol.* **2006**, *289*, 3–16.
10. Naqvi, A.; Sarwat, M.; Hasan, S.; Choudhury, N. Biogenesis, functions and fate of plant microRNAs. *J. Cell. Physiol.* **2012**, *227*, 3163–3168.
11. Mallory, A.C.; Bouche, N. MicroRNA-directed regulation: To cleave or not to cleave. *Trends Plant Sci.* **2008**, *13*, 359–367.
12. Mallory, A.C.; Elmayan, T.; Vaucheret, H. MicroRNA maturation and action—The expanding roles of ARGONAUTEs. *Curr. Opin. Plant Biol.* **2008**, *11*, 560–566.
13. Baulcombe, D. RNA silencing in plants. *Kent* **2004**, *431*, 356–363.
14. Wu, L.; Zhou, H.; Zhang, Q.; Zhang, J.; Ni, F.; Liu, C.; Qi, Y. DNA methylation mediated by a microRNA pathway. *Mol. Cell* **2010**, *38*, 465–475.
15. Chellappan, P.; Xia, J.; Zhou, X.; Gao, S.; Zhang, X.; Coutino, G.; Vazquez, F.; Zhang, W.; Jin, H. siRNAs from miRNA sites mediate DNA methylation of target genes. *Nucleic Acids Res.* **2010**, *38*, 6883–6894.
16. Khraiwesh, B.; Zhu, J.K.; Zhu, J. Role of miRNAs and siRNAs in biotic and abiotic stress responses of plants. *Biochim. Biophys. Acta* **2012**, *1819*, 137–148.
17. Rhoades, M.W.; Reinhart, B.J.; Lim, L.P.; Burge, C.B.; Bartel, B.; Bartel, D.P. Prediction of plant microRNA targets. *Cell* **2002**, *110*, 513–520.
18. Noctor, G.; Foyer, C.H. Ascorbate and glutathione: Keeping active oxygen under control. *Annu. Rev. Plant Physiol. Plant Mol. Biol.* **1998**, *49*, 249–279.
19. Dat, J.; Vandenabeele, S.; Vranova, E.; van Montagu, M.; Inze, D.; van Breusegem, F. Dual action of the active oxygen species during plant stress responses. *Cell. Mol. Life Sci.* **2000**, *57*, 779–795.

20. Foyer, C.H.; Noctor, G. Redox homeostasis and antioxidant signaling: A metabolic interface between stress perception and physiological responses. *Plant Cell* **2005**, *17*, 1866–1875.
21. Scandalios, J.G. Oxidative stress: Molecular perception and transduction of signals triggering antioxidant gene defenses. *Braz. J. Med. Biol. Res.* **2005**, *38*, 995–1014.
22. Sunkar, R.; Kapoor, A.; Zhu, J.K. Posttranscriptional induction of two Cu/Zn superoxide dismutase genes in *Arabidopsis* is mediated by downregulation of miR398 and important for oxidative stress tolerance. *Plant Cell* **2006**, *18*, 2051–2065.
23. Kozomara, A.; Griffiths-Jones, S. miRBase: Integrating microRNA annotation and deep-sequencing data. *Nucleic Acids Res.* **2011**, *39*, D152–D157.
24. Griffiths-Jones, S.; Grocock, R.J.; van Dongen, S.; Bateman, A.; Enright, A.J. miRBase: microRNA sequences, targets and gene nomenclature. *Nucleic Acids Res.* **2006**, *34*, D140–D144.
25. Griffiths-Jones, S.; Saini, H.K.; van Dongen, S.; Enright, A.J. miRBase: Tools for microRNA genomics. *Nucleic Acids Res.* **2008**, *36*, D154–D158.
26. Gébelin, V.; Argout, X.; Engchuan, W.; Pitollat, B.; Duan, C.; Montoro, P.; Leclercq, J. Identification of novel microRNAs in *Hevea brasiliensis* and computational prediction of their targets. *BMC Plant Biol.* **2012**, *12*, doi:10.1186/1471-2229-12-18.
27. Lertpanyasampatha, M.; Gao, L.; Kongsawadworakul, P.; Viboonjun, U.; Chrestin, H.; Liu, R.; Chen, X.; Narangajavana, J. Genome-wide analysis of microRNAs in rubber tree (*Hevea brasiliensis* L.) using high-throughput sequencing. *Planta* **2012**, *236*, 437–445.
28. Gébelin, V.; Leclercq, J.; Kuswanhadi; Argout, X.; Chaidamsari, T.; Hu, S.; Tang, C.; Sarah, G.; Yang, M.; Montoro, P. The small RNA profile in latex from *Hevea brasiliensis* trees is affected by Tapping Panel Dryness. *Tree Physiol.* **2013**, in press.
29. Voinnet, O. Origin, biogenesis, and activity of plant microRNAs. *Cell* **2009**, *136*, 669–687.
30. Duan, C.; Argout, X.; Gebelin, V.; Summo, M.; Dufayard, J.F.; Leclercq, J.; Hadi, K.; Piyatrakul, P.; Pirrello, J.; Rio, M.; *et al.* Identification of the *Hevea brasiliensis* AP2/ERF superfamily by RNA sequencing. *BMC Genomics* **2013**, *14*, doi:10.1186/1471-2164-14-30.
31. Duan, C.; Rio, M.; Leclercq, J.; Bonnot, F.; Oliver, G.; Montoro, P. Gene expression pattern in response to wounding, methyl jasmonate and ethylene in the bark of Hevea brasiliensis. *Tree Physiol.* **2010**, *30*, 1349–1359.
32. Reyes, J.L.; Chua, N.H. ABA induction of miR159 controls transcript levels of two MYB factors during *Arabidopsis* seed germination. *Plant J.* **2007**, *49*, 592–606.
33. Lu, S.; Sun, Y.H.; Shi, R.; Clark, C.; Li, L.; Chiang, V.L. Novel and mechanical stress-responsive microRNAs in *Populus trichocarpa* that are absent from *Arabidopsis*. *Plant Cell* **2005**, *17*, 2186–2203.
34. Sanan-Mishra, N.; Kumar, V.; Sopory, S.K.; Mukherjee, S.K. Cloning and validation of novel miRNA from basmati rice indicates cross talk between abiotic and biotic stresses. *Mol. Genet. Genomics* **2009**, *282*, 463–474.
35. Sunkar, R.; Zhu, J.K. Novel and stress-regulated microRNAs and other small RNAs from *Arabidopsis*. *Plant Cell* **2004**, *16*, 2001–2019.
36. Lu, S.; Sun, Y.H.; Chiang, V.L. Stress-responsive microRNAs in *Populus*. *Plant J.* **2008**, *55*, 131–151.

37. Liu, H.H.; Tian, X.; Li, Y.J.; Wu, C.A.; Zheng, C.C. Microarray-based analysis of stress-regulated microRNAs in *Arabidopsis thaliana. RNA* **2008**, *14*, 836–843.
38. Ding, D.; Zhang, L.; Wang, H.; Liu, Z.; Zhang, Z.; Zheng, Y. Differential expression of miRNAs in response to salt stress in maize roots. *Ann. Bot.* **2009**, *103*, 29–38.
39. Ben Chaabane, S.; Liu, R.; Chinnusamy, V.; Kwon, Y.; Park, J.-H.; Kim, S.Y.; Zhu, J.-K.; Yang, S.W.; Lee, B.-H. STA1, an *Arabidopsis* pre-mRNA processing factor 6 homolog, is a new player involved in miRNA biogenesis. *Nucleic Acids Res.* **2013**, *41*, 1984–1997.
40. Cuperus, J.T.; Fahlgren, N.; Carrington, J.C. Evolution and functional diversification of *MIRNA* genes. *Plant Cell* **2011**, *23*, 431–442.
41. Axtell, A. JPM patient information. Depression in palliative care. *J. Palliat. Med.* **2008**, *11*, 529–530.
42. Zhu, C.; Ding, Y.; Liu, H. MiR398 and plant stress responses. *Physiol. Plant* **2011**, *143*, 1–9.
43. Lardet, L.; Martin, F.; Dessailly, F.; Carron, M.P.; Montoro, P. Effect of exogenous calcium on post-thaw growth recovery and subsequent plant regeneration of cryopreserved embryogenic calli of *Hevea brasiliensis* (Mull. Arg.). *Plant Cell Rep.* **2007**, *26*, 559–569.
44. Piyatrakul, P.; Putranto, R.A.; Martin, F.; Rio, M.; Dessailly, F.; Leclercq, J.; Dufayard, J.F.; Lardet, L.; Montoro, P. Some ethylene biosynthesis and *AP2/ERF* genes reveal a specific pattern of expression during somatic embryogenesis in *Hevea brasiliensis. BMC Plant Biol.* **2012**, *12*, doi:10.1186/1471-2229-12-244.
45. German, M.A.; Luo, S.; Schroth, G.; Meyers, B.C.; Green, P.J. Construction of Parallel Analysis of RNA Ends (PARE) libraries for the study of cleaved miRNA targets and the RNA degradome. *Nat. Protoc.* **2009**, *4*, 356–362.

3

Fluctuation of Global Gene Expression by Endogenous miRNA Response to the Introduction of an Exogenous miRNA

Yoshiro Nagata [1,†], Eigo Shimizu [2,†], Naoki Hibio [1] and Kumiko Ui-Tei [1,2,*]

[1] Department of Computational Biology, Graduate School of Frontier Sciences, University of Tokyo, 5-1-5 Kashiwanoha, Kashiwa-shi, Chiba-ken 277-8561, Japan;
E-Mails: a5205075@gmail.com (Y.N.); lapcat.nh@gmail.com (N.H.)

[2] Department of Biophysics and Biochemistry, Graduate School of Science, University of Tokyo, 7-3-1 Hongo, Bunkyo-ku, Tokyo 113-0033, Japan; E-Mail: eigo0shimizu@gmail.com

† These authors contributed equally to this work.

* Author to whom correspondence should be addressed; E-Mail: ktei@bi.s.u-tokyo.ac.jp;

Abstract: Most of the intracellular endogenous microRNAs (endo-miRNAs) are considered to be saturated in Argonaute (Ago) proteins in the RNA-induced silencing complexes (RISCs). When exogenous miRNAs (exo-miRNAs) are introduced into cells, endo-miRNAs in the RISC may be replaced with exo-miRNAs or exo-miRNAs, and endo-miRNAs might also compete for the position in the newly synthesized RISC with each other. This would lead to the fluctuation of global gene expression not only by repression of exo-miRNA target gene expression, but also by the increase of the endo-miRNA target gene expression. In the present study, we quantified the changes in the expression levels of target genes of exo-miRNA and endo-miRNA in the cells transfected with fifteen different exo-miRNAs by microarray experiments. Different exo-miRNAs increased ratios of expression levels of target genes of a given endo-miRNA to different extents, suggesting that the replacement efficiencies might differ according to the exo-miRNA types. However, the increased ratios in the expression levels of each endo-miRNA target genes by the transfection of any particular exo-miRNA were mostly equivalent, suggesting that the endo-miRNAs present in the RISC might be replaced with excessive exo-miRNAs at similar levels, probably because they exist in single-stranded forms in the RISC.

Keywords: exogenous microRNA; endogenous microRNA; microarray; seed-matched target

1. Introduction

MicroRNAs (miRNAs) are an abundant class of non-coding RNAs, about ~22 nucleotides long, that are key posttranscriptional regulators of gene expression in various organisms, including animals, plants and protozoa [1–3]. Most endogenous miRNAs (endo-miRNAs) actively silence target genes mainly by contiguous and perfect Watson-Crick base-pairing between the miRNA 5'-proximal seed region (positions 2–8) and its complementary sequences in 3' untranslated regions (3' UTRs) of target genes [4–6]. However, transfection of the small interfering RNA (siRNA)/miRNA expression construct into cells relieves repression of the target genes of endo-miRNAs dose-dependently at low concentrations and reaches the saturation level at high concentration [7–10].

These interfering effects are shown to be caused by competition for RNA silencing components. One of the key components is the nuclear karyopherin Exportin-5 [11–13], which binds to siRNA/miRNA precursors to transport them from the nucleus to the cytoplasm in the presence of Ran-GTP. The overloading of Exportin-5 by the excessive production of hairpin-structured siRNAs/miRNAs transcribed from their expression constructs could result in a decrease of cellular miRNA function [7,14]. However, the silencing activities of siRNA/miRNA duplexes are not affected by Exportin-5, since they do not need to be transported from the nucleus to the cytoplasm for their function [8,10,11]. Nonetheless, siRNA/miRNA duplexes have been shown to cause upregulation of their non-target genes by competition with endo-miRNAs. They compete for another saturable component of RNA silencing machinery, the RNA-induced silencing complex (RISC) [8,10,15]. The displacement of endo-miRNAs from the RISC by the introduction of synthetic siRNAs/miRNA duplexes are shown to be observed as the increase of endo-miRNA target gene expression [10].

RISC is the cytoplasmic effector machine of the miRNA silencing pathway. RISC assembly is mediated by the RISC loading complex, which is a multi-protein complex composed of the core protein, Argonaute (Ago), the RNase Dicer and the double-stranded RNA-binding protein, TRBP (TAR RNA binding protein) [16–19]. Initially, miRNA duplexes and siRNAs are loaded into Ago protein contained in the RISC loading complex [20,21]. To form the active RISC that performs gene silencing, the small RNA duplex needs to be separated or unwound into the single stranded form guiding it to its target mRNAs, within Ago protein. Then Dicer and its interactor TRBP dissociate from the RISC. Generally, miRNA forms an imperfect duplex composed of a miRNA strand and an opposite-strand miRNA. Evolutionary pressure has selected one particular strand of the duplex as the main regulator, which is preferentially loaded onto RISC, with the opposite strand being less functional [22,23]. The strand choice is considered to be not random and is partly determined by the intrinsic sequence and/or mismatched base-pairing of the miRNA duplex. The major determinants of RISC loading have been shown to be the thermodynamic properties: the strand with the less stable 5' end is more often loaded onto active RISC [24–26]. Furthermore, central mismatches of miRNAs have also been shown to promote RISC loading [27]. In the cells, RISC is considered to be saturated with endo-miRNAs. Because these experiments were carried out by the transfection of exogenous miRNAs (exo-miRNAs), exo-miRNAs are anticipated to exclude the endo-miRNAs from RISC or compete with endo-miRNAs for RISC with each other.

In this study, to investigate the mechanism to perturb the endo-miRNA function caused by the introduction of exo-miRNA, we performed microarray profiling to quantify changes in the expression levels of endo-miRNA target genes following the transfection of fifteen different synthetic exo-miRNA duplexes. Such competition among cellular miRNAs is considered to have a role in normal biological and disease-related cellular processes. The exo-miRNAs, in addition to silencing their own target genes, clearly increased the expression of endo-miRNA target genes. The increased levels of endo-miRNA target gene expression were varied according to the types of the transfected exo-miRNA duplexes, suggesting each exo-miRNA duplex is presumed to have distinct characteristics, such as structures and sequences, which affect its incorporation into the RISC. However, the increased levels were mostly equivalent according to the types of endo-miRNAs after transfection of any particular exo-miRNA duplex. These results might propose the possibility that the endo-miRNAs present in the RISC in single-stranded form, but not in duplex form, nor in target-pairing form, were replaced with exo-miRNA duplexes, because double-stranded endo-miRNAs with different characteristics might not be replaced with exo-miRNAs in the similar levels. A part of exo-miRNAs transfected might compete with endo-miRNAs for newly synthesized Ago proteins, but such effects might be negligible, because the transfected exo-miRNAs are sufficiently abundant. Furthermore, the endo-miRNA target genes with a large number of target sites in their 3' UTRs are revealed to be strongly repressed by endo-miRNAs in normal conditions. So, the genes with short 3' UTRs, which have a small number of miRNA target sites, were weakly repressed by endo-miRNAs in the normal condition, then efficiently repressed when exo-miRNAs were transfected.

2. Results and Discussion

2.1. Microarray Profiling of the Reduced Expression of Exo-miRNA Target Genes

To quantify global changes in the expression levels of endo-miRNA target genes resulting from the transfection of various exo-miRNA duplexes, microarray experiments were carried out using fifteen different exo-miRNA duplexes (let-7b, miR-1, miR-21, miR-22, miR-28, miR-30c-1, miR-186, miR-199b, miR-200b, miR-330, miR-335, miR-346, miR-466, miR-574 and miR-3126), which were chemically synthesized to form the duplex structures the same as those shown in miRBase [28]. At first, 24 h following the transfection of each exo-miRNA duplexes into human HeLa cells, the effects on own target genes of either of both miRNA strand were examined. We chose this time point, because it is reported that RNA silencing is generally maximal ~24 h post-transfection and that protein silencing varies depending on the target, but is generally maximal ~48–72 h post-transfection [29]. Thus, the result at 24 h post-transfection may largely reflect the direct effects without downstream effects of exo-miRNAs, since major downstream effects should be observed after 48 h post-transfection. It has been reported that miRNA target genes mainly contain sequences complementary to miRNA seed regions at positions 2–8 from the 5' terminus [4–6], then the expression patterns of these genes were analyzed. The expression patterns of each miRNA target genes containing seed-complementary sequences in HeLa cells transfected with fifteen exo-miRNAs are shown as MA (M = intensity ratio, A = average intensity) plots (Figure S1A-1~O-1 and S1A-3~O-3) and cumulative distributions (Figure S1A-2~O-2 and S1A-4~O-4). The changes in expression were calculated as the difference values subtracting the average fold changes ($\log_2$) for exo-miRNA seed-matched target genes (blue line in Figure S1A-2~O-2 and S1A-4~O-4) from the average value for the genes without seed-matched sequence (black line in Figure S1A-2~O-2 and S1A-4~O-4). The differential fold change values were between 0.02 and −0.17, as summarized in Figure 1, indicating that all of the exo-miRNAs used in this study could decrease their own target genes at the different levels, due to the miRNA species. However, these decreases in exo-miRNA target gene expression, as a result of competition, with the corresponding endo-miRNAs might be underestimated in this experiment.

Figure 1. Microarray analysis of exo-miRNA target gene expression. HeLa cells transfected with each of exo-miRNA duplexes were subjected to microarray profiling. Differential fold changes in gene expression caused by the transfection of either strand of fifteen different exo-miRNAs. The results of the opposite strands of miR-1, miR-346 and miR-466 were not shown, because these miRNAs were not registered in miRBase. The MA plots and cumulative distribution patterns are shown in Figure S1.

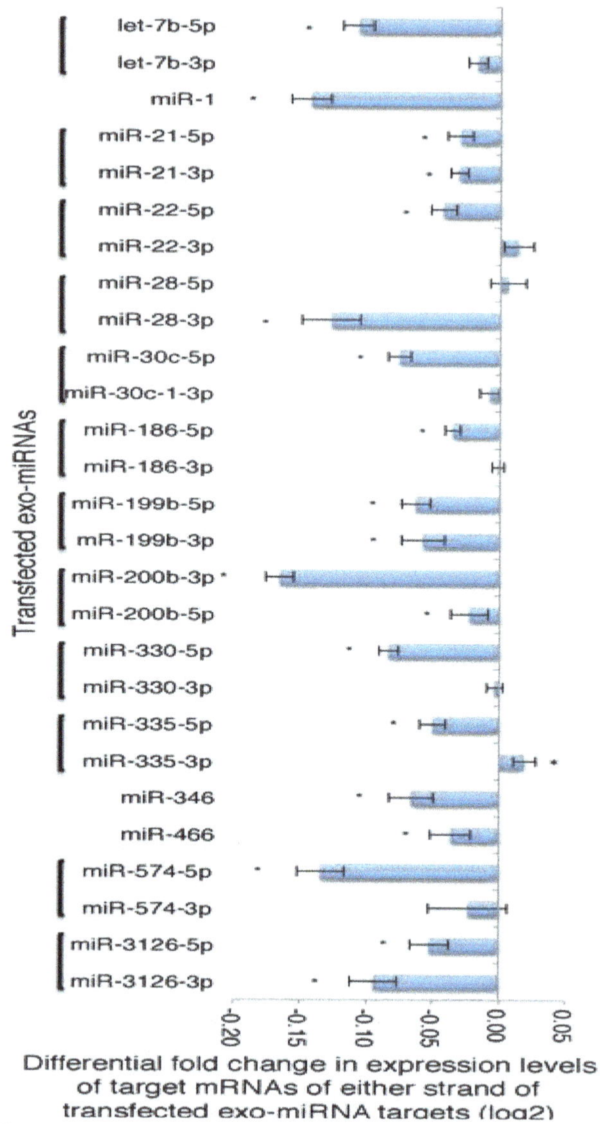

2.2. Microarray Profiling of the Increased Expression of Endo-miRNA Target Genes Resulting from the Introduction of Exo-miRNAs

Having found reduced expression of exo-miRNA target genes, we next analyzed the expression of endo-miRNA target genes. As let-7b, miR-21, miR-27a, miR-17, miR-26a, miR-24, miR-30a, miR-92a, miR-19a, miR-15a, miR-22, miR-29a, miR-125a, miR-93, miR-191, miR-103a, miR-143, miR-100, miR-23a and miR-186 are reported to be the top 20 most highly expressed miRNA families in HeLa cells [30], changes in the target gene expression of these representative miRNAs were calculated (Figures 2 and 3). In this analysis, "endo-miRNA target genes" were defined as genes with sequences complementary to endo-miRNA seed sequences (positions 2–8) in their 3' UTRs, but not to both strands of exo-miRNA seed sequences. The "endo-miRNA non-target genes" were genes with no sequences complementary to endo-miRNA or both strands of exo-miRNA seed sequences. The MA plots and cumulative distributions were shown in Figures S2–S21. The difference values were calculated by subtracting the average fold changes ($\log_2$) for endo-miRNA target genes (blue line in Figures S2–S21 A~O-2) from the average value for endo-miRNA non-target genes (black line in Figures S2–S21 A-2~O-2). The expression of most of the endo-miRNA target genes was increased by the transfection of exo-miRNAs (Figures 2 and 3, Figures S2–S21).

Next, we compared changes in the expression of endo-miRNA target genes according to the types of exo-miRNAs (Figure 2). Expression changes were determined by subtracting the average $\log_2$ fold change for endo-miRNA target genes containing more than one target site(s) from the average value for endo-miRNA non-target genes and shown as a differential fold change ($\log_2$). The increases in target gene expression of endogenous top 20 miRNAs varied, but showed similar tendencies according to the kind of exo-miRNA transfected (Figure 2). Of the fifteen transfected exo-miRNAs, the exogenously transfected miR-466 duplex increased the expression levels of target genes of at least 16 endo-miRNAs (let-7b-5p, miR-21-5p, miR-27a-3p, miR-17-5p, miR-26a-5p, miR-24-3p, miR-30a-5p, miR-92a-3p, miR-19a-3p, miR-15a-5p, miR-22a-3p, miR-29a-3p, miR-93a-5p, miR-143-3p, miR-23a-3p and miR-186-5p) significantly (Figure 2A–T), and their averaged expression level was the highest (Figure 2U). The exo-miRNA duplexes, miR-21, miR-22, miR-28, miR-30c-1, miR-200b, miR-346, miR574 and miR-3126, produced significant, but modest levels of increases in the endo-miRNA target genes (Figure 2U), while exo-miRNA duplexes, let-7b, miR-1, miR-28, miR-199b and miR-335, showed low values on average (Figure 2U).

Considering another dimension of the results, changes in the target gene expression of top 20 endo-miRNAs were examined. The results clearly show that the increased ratios of target genes of any endo-miRNAs were roughly equivalent when a particular exo-miRNA was transfected (Figure 3). The average expression of seed-matched target genes of these 20 endo-miRNAs for each of 15 transfected exo-miRNA are shown in Figure 3. These results indicated that the increased fold changes of expression levels of endo-miRNA target genes were essentially determined by the kind of exo-miRNA and not by the type of endo-miRNA, although the increase of target genes of a few types of endo-miRNAs, such as miR-1 (Figure 3B), miR-28 (Figure 3E), miR-199b (Figure 3H) and miR-335 (Figure 3K), by the transfection of a given miRNA were not equivalent, and averaged values after transfection of these exo-miRNAs became low due to combining the up- and down-regulated targets (Figure 2U). The similar results, indicating the equivalent expression levels of endo-miRNA targets by most of exo-miRNAs, except for miR-1, miR-28, miR-199b and miR-335, were also observed when the increased levels were calculated for top 383 endo-miRNAs (Figure S22).

Figure 2. Mean fold changes of the expression levels of endo-miRNA target transcripts by the transfection of different exo-miRNAs. HeLa cells were transfected with each of fifteen exo-miRNA duplexes, $\log_2$ mean differential fold changes of the expression levels of seed-matched target genes of each of the top 20 endo-miRNAs were calculated. Data were shown with respect to endo-miRNAs, let-7b-5p (**A**); miR-21 (**B**); mR-27a-3p (**C**); miR-17-5p (**D**); miR-26a-5p (**E**); miR-24-3p (**F**); miR-30a-5p (**G**); miR-92a-5p (**H**); miR-19a-5p (**I**); miR-15a-5p (**J**); miR-22-3p (**K**); miR-29a-3p (**L**); miR-125a-5p (**M**); miR-93-5p (**N**); miR-191-5p (**O**); miR-103a-3p (**P**); miR-143-3p (**Q**); miR-100-5p (**R**); miR-23a-3p (**S**); and mR-186-5p (**T**); and their averaged values were shown in (**U**). Note that each exo-miRNA increased the endo-miRNA target genes in different degrees. The individual data are shown in Figures S2–S21. Data represent the mean ± SE (* $p < 0.05$).

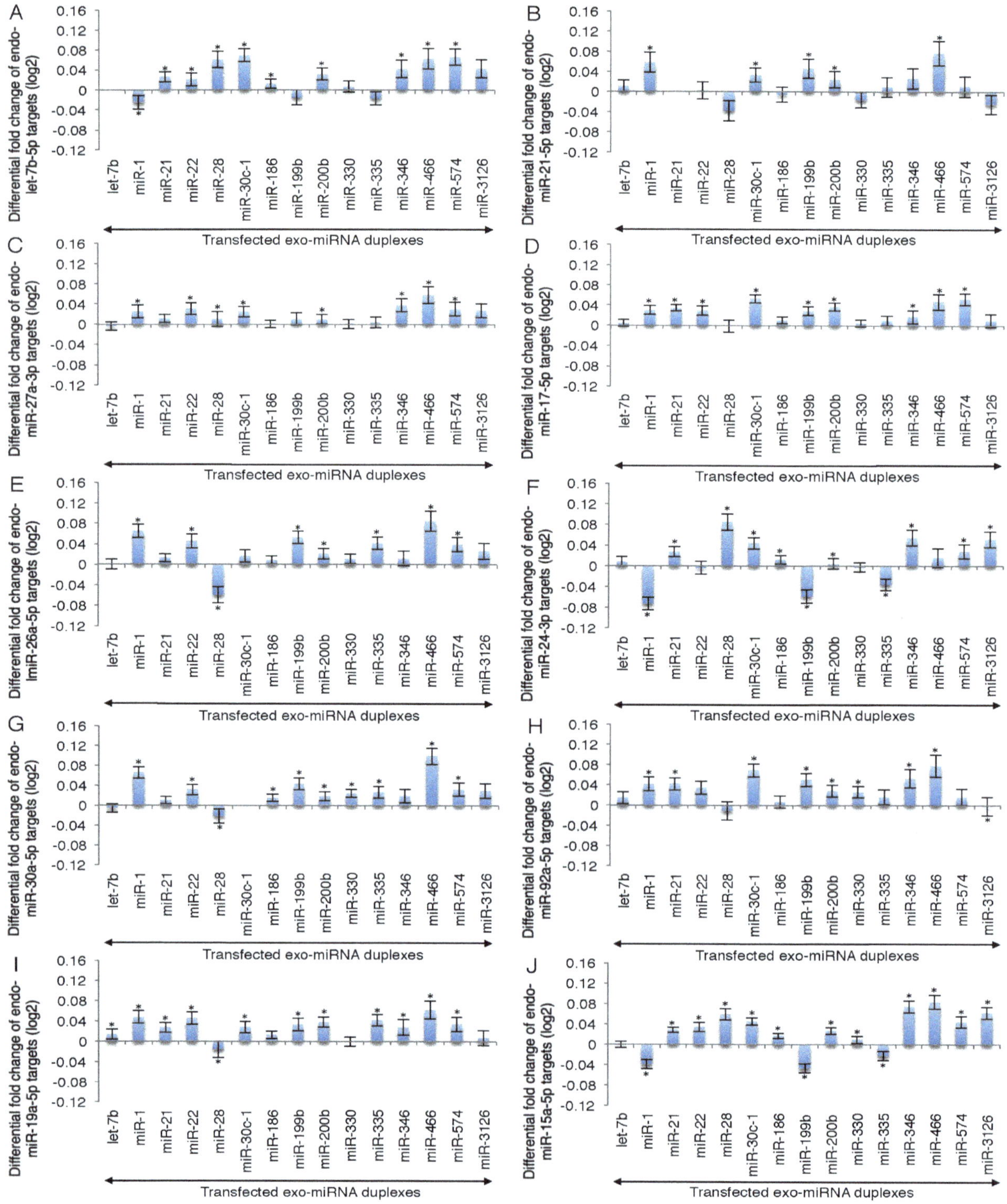

Figure 2. *Cont.*

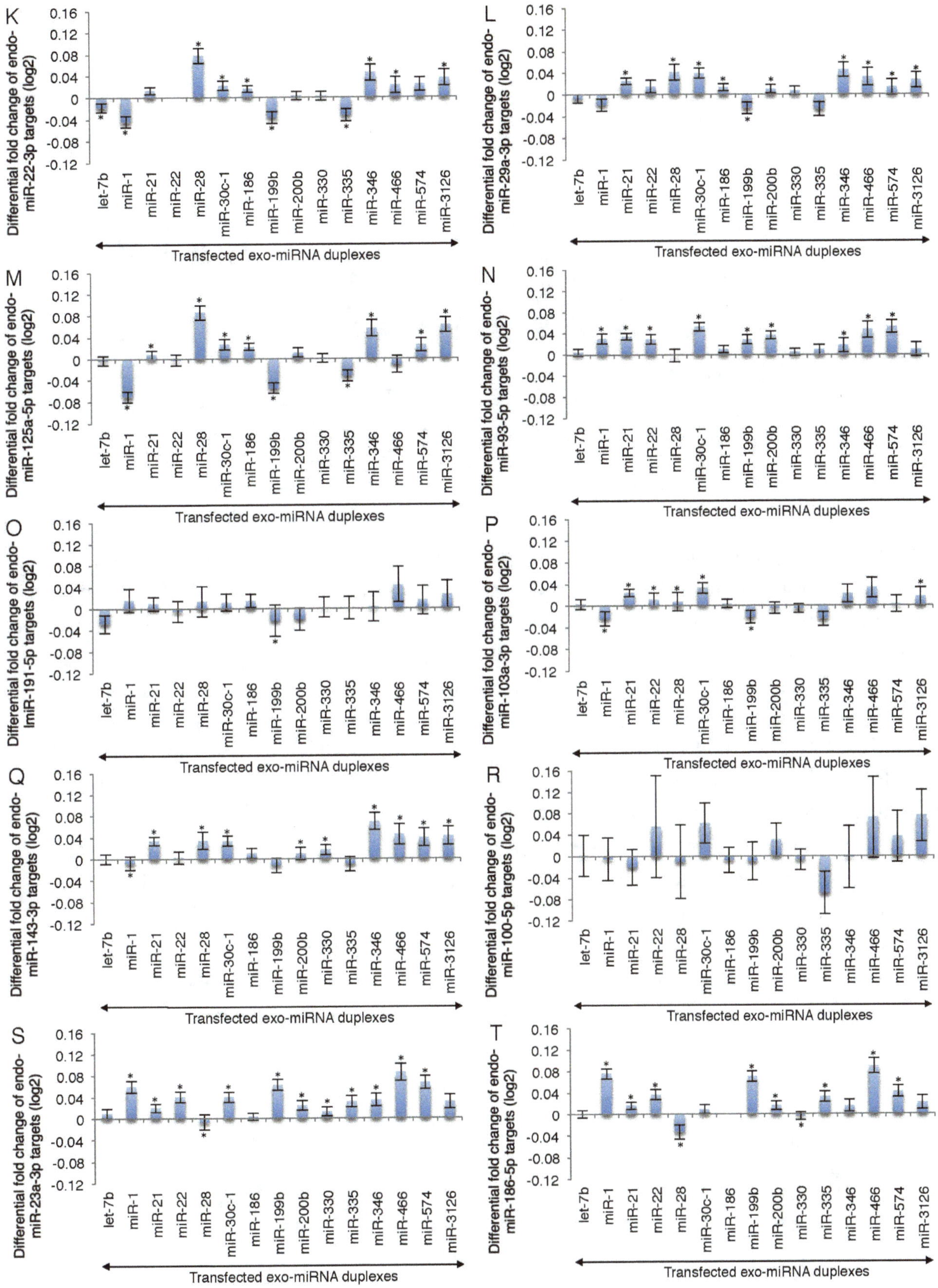
K
Differential fold change of endo-miR-22-3p targets (log2)
L
Differential fold change of endo-miR-29a-3p targets (log2)
M
Differential fold change of endo-miR-125a-5p targets (log2)
N
Differential fold change of endo-miR-93-5p targets (log2)
O
Differential fold change of endo-lmiR-191-5p targets (log2)
P
Differential fold change of endo-miR-103a-3p targets (log2)
Q
Differential fold change of endo-miR-143-3p targets (log2)
R
Differential fold change of endo-miR-100-5p targets (log2)
S
Differential fold change of endo-miR-23a-3p targets (log2)
T
Differential fold change of endo-miR-186-5p targets (log2)
0.16
0.12
0.08
0.04
0
-0.04
-0.08
-0.12
let-7b
miR-1
miR-21
miR-22
miR-28
miR-30c-1
miR-186
miR-199b
miR-200b
miR-330
miR-335
miR-346
miR-466
miR-574
miR-3126
Transfected exo-miRNA duplexes

Figure 2. *Cont.*

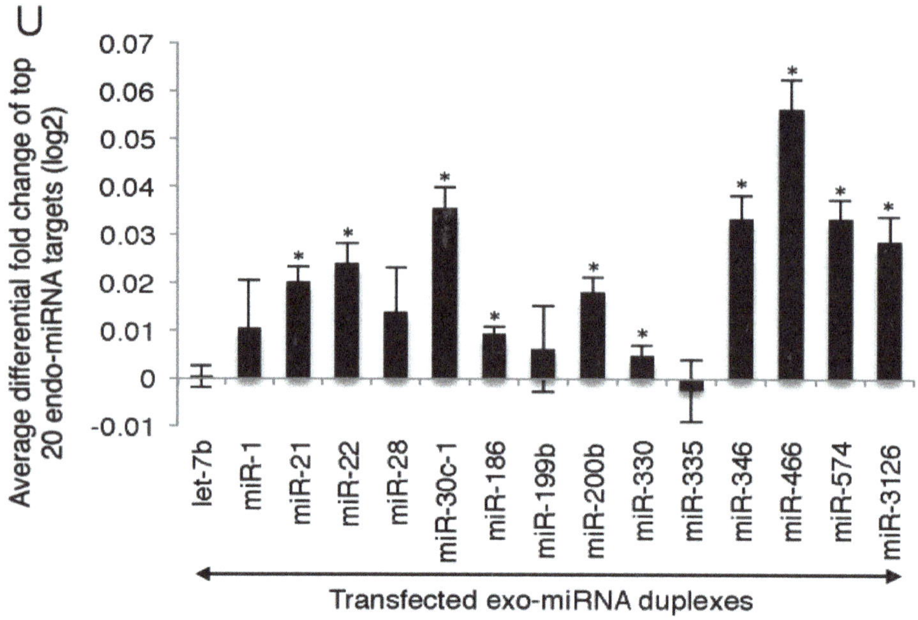

Our results were obtained from our own microarray experiments using one human cell line; it should reduce variables compared to the previous studies. To confirm the reliability of the microarray data, we analyzed 28 transcripts by quantitative RT-PCR (Figure S23). The expression levels estimated by quantitative RT-PCR were essentially identical to those obtained in the microarray analysis, with an estimated correlation coefficient of 0.83.

Figure 3. Mean fold changes of the expression levels of endo-miRNA target transcripts following the transfection of exo-miRNAs. HeLa cells were transfected with each of fifteen exo-miRNA duplexes, $\log_2$ mean differential fold changes of the expression levels of seed-matched target genes of each of the top 20 endo-miRNAs were calculated. Data were shown with respect to transfected exo-miRNAs, miR-7b (**A**); miR-1 (**B**); miR-21 (**C**); miR-22 (**D**); miR-28 (**E**); miR-30c-1 (**F**); miR-186 (**G**); miR-199b (**H**); miR-200b (**I**); miR-330 (**J**); miR-335 (**K**); miR-346 (**L**); miR-466 (**M**); miR-574 (**N**); miR-3126 (**O**); and their averaged values were shown in (**P**). Target genes of different endo-miRNAs are increased by exo-miRNAs in approximately comparable levels. The individual data are shown in Figures S2–S21. Data represent the mean ± SE (* $p < 0.05$).

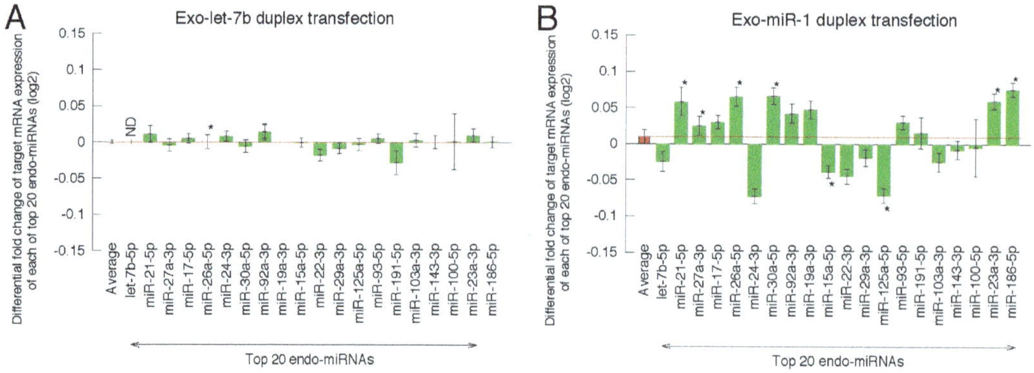

Figure 3. *Cont.*

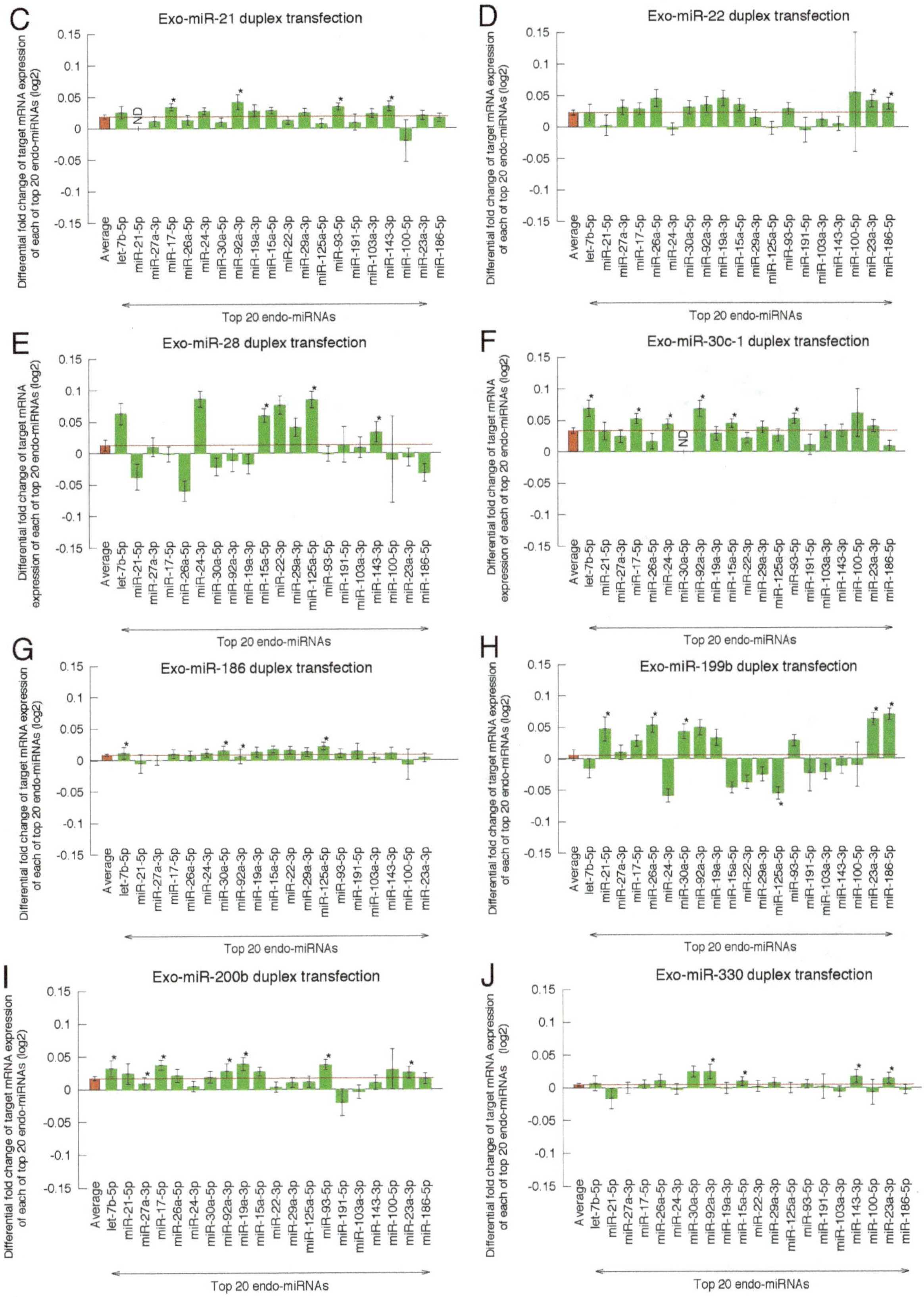

Figure 3. *Cont.*

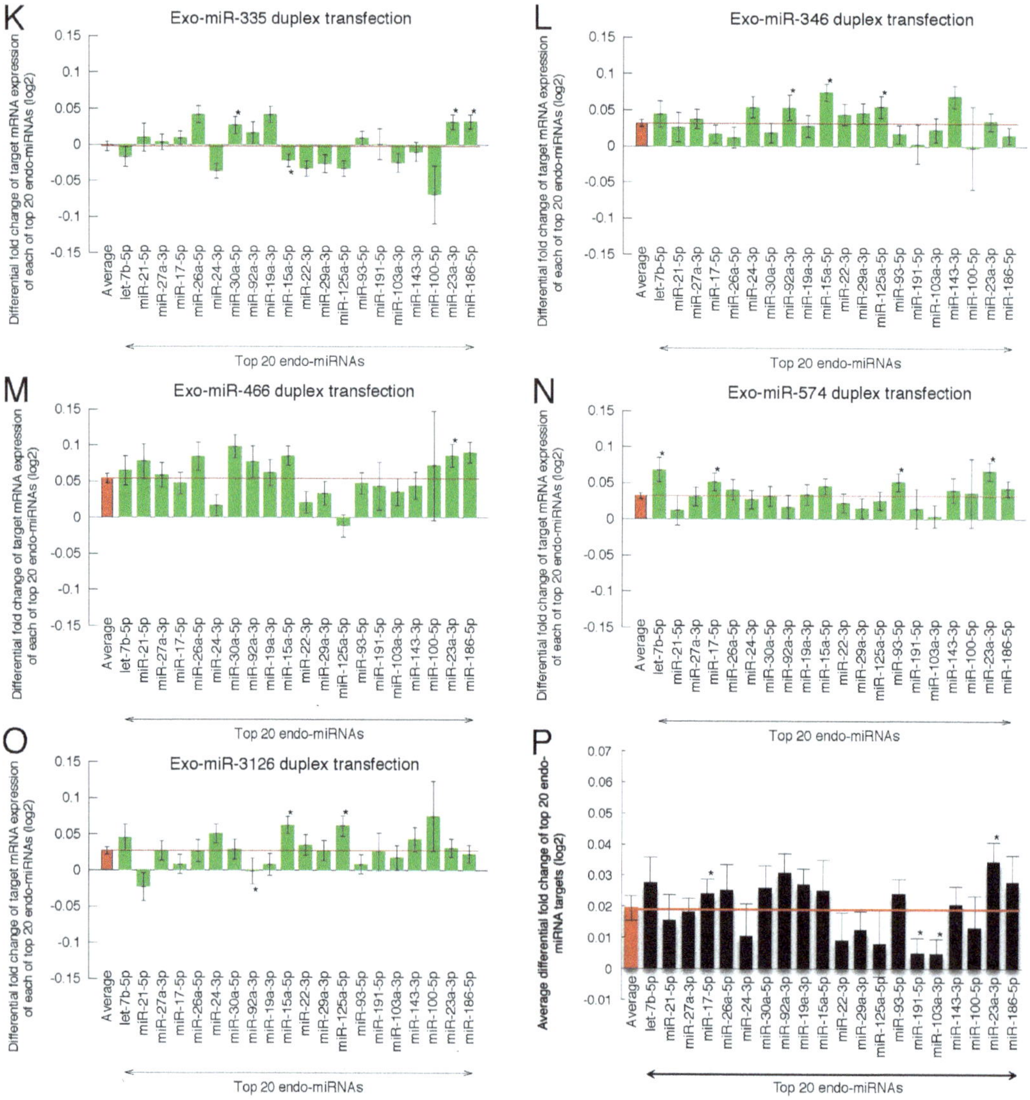

2.3. Reporter Analysis of Exo-miRNA and Endo-miRNA Target Expression in Cells Transfected with Exo-miRNA

A part of the target genes with 3' UTRs complementary to endo-miRNA seed sequences is not necessarily downregulated by endo-miRNAs in normal conditions. The accessibilities of endo-miRNAs to such genes are speculated to be interfered by some causes, such as binding of specific RNA binding proteins or RNA secondary structures. Thus, the changes in the expression levels of seed-complementary target genes of endo-miRNAs shown in Figures 2 and 3 might be modest or low (generally ~2 to 10%). To confirm the downregulation of exo-miRNA target gene expression and upregulation of endo-miRNA target gene expression resulting from transfection of exo-miRNA, we carried out reporter assays using luciferase expression constructs carrying sequences perfectly complementary to exo-miRNAs or endo-miRNAs in the 3' UTR of the *Renilla* luciferase gene in psiCHECK-1 (Figure 4A). Even when the changes in the expression levels of seed-complementary endogenous target genes of exo-miRNAs or endo-miRNAs are small, the expression levels of luciferase reporter with perfect complementary sequences of miRNAs are expected to show remarkable effects, although the effect is not biologically relevant. Exo-miR-200b or exo-miR-330 was transfected into HeLa cells with each reporter construct and a firefly luciferase expression construct (pGL3-Cont) (internal control). A double-stranded DNA (miDNA), which mimics the miRNA structure, was used as a miRNA control. At 24 h post-transfection, the relative luciferase activity was calculated (Figure 4). Both exo-miRNAs silenced their own targets (Figure 4B,E). In contrast, the luciferase activities of psiCHECK-1 constructs containing the endo-miRNA target sequences apparently increased (Figure 4C,D,F,G). These results suggest that exo-miRNAs interfered with endo-miRNA silencing activity by competing for the RNA silencing machinery downstream of Exportin-5 probably replacing with endo-miRNAs in the RISC, thereby leading to increased expression of endo-miRNA target genes.

Figure 4. Luciferase reporter analyses of changes in the expression of exo-miRNA and endo-miRNA targets. (**A**) Schematic structure of a luciferase reporter containing a miRNA completely matched target sequence in the 3' UTR and mRNA transcribed from the reporter. The exo-miRNAs, miR-200b (**B–D**) and miR-330 (**E–G**), respectively, were transfected into HeLa cells at 50 nM. At 24 h post-transfection, luciferase activities were measured. The luciferase activities derived from the constructs carrying exo-miRNA target sequences were decreased by the transfection of respective exo-miRNA (**B**,**E**). In contrast, the luciferase activities from the constructs carrying endo-let-7b-5p (**C**), endo-miR-21-5p (**D**,**G**) and endo-miR-7b-5p (**F**) target sequences were increased, indicating that exo-miRNAs repressed endo-miRNA silencing activities.

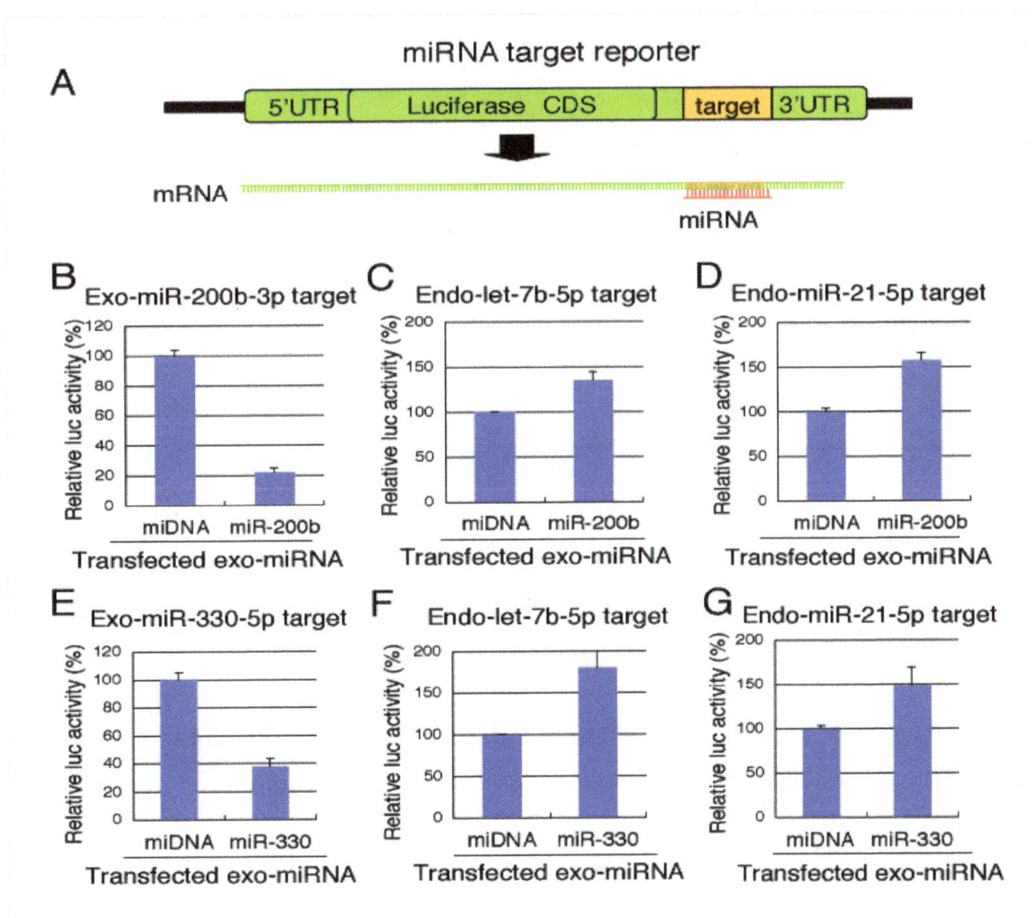

Figure 5. Mean fold changes of target genes of exo-miRNAs and endo-miRNAs according to the 3' UTR lengths. Differential fold changes (log_2) of expression levels of exo-miRNA targets (**A**); and endo-miRNA targets (**B**); and the number of top 20 endo-miRNA target sites (**C**) according to the lengths of 3' UTRs. The averaged numbers of exo-miRNA target sites are 53, 137, 210, 250 and 149 (**A**), and those of endo-miRNA target sites are 53, 134, 185, 184 and 84 (**B**) in the 3' UTRs of 0–284, 285–647, 648–1236, 1237–2392 and 2393~nucleotides (nts). Each fraction contains the same number of mRNAs registered in the RefSeq database.

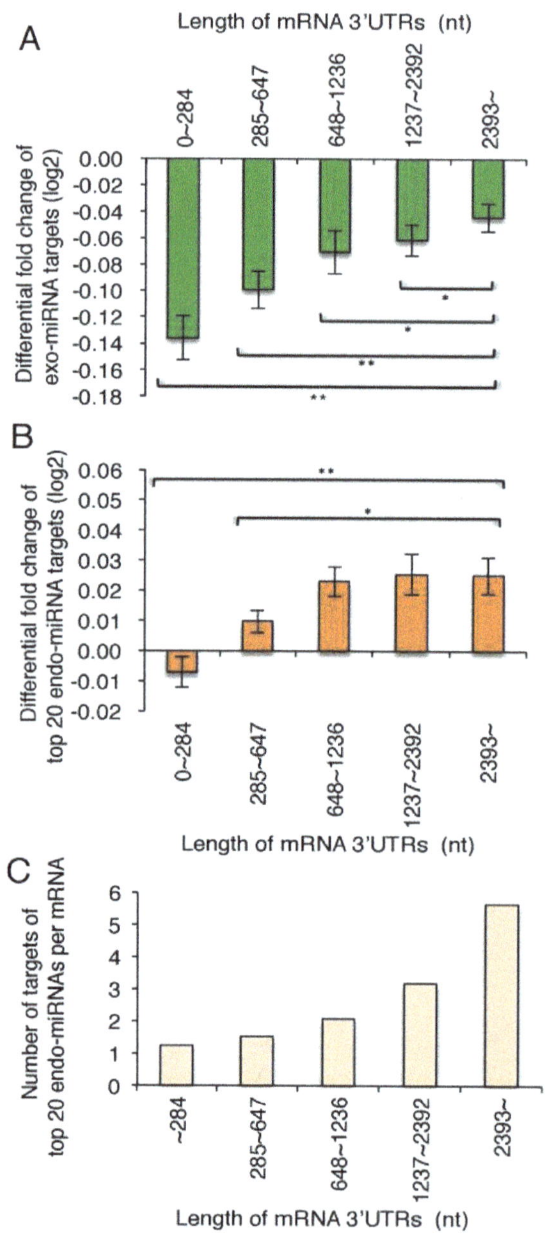

2.4. Exo-miRNA Targets with Short 3' UTRs Are Efficiently Downregulated and Endo-miRNA Targets with Long 3'UTRs Are Efficiently Upregulated by the Introduction of Exo-miRNA Duplex

Among exo-miRNA target genes, the genes with short 3' UTRs were more efficiently downregulated than those with long 3' UTRs by the transfection of exo-miRNAs (Figure 5A), consistent with the previous report [31]. In this analysis, we used the fold change values of exo-miRNA targets with 3' UTRs complementary to seed sequence of either one strand of exo-miRNA duplex: let-7b-5p, miR-1, miR-21-5p, miR-22-5p, miR-28-5p, miR-30c-5p, miR-186-5p, miR-199b-5p, miR-200b-3p, miR-330-5p, miR-335-5p, miR-346, miR-466, miR-574-5p and miR-3126-5p. In contrast, the top 20 endo-miRNA targets, which have no exo-miRNA target sites, were found to be more strongly upregulated, according to the increase of the 3' UTR lengths (Figure 5B). It was presumed that a large number of endo-miRNA target sites are situated in the long 3' UTRs, but small in the short 3' UTRs; then, the genes with long 3' UTRs might be strongly repressed by endo-miRNAs in the normal condition, but those with short 3' UTRs might be repressed weakly. So, we calculated the number of top 20 endo-miRNA target sites in the various length of mRNA 3' UTRs. As expected, a large number of target sites of top 20 endo-miRNAs were found in the long 3' UTRs, but a small number of them were in the short 3' UTRs (Figure 5C), indicating that the genes with long 3' UTRs should be strongly repressed by endo-miRNAs compared to those with short 3' UTRs in the normal condition. However, once exo-miRNAs are transfected into the cells, a part of the endo-miRNAs loaded on the RISCs should be replaced with exo-miRNAs. As a result, the constant downregulation of endo-miRNA targets by endo-miRNAs might be cancelled, and the expression of these genes are upregulated. Thus, the changes of expression levels of the exo-miRNA target genes might be elaborately regulated by the endo-miRNAs pre-situated in the RISCs.

2.5. Highly Expressed Exo-miRNA Targets Are Efficiently Downregulated and Endo-miRNA Targets with Low Expression Levels Are Efficiently Upregulated by the Transfection of the Exo-miRNA Duplex

Among exo-miRNA targets, the genes with high expression levels in the normal conditions were more efficiently downregulated by the transfection of exo-miRNAs compared to those with low expression levels (Figure 6A). The results showed good agreement with the previous study [31]. In contrast, the endo-miRNA targets with low expression levels in the normal conditions were uncovered to be intensively upregulated by the exo-miRNA transfection, and those with high expression levels were not upregulated efficiently (Figure 6B). So, we investigated the number of endo-miRNA target sites in gene with different expression levels. As a result, a large number of target sites were found in the mRNAs with low expression levels, but a small number of the sites were detected in the mRNAs with high expression levels (Figure 6C), suggesting that many genes with low expression levels might be downregulated by endo-miRNAs, but those with high expression levels are not repressed by endo-miRNAs in the normal condition. Thus, the changes of expression levels of the exo-miRNA target genes might be also suitably regulated by the endo-miRNAs pre-situated in the RISCs.

Figure 6. Mean fold changes of target genes of exo-miRNAs and endo-miRNAs, according to the expression levels. Differential fold changes (log_2) of target gene expression levels of exo-miRNAs (**A**) and endo-miRNAs (**B**) and the number of the top 20 endo-miRNA target sites (**C**), according to the expression levels of exo-miRNA targets. The averaged numbers of exo-miRNA target sites are 130, 217, 237 and 215 (**A**), and those of endo-miRNA target sites are 92, 164, 176 and 202 (**B**) in the 3' UTRs of genes with differential fold changes of 0~7, 7~9, 9~11 and 11~.

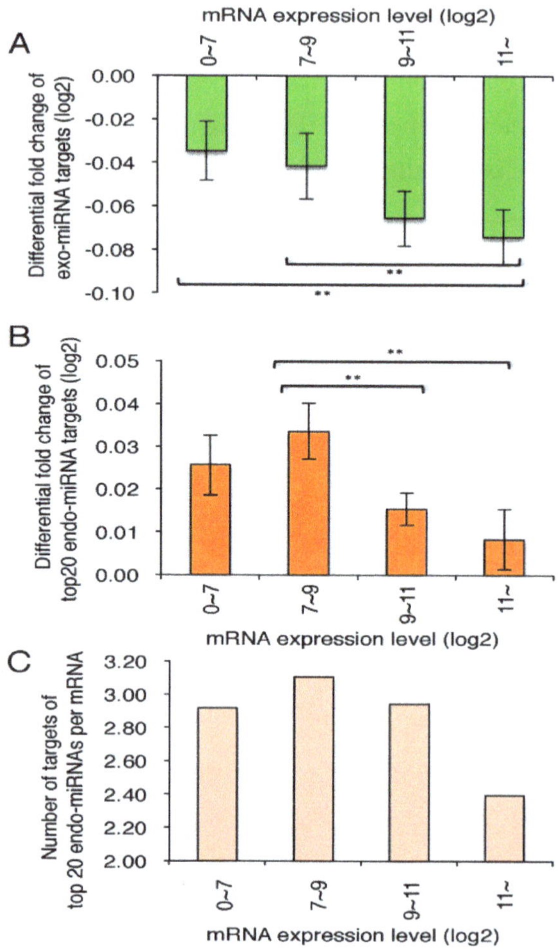

3. Experimental Section

3.1. Cell Culture and miRNA Synthesis

Human HeLa cells were cultured and used in reporter assays and microarray analyses. Cells were cultured at 37 °C in Dulbecco's modified Eagle's Medium (Invitrogen, Carlsbad, NM, USA), supplemented with 10% heat-inactivated fetal bovine serum (Sigma, St. Louis, MO, USA). They were plated on 24-well culture plates (1×10^5 cells/mL/well) 24 h prior to transfection. Transfection was carried out using Lipofectamine 2000 (Invitrogen, Carlsbad, NM, USA). Each RNA strand of the miRNA duplex was chemically synthesized (Sigma, St. Louis, MO, USA) and annealed to form the duplex structures the same as those shown in miRBase [28]. The sequences of the synthetic miRNAs (let-7b, miR-1, miR-21, miR-22, miR-28, miR-30c-1, miR-186, miR-199b, miR-200b, miR-330, miR-335, miR-346, miR-466, miR-574 and miR-3126) are listed in Table S1.

3.2. Construction of Luciferase Reporters

All of the reporter plasmids constructed were derivatives of psiCHECK-1 (Promega, Fitchburg, WI, USA). Oligonucleotides with target sequences completely matched to each miRNA strand (cm-target) were chemically synthesized with cohesive *Xho*I/*Eco*RI ends (Table S2). They then were inserted into the corresponding restriction sites of psiCHECK-1 to generate miRNA cm-targets (miR-200b-3p target, miR-7b-5p target, miR-21-5p target and miR-330-5p target). Each of the inserted targets was expressed as part of the 3'_UTR region of *Renilla* luciferase mRNA in transfected cells.

HeLa cells growing in 24-well plates were transfected simultaneously with miRNA target (100 ng), pGL3-Control (Promega, 0.5 μg) and miRNA (50 nM). The cells were harvested 24 h post-transfection and the relative luciferase activity (*Renilla* luc activity/firefly luc activity) was determined using a Dual-Luciferase Reporter Assay System (Promega, Fitchburg, WI, USA). The pGL3-Control encoding firefly luciferase served as a control for the calculation of relative luciferase activity for miRNAs.

3.3. Microarray Analysis

HeLa cells (1×10^5 cells/mL) were transfected with 50 nM of each of 15 miRNA duplexes. At 24 h post-transfection, total RNA was purified using an RNeasy Kit (Qiagen, Hilden, Germany). The steps were repeated four times and RNA quality assessed using a NanoDrop 2000 spectrophotometer (Thermo Scientific, Waltham, MA, USA) and a Bioanalyzer (Agilent, Santa Clara, CA, USA). RNAs recovered independently were mixed equally for cDNA synthesis using an Agilent One Color Spike Mix Kit (Agilent, Santa Clara, CA, USA). Cy3-labeled cRNA was synthesized using a Quick Amp Labeling Kit (Agilent, Santa Clara, CA, USA) and was hybridized to an Agilent Whole Human Genome Microarray (4 × 44 K multi-pack format), according to the manufacturer's protocol. RNA from mock-transfected cells treated with transfection reagent in the absence of miRNA was used as a control. Transcript expression values were calculated using Microarray Suite 5.0 (MAS5: Affymetrix, Santa Clara, CA, USA) [32] with quantile normalization [33]. To identify transcripts whose expression was upregulated or downregulated, the cumulative distribution of expression changes for transcripts containing the site was compared with that for transcripts with no canonical site. NCBI's Reference Sequence (RefSeq) was used to identify mRNAs with sequences complementary to the seed regions of the transfected miRNAs. Data are presented as an MA plot (M = intensity ratio, A = average intensity) and a cumulative frequency distribution. Changes in expression are shown as fold changes ($\log_2$).

3.4. Quantitative RT-PCR

Total RNA was reverse-transcribed using a Transcriptor High Fidelity cDNA Synthesis kit (Roche, Basel, Switzerland). The resultant cDNA samples were incubated with FastStart Universal SYBR Green Master (Roche, Basel, Switzerland) at 95 °C for 10 min, followed by PCR amplification. PCR product levels were monitored using an ABI PRISM 7000 sequence detection system and analyzed with ABI PRISM 7000 SDS software (Applied Biosystems). The expression of each target gene was first normalized to that of β-actin and then to the mock-transfection control. The primer sets used are listed in Table S3.

4. Conclusions

In this study, we quantified the changes in expression levels of endo-miRNA target genes resulting from the transfection of exo-miRNA duplexes. The expression levels of endo-miRNA target genes with seed-complementary sequences were increased by the transfection of exo-miRNA duplexes, while exo-miRNA target gene expression was reduced (Figures 1–3 and Figures S1–S3). These results suggest that exo-miRNA duplex transfected into the cells may compete with endo-miRNAs for the RISC, which may be saturated with endo-miRNAs under normal conditions in HeLa cells.

In miRNA-mediated gene silencing, the structure of the RNA-protein complex is known to be altered [23]. The miRNA duplex in RISC loading complex is unwound, yielding single-stranded RNA, which is loaded onto the RISC and recognizes target mRNAs through base-pairing in the seed region [4–6]. Our results indicate that the expression of the target genes of a given endo-miRNA differed according to the exo-miRNA duplex that was transfected (Figures 2 and 7A), whereas with a given exo-miRNA duplex, the fold changes in the target gene expression of the endo-miRNAs examined were mostly equivalent, except for a limited types of exo-miRNAs (miR-1, miR-28, miR-199b and miR-335), despite differences in the structures and sequences of the endo-miRNA duplexes (Figure 3 and Figure 7B). One of the possible explanations of these results is that the RISC exchange reaction might be occurred associated with single-stranded endo-miRNAs and not endo-miRNA duplexes or target-paired endo-miRNAs (Figure 7), because double-stranded endo-miRNAs in the RISC might not be replaced with exo-miRNAs at similar levels, due to their different structures and sequences. Most miRNA-RISCs might be in the form of single-stranded miRNAs in RISCs, so as to be readily replaced by double-stranded miRNAs.

Figure 7. Predicted model for the RNA-induced silencing complexes (RISCs) replacement. Different types of exo-miRNAs transfected into cells may be replaced with single-stranded endo-miRNAs loaded on the RISC with different efficiencies (**A**); however, different types of endo-miRNAs may be replaced with a given exo-miRNAs with similar efficiencies (**B**).

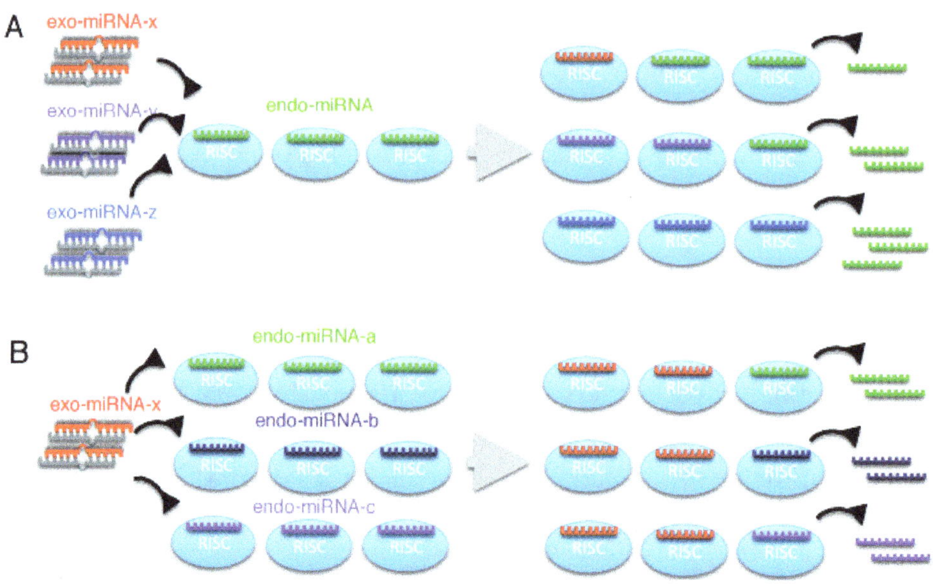

Furthermore, it was apparently revealed that endo-miRNAs constantly repress the expression of endogenous mRNAs with endo-miRNA target sites, and their repression is probably relieved by the replacement of endo-miRNAs on the RISC by the exo-miRNAs transfected (Figures 5 and 6). Thus, global gene expression by endogenous miRNAs might be fluctuated by the transfection of exo-miRNAs or the increase of expression levels of endo-miRNAs. Competition similar to that shown here between exo-miRNA duplexes and endo-miRNAs may also occur among newly transcribed endo-miRNAs and may provide a mechanism for orchestrating cellular programs.

Acknowledgments

We thank Yuki Naito for helpful discussions about the bioinformatics analysis and Kenji Nishi for technical advises on microarray experiments. This work was supported by a Grant-in-Aid for Scientific Research (grant numbers 21115004, 21310123), the Cell Innovation Project from the Ministry of Education, Culture, Sports, Science and Technology of Japan and the Core Research Project for Private University matching fund subsidy to K.U.-T.

References

1. Rana, T.M. Illuminating the silence: Understanding the structure and function of small RNAs. *Nat. Rev. Mol. Cell Biol.* **2007**, *8*, 23–36.
2. Filipowicz, W.; Bhattacharyya, S.N.; Sonenberg, N. Mechanisms of post-transcriptional regulation by microRNAs: Are the answers in sight? *Nat. Rev. Genet.* **2008**, *9*, 102–114.
3. Bartel, D.P. MicroRNAs: Genomics, biogenesis, mechanism, and function. *Cell* **2004**, *116*, 281–297.
4. Grimson, A.; Farh, K.K.; Johnston, W.K.; Garrett-Engele, P.; Lim, L.P.; Bartel, D.P. MicroRNA targeting specificity in mammals: Determinants beyond seed pairing. *Mol. Cell* **2007**, *27*, 91–105.
5. Lewis, B.P.; Burge, C.B.; Bartel, D.P. Conserved seed pairing, often flanked by adenosines, indicates that thousands of human genes are microRNA targets. *Cell* **2005**, *120*, 15–20.
6. Lim, L.P.; Lau, N.C.; Garrett-Engele, P.; Grimson, A.; Schelter, J.M.; Castle, J.; Bartel, D.P.; Linsley, P.S.; Johnson, J.M. Microarray analysis shows that some microRNAs downregulate large numbers of target mRNAs. *Nature* **2005**, *433*, 769–773.
7. Grimm, D.; Streetz, K.L.; Jopling, C.L.; Storm, T.A.; Pandey, K.; Davis, C.R.; Marion, P.; Salazar, F.; Kay, M.A. Fatality in mice due to oversaturation of cellular microRNA/short hairpin RNA pathways. *Nature* **2006**, *441*, 537–541.
8. Castanotto, D.; Sakurai, K.; Lingeman, R.; Li, H.; Shively, L.; Aagaard, L.; Soifer, H.; Gatignol, A.; Riggs, A.; Rossi, J.J. Combinatorial delivery of small interfering RNAs reduces RNAi efficacy by selective incorporation into RISC. *Nucleic Acids Res.* **2007**, *35*, 5154–5164.
9. Stewart, C.K.; Li, J.; Golovan, S.P. Adverse effects induced by short hairpin RNA expression in porcine fetal fibroblasts. *Biochem. Biophys. Res. Commun.* **2008**, *370*, 113–117.

10. Khan, A.A.; Betel, D.; Miller, M.L.; Sander, C.; Leslie, C.S.; Marks, D.S. Transfection of small RNAs globally perturbs gene regulation by endogenous microRNAs. *Nat. Biotechnol.* **2009**, *27*, 549–555.
11. Yi, R.; Qin, Y.; Macara, J.G.; Cullen, B.R. Exportin-5 mediates the nuclear export of pre-microRNAs and short hairpin RNAs. *Genes Dev.* **2003**, *17*, 3011–3016.
12. Lund, E.; Guttinger, S.; Calado, A.; Dahlberg, J.E.; Kutay, U. Nuclear export of microRNA precursors. *Science* **2004**, *303*, 95–98.
13. Bohnsack, M.T.; Czaplinski, K.; Gorlich, D. Exportin-5 is a RanGTP-dependent dsRNA-binding protein that mediates nuclear export of pre-miRNAs. *RNA* **2004**, *10*, 185–191.
14. Kim, V.N. MicroRNA precursors in motion; exportin-5 mediates their nuclear export. *Trends Cell Biol.* **2004**, *14*, 156–159.
15. Vankoningsloo, S.; de Longueville, F.; Evrard, S.; Rahier, P.; Houbion, A.; Fattaccioli, A.; Gastellier, M.; Remacle, J.; Raes, M.; Ranard, P.; *et al.* Gene expression silencing with "specific" small interfering RNA goes beyond specificity–a study of key parameters to take into account in the onset of small interfering RNA off-target effects. *FEBS J.* **2008**, *275*, 2738–2753.
16. Gregory, R.I.; Chendrimada, T.P.; Cooch, N.; Shienkhattar, R. Human RISC couples microRNA biogenesis and posttranscriptional gene silencing. *Cell* **2005**, *123*, 631–540.
17. Haase, A.D.; Jaskiewicz, L.; Zhang, H.; Lainé, S.; Sack, R.; Gatignol, A.; Filipowicz, S. TRBP, a regulator of cellular PKR and HIV-1 virus expression, interacts with Dicer and functionas in RNA silencing. *EMBO J.* **2005**, *6*, 961–967.
18. Macrae, I.J.; Ma, E.; Zhou, M.; Robinson, C.V.; Doudna, J.A. *In vitro* reconstitution of the human RISC-loading complex. *Proc. Natl. Acad. Sci. USA* **2008**, *105*, 512–517.
19. Winter, J.; Jung, S.; Keller, S.; Gregory, R.; Diederichs, S. Many roads to maturity: microRNA biogenesis pathways and their regulation. *Nat. Cell Biol.* **2009**, *11*, 228–234.
20. Matranga, C.; Tomari, Y.; Shin, C.; Bartal, D.P.; Zamore, P.D. Passenger-strand cleavage facilitates assembly of siRNA into Ago2-containing RNAi enzyme complexes. *Cell* **2005**, *123*, 607–620.
21. Leuschner, P.J.; Ameres, S.L.; Kueng, S.; Martinez, J. Cleavage of the siRNA passenger strand during RISC assembly in human cells. *EMBO Rep.* **2006**, *7*, 314–320.
22. Tolia, N.H.; Joshua-Tor, L. Slicer and the Argonaute. *Nat. Chem. Biol.* **2006**, *3*, 36–43.
23. Czech, B.; Hannon, G.J. Small RNA sorting: Matchmaking for Argonautes. *Nat. Rev. Genetics* **2011**, *12*, 19–31.
24. Khvorova, A.; Reynolds, A.; Jayasena, S.D. Functional siRNAs and miRNAs exhibit strand bias. *Cell* **2003**, *115*, 209–215.
25. Schwarz, D.S.; Hutvágner, G.; Du, T.; Xu, A.; Aronin, N.; Zamore, P.D. Asymmetry in the assembly of the RNAi enzyme complex. *Cell* **2003**, *115*, 199–205.
26. Ui-Tei, K.; Naito, Y.; Takahashi, F.; Haraguchi, T.; Ohki-Hamazaki, H.; Juni, A.; Ueda, R.; Saigo, K. Guidelines for the selection of highly effective siRNA sequences for mammalian and chick RNA interference. *Nucleic Acids Res.* **2004**, *32*, 936–948.
27. Yoda, M.; Kawamata, T.; Paroo, Z.; Ye, X.; Iwasaki, S.; Liu, Q.; Tomari, Y. ATP-dependent human RISC assembly pathways. *Nat. Struct. Mol. Biol.* **2010**, *17*, 17–23.

28. Griffiths-Jones, S.; Saini, H.K.; van Dongen, S.; Enright, A.J. MiRBase tools for microRNA genomics. *Nucleic Acids Res.* **2008**, *36*, D154–D158.
29. Jackson, A.L.; Burchard, J.; Schelter, J.; Chau, B.N.; Cleary, M.; Lim, L.; Linsley, P.S. Widespread siRNA "off-target" transcript silencing mediated by seed region sequence complementarity. *RNA* **2006**, *12*, 1179–1187.
30. Mayr, C.; Bartel, D.P. Widespread shortening of 3'UTRs by alternative cleavage and polyadenylation activates oncogenes in cancer cells. *Cell* **2009**, *138*, 673–684.
31. Saito, T.; Sætrom, P. Target gene expression levels and competition between transfected and endogenous microRNAs are strong confounding factors in microRNA high-throughput experiments. *Silence* **2012**, *3*, 3.
32. Hubbell, E.; Liu, W.M.; Mei, R. Robust estimation for expression analysis. *Bioinformatics* **2002**, *18*, 1585–1592.
33. Bolstad, B.M.; Irizarry, R.A.; Gautier, L.; Wu, Z. Preprocessing High-Density Oligonucleotide Arrays. In *Bioinformatics and Computational Biology Solutions Using R and Bioconductor*; Gentleman, R., Carey, V.J., Huber, W., Irizarry, R.A., Dudoit, S., Eds.; Springer: New York, NY, USA, 2005; pp. 13–32.

4

MicroRNA Regulation in Renal Pathophysiology

Jianghui Hou [1,*] and Dan Zhao [2]

1 Renal Division and Center for Investigation of Membrane Excitability Diseases, Washington University in St. Louis, 660 South Euclid Avenue, St. Louis, MO 63110, USA

2 Division of Pharmacology, PLA 85th Hospital, 1328 Hua Shan Road, Shanghai 20052, China; E-Mail: zdandzjw@sina.com

* Author to whom correspondence should be addressed; E-Mail: jhou@wustl.edu;

Abstract: MicroRNAs are small, noncoding RNA molecules that regulate a considerable amount of human genes on the post-transcriptional level, and participate in many key biological processes. MicroRNA deregulation has been found associated with major kidney diseases. Here, we summarize current knowledge on the role of microRNAs in renal glomerular and tubular pathologies, with emphasis on the mesangial cell and podocyte dysfunction in diabetic nephropathy, the proximal tubular cell survival in acute kidney injury, the transport function of the thick ascending limb in Ca^{++} imbalance diseases, and the regulation of salt, K^+ and blood pressure in the distal tubules. Identification of microRNAs and their target genes provides novel therapeutic candidates for treating these diseases. Manipulation of microRNA function with its sense or antisense oligonucleotide enables coordinated regulation of the entire downstream gene network, which has effectively ameliorated several renal disease phenotypes. The therapeutic potentials of microRNA based treatments, though promising, are confounded by major safety issues related to its target specificity, which remain to be fully elucidated.

Keywords: microRNA; kidney; diabetic nephropathy; hypercalciuria; hypertension

---

1. Introduction

MicroRNAs comprise a large family of 21–22-nucleotide-long RNAs that have emerged as key post-transcriptional regulators of gene expression in animals and plants [1,2]. In animals, microRNAs are predicted to control the activity of ~50% of all protein-coding genes [3]. Functional studies indicate that microRNAs participate in the regulation of almost every cellular process, and are intrinsically associated with many human pathologies.

In animals, microRNAs are processed from longer hairpin transcripts, known as pre-microRNA, by the RNase III-like enzymes Drosha and Dicer, whereas in plants only Dicer is responsible for microRNA processing [4,5]. One strand of the hairpin duplex is loaded into an Argonaute family protein (AGO) to form the core of microRNA-induced silencing complexes (miRISCs). miRISCs silence the expression of target genes through mRNA decay and translational repression. The target recognitions are achieved through base-pairing complementarity between the loaded microRNA and the target mRNA that contains a partially or fully complementary sequence [4,5]. Unlike plant microRNAs, that recognize fully complementary binding sites within the open reading frame (ORF), animal microRNAs recognize partially complementary binding sites generally located in the 3'-untranslated region (UTR) (Figure 1). For most microRNA binding sites, the complementarity is limited to the seed sequence found in the 5'-end of microRNA from nucleotide 2 to 7. The partial recognition between microRNA and its target is sufficient to trigger silencing.

Figure 1. MicroRNA target recognition mechanism. Plant microRNAs recognize fully complementary binding sites within the open reading frame (ORF) of the target mRNA. Animal microRNAs recognize partially complementary binding sites located in the 3'-untranslated region (3'-UTR). For most microRNA binding sites, the complementarity is limited to the seed sequence containing nucleotides #2–7 on the microRNA molecule. Note that in both plant and animal microRNAs the 5'-terminal nucleotide (#1) is not involved in target recognition.

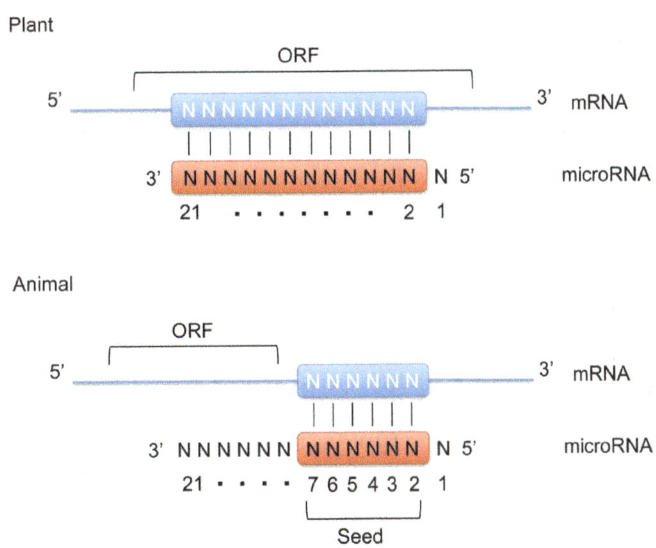

The significance of microRNA in renal pathophysiology has been demonstrated in Dicer knockout animal models. During kidney development, the global knockout of Dicer in nephron progenitor cells results in a marked decrease in nephron number [6]. Conditional removal of Dicer from the ureteric lineage results in cystic kidney disease [7]. Podocyte-specific loss of Dicer function causes proteinuria, foot process effacement, and glomerulosclerosis [8–10]. The inducible deletion of another microRNA processing enzyme, Drosha, in mature podocytes from two- to three-month-old mice, results in a similar phenotype, demonstrating a post-developmental need for microRNA activity in podocytes [11]. Deletion of Dicer in renin-secreting juxtaglomerular cells results in a selective loss of these juxtaglomerular cells, suggesting a role in cell fate determination [12]. In the proximal tubule, microRNA appear to promote cellular injury because a selective loss of Dicer in animals after three weeks of age confers resistance to ischemia-reperfusion injury [13].

2. MicroRNA in Glomerular Diseases

Hypertrophy and expansion in the glomerular mesangium, along with podocyte dysfunction and accumulation of extracellular matrix proteins, are major features of diabetic nephropathy, glomerulonephritis, glomerulosclerosis, and many other types of glomerular pathologies. Studies of microRNAs and their targets in glomeruli will provide critical insights of the pathogenesis of glomerular diseases and reveal new therapeutic targets for pharmacological intervention.

2.1. MicroRNA in Glomerular Mesangial Cell

Among the microRNAs highly expressed in the kidney [14,15], several key microRNAs (miR-192, miR-200b, miR-200c, miR-216a, and miR-217) were upregulated in glomerular mesangium of diabetic mouse models (type I (streptozotocin (STZ)-induced) and type2 (db/db)) (*vide infra*). *In vitro*, TGF-β-induced miR-192 was shown to increase the gene expression of collagen 1α2 by reducing the expression of two E-box repressors (Zeb1 and Zeb2) that control collagen 1α2 gene activation [16]. *In vivo*, the miR-192 and Collagen 1α2 levels were substantially increased in the mesangial cells of STZ-induced diabetic mice, as well as of db/db diabetic mice, suggesting a role in glomerular basement membrane thickening (Figure 2A) [16]. miR-216a and miR-217 are downstream targets of miR-192 through Zeb1/2 mediated mechanisms [17]. In diabetic mesangium, the cellular levels of both microRNAs were increased, resulting in the silencing of their target—PTEN (phosphatase and tensin homolog). PTEN is a phosphatidylinositol-3,4,5-triphosphate (PIP3)-phosphatase that inhibits the phosphoinositide-3-kinase pathway (PI3K) and thus prevents Akt activation [18]. Akt activation by miR-216a and miR-217 led to glomerular mesangial cell expansion and hypertrophy, another hallmark of diabetic nephropathy (Figure 2A). MiR-377 was also upregulated in cultured human and mouse mesangial cells by glucose or TGF-β treatment or in type I diabetes [19]. Overexpression of miR-377 in mesangial cells *in vitro* increased fibronectin protein production, another component of the glomerular extracellular matrix. Mechanistically, miR-377 silenced the expression of serine/threonine protein kinase PAK1, also known as p21 protein (Cdc42/Rac)-activated kinase 1, and superoxide dismutase, both of which negatively controlled fibronectin protein production (Figure 2A) [19]. The miR-200 family members are separated into two clusters located in different genomic loci. Among them, miR-200b and c are regulated by the miR-192 targets—Zeb1/2 through E-boxes in the promoters of their host genes [20,21]. miR-200b and c also target the transcripts of Zeb1/2 to auto-regulate their own expression and amplify the signaling response leading to collagen expression and glomerular fibrosis (Figure 2A) [22].

Figure 2. The role of microRNA in glomerular diseases. (**A**) In the mesangial cell, diabetic conditions increases the levels of miR-377 and miR-192, both of which promote fibrosis and hypertrophy through the signaling cascades involving PAK1 and Zeb1/2; (**B**) In the podocyte, focal segmental glomerulosclerosis (FSGS) induces the expression of miR-193a, which in turn inhibits WT1, a master regulator of podocyte homeostasis through podocalyxin (PODXL) and nephrin (NPHS1).

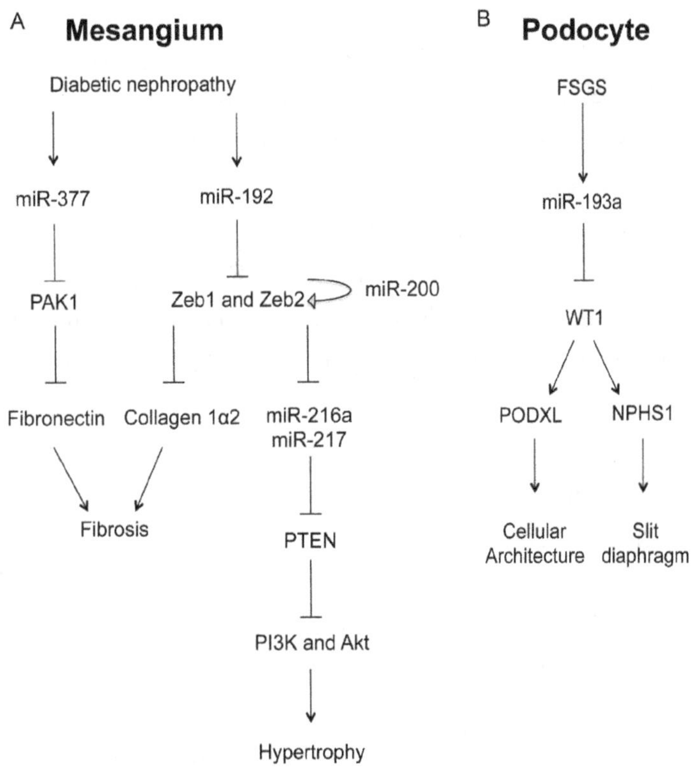

IgA nephropathy is a leading cause of idiopathic glomerulonephritis and is characterized by mesangial deposition of IgA. The levels of glomerular microRNAs were deranged in patients with IgA nephropathy [23]. Among the analyzed microRNAs, the expression level of miR-200c was profoundly downregulated in these patients and negatively correlated with proteinuria, while the level of miR-192 was significantly upregulated and positively correlated with glomerulosclerosis [23].

2.2. MicroRNA in Glomerular Podocyte

Podocytes in the glomerular basement membrane are critical in the maintenance of structure and function of the glomerular filtration barrier. To study an overall role of miRNAs in podocyte biology, two independent lines of Dicer KO mice were generated for podocytes [9,10]. Mutant mice developed proteinuria by three weeks after birth and progressed rapidly to end-stage kidney disease. Multiple abnormalities were observed in glomeruli of mutant mice, including foot process effacement, irregular and split areas of the glomerular basement membrane, podocyte apoptosis and depletion, mesangial expansion, capillary dilation, and glomerulosclerosis [10]. Cytoskeletal dynamics was significantly altered in mutant animals, including early loss of synaptopodin and downregulation of the ERM protein family (ezrin-radixin-moesin) at three weeks [9]. Gene profiling revealed upregulation of 190 genes in glomeruli isolated from mutant mice at the onset of proteinuria. Target sequences for 16 microRNA were significantly enriched in the 3'-untranslated regions of 190 upregulated genes [10]. Bioinformatic approaches were used to validate six of the eight top-candidate microRNAs, which were miR-28, miR-34a, and four members of the miR-30 family (miR-30c-1, miR-30b, miR-30d, and miR-30c-2) [10]. The miR-30 family was shown to target four genes known to be functional in podocytes, including genes that mediate podocyte apoptosis (receptor for advanced glycation end product and immediate early response 3 protein) and cytoskeletal disruption (vimentin and heat shock protein 20).

Focal segmental glomerulosclerosis (FSGS) is a devastating glomerular diseases caused by podocyte dysfunction. Deranged expression of several podocyte specific genes (WT1, NPHS1, ACTN4, and TRPC6), accompanied by collapse of normal podocyte shape and podocyte foot process effacement, presents as major pathogenic origins for FSGS. The importance of microRNAs in FSGS has been demonstrated by Gebeshuber and colleagues in a recent study [24]. Through transgenic screening in mice, Gebeshuber *et al.* have identified miR-193a as a powerful inducer of FSGS. Mechanistically, miR-193a silences the Wilms' tumor (WT1) gene, which encodes a transcriptional factor and acts as a master regulator for podocye homeostasis [24]. In normal podocytes, WT1 positively regulates the expression of several key genes crucial for podocyte architecture, e.g., podocalyxin (PODXL) and for slit diaphragm formation, e.g., nephrin (NPHS1). The level of miR-193a was consistently higher in isolated glomeruli from FSGS patients compared to normal kidneys, which provides an important mechanism for FSGS pathogenesis (Figure 2B) [24]. Deranged expression of selected microRNAs also causes podocyte abnormalities under diabetic conditions such as apoptosis and fibrosis [25,26]. In diabetic podocytes, the level of miR-195 is significantly elevated [25] while miR-29 is reduced [26]. MiR-195 targets the BCL2 gene and contributes to podocyte apoptosis via an increase in caspase-3 protein [25]. MiR-29, on the other hand, represses the expression of collagens I and IV, at both the mRNA and protein levels, by targeting the 3'-untranslated region of these genes [26].

3. MicroRNA in the Proximal Nephron

Renal ischemia-reperfusion injury (IRI) is a major cause of acute kidney injury (AKI), which is often associated with renal failure and high mortality rates [27]. The pathogenic role of microRNAs in AKI was interrogated by Wei *et al.* [13] using a proximal tubule specific Dicer KO model. Despite normal development, histology, and function of the kidney, these conditional KO mice are remarkably resistant to renal IRI, showing significantly better renal function, less tissue damage, lower tubular apoptosis rate, and higher survival rates. Microarray analyses in wildtype animals undergoing the same IRI procedure revealed changes in microRNA expression levels in the proximal tubule. Among the 173 microRNAs detected in renal cortex, miRNA-132, -362, -379, -668, and -687 showed continuous change during 12–48 h of reperfusion [13]. Another study by Godwin *et al.* demonstrated similar changes in microRNA expression during renal IRI in laboratory mice [28]. Consistent with the pathogenic hypothesis of microRNA in AKI, miR-34a was upregulated in mouse proximal tubular cells within a few hours of cisplatin-induced nephrotoxicity [29]. Inhibition of p53 with pifithrin-α abrogated the induction of miR-34a during cisplatin treatment. Ablation of miR-34a with antisense oligonucleotides led to increased apoptosis and reduced cell survival [29]. MiR-34a is therefore a novel downstream target of p53 in response to cisplatin induced DNA damage.

In contrast to its role in the glomerular mesangium, miR-192 appears to play a protective role against tubulointerstitial fibrosis in the proximal tubule of diabetic patients [30]. Krupa and colleagues performed a systematic study to profile microRNA expression levels in renal biopsies from patients with established diabetic nephropathy and identified 12 microRNAs showing significant differences from normal kidneys [30]. Among them, miR-192 showed the greatest change, the level of which was consistently lower in diabetic patients. In individual biopsies, miR-192 expression was inversely correlated with the progression of tubulointerstitial fibrosis and the loss of GFR. Mechanistically, overexpression of miR-192 in cultured proximal tubular cells suppressed the expression of Zeb1 and 2, opposing TGF-β-mediated epithelial-to-mesenchymal transition (EMT) [30].

4. MicroRNA in the Loop of Henle

The thick ascending limb of Henle's loop (TALH) is responsible for extracellular Ca^{++} and Mg^{++} homeostasis. A signaling cascade made of CaSR and claudins senses extracellular Ca^{++}/Mg^{++} differences and makes corresponding changes in renal excretion rates. CaSR—the Ca^{++} sensing receptor is a member of the G protein-coupled receptors (GPCRs). Mutations in the CaSR gene cause familial hypocalciuric hypercalcemia (FHH) and neonatal severe hyperparathyroidism (NSHPT), two inherited conditions characterized by altered calcium homeostasis [31]. Claudins are integral membrane proteins of the tight junction that is responsible for the paracellular transport of ions and solutes between apical and basolateral membranes. Mutations in the claudin genes, claudin-16 and claudin-19, cause familial hypomagnesemia with hypercalciuria and nephrocalcinosis (FHHNC) [32,33]. Synonymous sequence variants in the claudin-14 gene are associated with hypercalciuric kidney stones and reduced bone mineral density [34]. Several *in vitro* studies have revealed that that claudin-16 and -19 form a heteromeric cation channel in the TALH [35], whereas claudin-14 interacts with the claudin-16 and inhibits its cation permeability as a regulatory subunit [36]. Although the promoter activity and the mRNA level of claudin-14 are very high in the kidney, its protein level is surprisingly low. Gong *et al.* have identified two microRNA molecules, miR-9 and miR-374 from TALH cells, both of which recognize partially complementary binding sites located in 3'-UTRs of claudin-14 mRNA (Figure 3) [36]. Treatments with antisense oligonucleotide against miR-9 or miR-374 revealed that both microRNAs suppressed claudin-14 translation and induced its mRNA decay in a synergistic manner [36]. High Ca^{++} intake significantly downregulated the expression levels of miR-9 and miR-374 in TALH cells, which in turn causes a reciprocal increase in claudin-14 expression level. Deletion of CaSR from TALH cells abolished extracellular Ca^{++} induced changes in microRNA and claudin-14 [36]. The dietary regulation of microRNA suggests a physiological role for microRNA based signaling in the TALH of the kidney. The observed association between claudin-14 and hypercalciuric nephrolithiasis [34] can be explained by claudin-14 deregulation that escapes microRNA suppression, inhibits claudin-16/-19 channel permeabilities and phenocopies FHHNC to variable degrees. FHHNC patients [33,37] and animal models [38,39] with claudin-16 or claudin-19 mutations are known to have hypercalciuria, nephrocalcinosis, and nephrolithiasis. The regulation of microRNA by CaSR may occur on several layers: microRNA transcription, processing, or degradation [4]. Transcriptional regulation is undoubtedly the most specific way for individual microRNA. The promoters of both miR-9 (miR-9-3 locus) and miR-374 genes contain a canonical myc-binding site (E-box: CACGTG). The transcription of miR-9-3 is upregulated by myc in human breast cancer cells [40]; miR-421/-374 cluster is upregulated by myc in HeLa cells [41]. Although these studies did not prove a role for myc in CaSR signaling, both CaSR and claudin have been implicated in tumorigenesis and metastasis [42–44].

Figure 3. The role of microRNA in renal CaSR signaling. A feedback loop of CaSR signaling in the thick ascending limb of Henle's loop (TALH) regulates urinary excretion of Ca^{++} through transcriptional regulation of miR-9 and miR-374. MicroRNAs in turn negatively regulate the expression level of claudin-14 through mRNA decay and translational repression. Claudin-14 directly binds to claudin-16 and inhibits its cation permeability to urinary Ca^{++}.

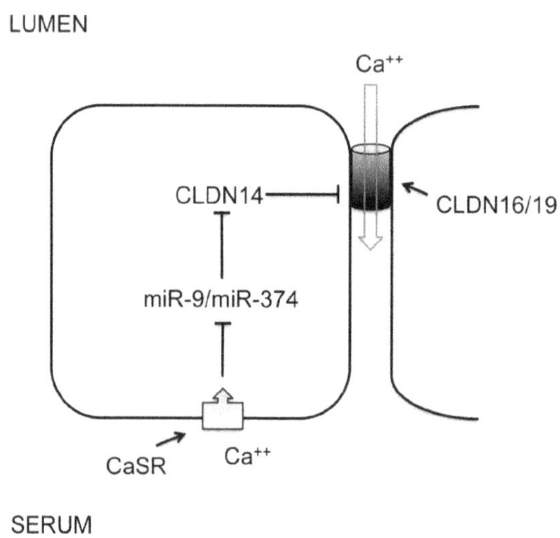

In addition to Ca^{++} metabolism, microRNAs also play a role in salt and fluid handling in the TALH. Mladinov *et al.* performed microarray assays in microdissected nephron segments and identified 12 microRNAs that were highly enriched in the TALH compared to the glomerulus and the proximal tubule [45]. Among them, miR-192 was found to target the Na^+/K^+-ATPase β1 subunit gene (Atp1b1). High salt diet increased the expression level of miR-192 in the TALH, which in turn suppressed Atp1b1 gene expression [45]. Knockdown of miR-192, *in vivo* with antisense oligonucleotide, upregulated the Atp1b1 protein level in the kidney, causing antidiuresis under high salt dietary condition. Contrasting with common microRNA target areas, miR-192 appeared to target Atp1b1 through the 5'-rather than 3'-untranslated region, although its binding sites were present within both regions [45].

5. MicroRNA in the Distal Nephron

The aldosterone-sensitive distal nephron (ASDN) encompasses the distal convoluted tubule (DCT), the connecting tubule, and the collecting duct. It is collectively responsible for the reabsorption of approximately 5% of filtered NaCl, and plays a vital role in the regulation of extracellular fluid volume (ECFV) and blood pressure [46]. The involvement of microRNA in hypertensive kidney disease was first studied by Sequeira-Lopez *et al.* in the juxtaglomerular cell [12]. The juxtaglomerular cell is found between the vascular pole of the renal corpuscle and the macula densa of the DCT. Its primary role in blood pressure control is the synthesis of renin. Renin is the key regulated step that initiates an enzymatic cascade that leads to angiotensin and aldosterone generation (collectively known as the rennin-angiotensin-aldosterone system). Conditional knockout of Dicer in juxtaglomerular cells resulted in a pronounced reduction in the number of juxtaglomerular cells accompanied by decreased expression of *Ren1* and *Ren2*, decreased plasma renin concentration, decreased blood pressure, and striking nephrovascular abnormalities, including striped corticomedullary fibrosis [12].

Two studies reported the role of microRNA in the regulation of ion transport by the ASDN. Elvira-Matelot *et al.* have shown that miR-192 expression is strongly reduced in the kidneys of mice treated with salt depletion, potassium load, or chronic aldosterone infusion, whereas its level is not modified by a high salt diet (Figure 4) [47]. The serine-threonine kinase WNK1 is the target of miR-192 when assayed *in vitro* and *ex vivo*. Its gene expression was reciprocally regulated by aldosterone, potassium and salt as to miR-192 in the kidney [47]. WNK1, belonging to the WNK (with no lysine-K) serine-threonine kinase subfamily, is essential for the coordinated regulation of Na^+ and K^+ transport in the kidney [48]. Mutations in WNK1 cause familial hyperkalemic hypertension (FHHt), a rare mendelian form of human hypertension [49].

Figure 4. The role of microRNA in Na^+ and K^+ handling by the distal nephron. Na^+ depletion or K^+ load increases serum aldosterone level, which decreases miR-192 levels in the aldosterone-sensitive distal nephron (ASDN). MiR-192 suppresses WNK1 gene expression, which in turn regulates Na^+ reabsorption through the NCC and ENaC channels. K^+ load also increases miR-802 levels in the ASDN. MiR-802 suppresses caveolin-1 gene expression, which regulates renal outer medullary potassium (ROMK) channel surface expression through endocytosis. ROMK mediates K^+ secretion to urine.

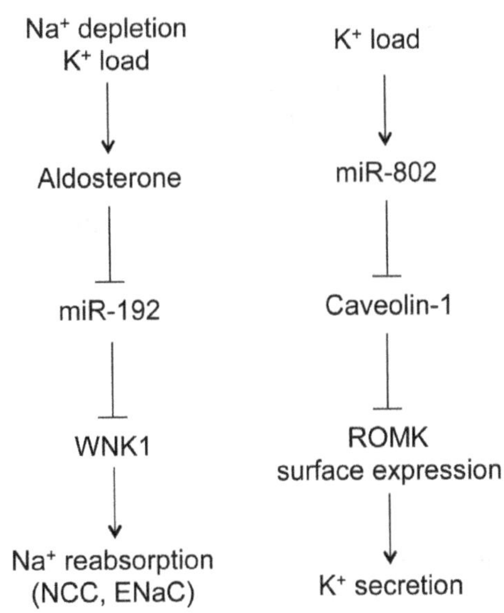

The renal outer medullary potassium channel (ROMK) is an ATP-dependent potassium channel encoded by the gene Kir1.1 and responsible for K^+ secretion along the ASDN. Its membrane localization is regulated by dietary K^+ intake through endocytosis-mediated mechanisms. Lin *et al.*

have revealed a physiological role for microRNA in ROMK endocytosis [50]. The ROMK channel is physically associated with caveolin-1, the principal protein component of caveolae. Expression of caveolin-1 varied inversely with the expression of ROMK in the plasma membrane, and caveolin-1 inhibited ROMK channel activity [50]. Caveolin-1 is the molecular target of miR-802 in the collecting duct. *In vitro*, expression of miR-802 suppressed the expression of caveolin-1; *in vivo*, the level of miR-802 was inversely correlated with that of caveolin-1 [50]. High K^+ intake stimulated the transcription of miR-802, which in turn decreased the expression of caveolin-1 and increased the membrane localization of ROMK, leading to higher K^+ excretion (Figure 4) [50].

MicroRNAs are also regulated by tonicity in the ASDN of the kidney. Exposure of the inner medullary collecting duct cells (mIMCD3) to a hypertonic solution induces a decrease in miR-200b and miR-717 expression level as early as two hours [51]. A common target for both microRNAs is the gene encoding the tonicity responsive element binding protein (TonEBP), also known as the osmotic response element binding protein (OREBP) [51]. TonEBP regulates many aspects of tonicity induced cellular responses, such as accumulation of inorganic osmolytes and expression of the HSP70 osmoprotective chaperone protein [52]. Depletion of microRNAs by knocking-down Dicer significantly increases TonEBP protein expression, while overexpression of miR-200b and miR-717 in mIMCD3 cells suppresses TonEBP expression on both the transcriptional and post-transcriptional levels [51]. *In vivo* in the mouse kidney, furosemide induced diuresis significantly upregulates miR-200b and miR-717, but downregulates TonEBP in the medulla [51]. The inverse correlation between the expression of microRNA and TonEBP suggests a signaling mechanism for renal cell adaptation to urinary osmolality.

6. MicroRNA as Therapeutic Candidates

Antisense oligonucleotides targeting specific microRNAs, termed "antagomirs", were successfully used to silence endogenous microRNAs *in vivo* in experimental animals [53]. Antagomirs can be modified with methoxyethyl group (MOE) [53] or locked nucleic acid (LNA) [54] to enhance their *in vivo* stability. Once administered systemically, antagomirs are absorbed by major organs such as the liver, lungs, skin, spleen, lymph nodes, bone marrow, and kidney, excepting the brain [53–56]. The effect of a single intravenous dose of antagomir can last up to 21 d in liver and kidney [54,55]. MiR-21 regulates the ERK-MAPK signaling pathway to stimulate fibroblast survival and growth factor secretion [57]. Antagomirs against miR-21 was used to block fibrosis in cardiovascular diseases [57] and in pulmonary diseases [58]. The same strategy was also proven effective for treating renal fibrosis in chronic kidney diseases [57]. The liver-expressed miR-122 is essential for hepatitis C virus (HCV) RNA accumulation [59]. Antagomir treatments in nonhuman primates with chronic HCV infection generated long-lasting suppression of HCV viremia with no evidence of viral resistance or other adverse effects in the treated animals [60]. The most effective treatment of kidney diseases with antagomirs was demonstrated in the case of miR-192. Anti-miR-192 treatments ameliorated glomerular fibrosis in mouse models of diabetic nephropathy through a concomitant repression of collagen and fibronectin levels in the mesangial cells [61]. Another approach to interfere with the binding of a microRNA to its cognate target mRNA is the use of "microRNA sponge". A "microRNA sponge" is a construct encoding an mRNA (e.g., the GFP mRNA) that contains a tandem repeat of microRNA binding sites in the 3'-UTR [62]. The sponges can be introduced *in vivo* to target organs via lentivirus or retrovirus mediated transgenesis.

The major limitation to microRNA therapeutics is the off-target effect, which occurs on two distinct levels. Most current miRNA modulators display a high degree of target specificity, as point mutations made into an antagomir can completely abolish its regulatory effect both *in vitro* and *in vivo* [63]. Nevertheless, the antagomir may still target a class of related microRNAs that differ only at one or two nucleotide loci. Additional ambiguity derives from the relatively modest inhibitory effects of individual microRNAs on mRNA targets. For example, a complete depletion of a microRNA often results in maximal increases (<1.5-fold) in the expression levels of its mRNA targets, suggesting that it is the cumulative impact of small changes in the expression of myriad targets—rather than pronounced changes in a single target—that mediates the biological actions of microRNAs. A more perplexing matter is the undesired effect of systemically administrated antagomirs in off-target organs or off-target cells in the target organ. Local administration of antagomirs represents a potential solution. For example, intracerebral injection of anti-miR-16 has been demonstrated to reduce miR-16 levels in the brain, which was, however, immune to any systemic manipulation [63]. Coupling antagomirs to tissue specific antibodies provides another means to manipulating microRNA levels locally. Upstream signals intervening with microRNA expression or maturation may be unique to each target organ, thus providing new layers of specificity in microRNA-based therapeutics. Despite these potential challenges, microRNA therapeutics have recently entered Phase I and Phase II clinical trials of Santaris Pharma's antagomir anti-miR-122, *miravirsen*, for the treatment of HCV.

In some cases, a gain-of-microRNA function is beneficial to treating diseases. This can be achieved by use of a synthetic double-stranded precursor microRNA molecule, known as the "miRNA mimic". In acute myocardial infarction, for example, miR-29 is significantly downregulated in fibroblast cells of the heart, resulting in cardiac fibrosis. Introducing miR-29 mimics profoundly reduced collagen expression and ameliorated fibrotic phenotypes in the heart [64].

MicroRNAs have also been found in blood and urine samples and have emerged as potential biomarkers for various diseases including kidney diseases [65]. Nevertheless, no association study is currently available to link the expression levels of these markers with disease progression and their power for predicting disease susceptibility is still debated.

7. Conclusions

MicroRNA has emerged to be a critical regulator underlying a diverse range of renal pathophysiologies. Owing to its short seed sequence, a cognate microRNA regulates multiple gene targets, making it a powerful signaling molecule to coordinate various cellular functions. Manipulation of tissue microRNA level as a novel therapeutic approach has been proven effective in several renal disease models. Systematic identification of downstream microRNA target genes will improve our knowledge in renal pathology and supply additional candidates for therapeutic intervention. What remain largely unknown are the mechanisms that endogenously control microRNA metabolism and the clinical evidence that derangement of such mechanisms causes human diseases.

Acknowledgments

This work was supported by the National Institutes of Health Grants RO1DK084059 and P30 DK079333, and American Heart Association Grant 0930050N to JH.

References

1. Bartel, D.P. MicroRNAs: Target recognition and regulatory functions. *Cell* **2009**, *136*, 215–233.
2. Carthew, R.W.; Sontheimer, E.J. Origins and mechanisms of miRNAs and siRNAs. *Cell* **2009**, *136*, 642–655.
3. Rajewsky, N. microRNA target predictions in animals. *Nat. Genet.* **2006**, *38*, S8–S13.
4. Krol, J.; Loedige, I.; Filipowicz, W. The widespread regulation of microRNA biogenesis, function and decay. *Nat. Rev. Genet.* **2010**, *11*, 597–610.
5. Voinnet, O. Origin, biogenesis, and activity of plant microRNA. *Cell* **2009**, *136*, 669–687.
6. Ho, J.; Pandey, P.; Schatton, T.; Sims-Lucas, S.; Khalid, M.; Frank, M.H.; Hartwig, S.; Kreidberg, J.A. The pro-apoptotic protein Bim is a microRNA target in kidney progenitors. *J. Am. Soc. Nephrol.* **2011**, *22*, 1053–1063.
7. Nagalakshmi, V.K.; Ren, Q.; Pugh, M.M.; Valerius, M.T.; McMahon, A.P.; Yu, J. Dicer regulates the development of nephrogenic and ureteric compartments in the mammalian kidney. *Kidney Int.* **2011**, *79*, 317–330.
8. Ho, J.; Ng, K.H.; Rosen, S.; Dostal, A.; Gregory, R.I.; Kreidberg, J.A. Podocyte-specific loss of functional microRNAs leads to rapid glomerular and tubular injury. *J. Am. Soc. Nephrol.* **2008**, *19*, 2069–2075.
9. Harvey, S.J.; Jarad, G.; Cunningham, J.; Goldberg, S.; Schermer, B.; Harfe, B.D.; McManus, M.T.; Benzing, T.; Miner, J.H. Podocyte-specificdeletionofdicer alters cytoskeletal dynamics and causes glomerular disease. *J. Am. Soc. Nephrol.* **2008**, *19*, 2150–2158.
10. Shi, S.; Yu, L.; Chiu, C.; Sun, Y.; Chen, J.; Khitrov, G.; Merkenschlager, M.; Holzman, L.B.; Zhang, W.; Mundel, P.; *et al.* Podocyte-selective deletion of dicer induces proteinuria and glomerulosclerosis. *J. Am. Soc. Nephrol.* **2008**, *19*, 2159–2169.
11. Zhdanova, O.; Srivastava, S.; Di, L.; Li, Z.; Tchelebi, L.; Dworkin, S.; Johnstone, D.B.; Zavadil, J.; Chong, M.M.; Littman, D.R.; *et al.* The inducible deletion of Drosha and microRNAs in mature podocytes results in a collapsing glomerulopathy. *Kidney Int.* **2011**, *80*, 719–730.
12. Sequeira-Lopez, M.L.; Weatherford, E.T.; Borges, G.R.; Monteagudo, M.C.; Pentz, E.S.; Harfe, B.D.; Carretero, O.; Sigmund, C.D.; Gomez, R.A. The microRNA-processing enzyme dicer maintains juxtaglomerular cells. *J. Am. Soc. Nephrol.* **2010**, *21*, 460–467.
13. Wei, Q.; Bhatt, K.; He, H.Z.; Mi, Q.S.; Haase, V.H.; Dong, Z. Targeted deletion of Dicer from proximal tubules protects against renal ischemia reperfusion injury. *J. Am. Soc. Nephrol.* **2010**, *21*, 756–761.
14. Sun, Y.; Koo, S.; White, N.; Peralta, E.; Esau, C.; Dean, N.M.; Perera, R.J. Development of a micro-array to detect human and mouse microRNAs and characterization of expression in human organs. *Nucleic Acids Res.* **2004**, *32*, e188.

15. Tian, Z.; Greene, A.S.; Pietrusz, J.L.; Matus, I.R.; Liang, M. MicroRNA-target pairs in the rat kidney identified by microRNA microarray, proteomic, and bioinformatics analysis. *Genome Res.* **2008**, *18*, 404–411.
16. Kato, M.; Zhang, J.; Wang, M.; Lanting, L.; Yuan, H.; Rossi, J.J.; Natarajan, R. MicroRNA-192 in diabetic kidney glomeruli and its function in TGF-beta-induced collagen expression via inhibition of E-box repressors. *Proc. Natl. Acad. Sci. USA* **2007**, *104*, 3432–3437.
17. Kato, M.; Putta, S.; Wang, M.; Yuan, H.; Lanting, L.; Nair, I.; Gunn, A.; Nakagawa, Y.; Shimano, H.; Todorov, I.; *et al.* TGF-beta activates Akt kinase through a microRNA-dependent amplifying circuit targeting PTEN. *Nat. Cell Biol.* **2009**, *11*, 881–889.
18. Jiang, B.H.; Liu, L.Z. PI3K/PTEN signaling in angiogenesis and tumorigenesis. *Adv. Cancer Res.* **2009**, *102*, 19–65.
19. Wang, Q.; Wang, Y.; Minto, A.W.; Wang, J.; Shi, Q.; Li, X.; Quigg, R.J. MicroRNA-377 is up-regulated and can lead to increased fibronectin production in diabetic nephropathy. *FASEB J.* **2008**, *22*, 4126–4135.
20. Bracken, C.P.; Gregory, P.A.; Kolesnikoff, N.; Bert, A.G.; Wang, J.; Shannon, M.F.; Goodall, G.J. A double-negative feedback loop between ZEB1-SIP1 and the microRNA-200 family regulates epithelial-mesenchymal transition. *Cancer Res.* **2008**, *68*, 7846–7854.
21. Burk, U.; Schubert, J.; Wellner, U.; Schmalhofer, O.; Vincan, E.; Spaderna, S.; Brabletz, T. A reciprocal repression between ZEB1 and members of the miR-200 family promotes EMT and invasion in cancer cells. *EMBO Rep.* **2008**, *9*, 582–589.
22. Kato, M.; Arce, L.; Wang, M.; Putta, S.; Lanting, L.; Natarajan, R. A microRNA circuit mediates transforming growth factor-β1 autoregulation in renal glomerular mesangial cells. *Kidney Int.* **2011**, *80*, 358–368.
23. Wang, G.; Kwan, B.C.; Lai, F.M.; Choi, P.C.; Chow, K.M.; Li, P.K.; Szeto, C.C. Intrarenal expression of microRNAs in patients with IgA nephropathy. *Lab. Invest.* **2010**, *90*, 98–103.
24. Gebeshuber, C.A.; Kornauth, C.; Dong, L.; Sierig, R.; Seibler, J.; Reiss, M.; Tauber, S.; Bilban, M.; Wang, S.; Kain, R.; *et al.* Focal segmental glomerulosclerosis is induced by microRNA-193a and its downregulation of WT1. *Nat. Med.* **2013**, *19*, 481–487.
25. Chen, Y.Q.; Wang, X.X.; Yao, X.M.; Zhang, D.L.; Yang, X.F.; Tian, S.F.; Wang, N.S. MicroRNA-195 promotes apoptosis in mouse podocytes via enhanced caspase activity driven by BCL2 insufficiency. *Am. J. Nephrol.* **2011**, *34*, 549–559.
26. Wang, B.; Komers, R.; Carew, R.; Winbanks, C.E.; Xu, B.; Herman-Edelstein, M.; Koh, P.; Thomas, M.; Jandeleit-Dahm, K.; Gregorevic, P.; *et al.* Suppression of microRNA-29 expression by TGF-β1 promotes collagen expression and renal fibrosis. *J. Am. Soc. Nephrol.* **2012**, *23*, 252–265.
27. Schrier, R.W.; Wang, W.; Poole, B.; Mitra, A. Acute renal failure: Definitions, diagnosis, pathogenesis, and therapy. *J. Clin. Invest.* **2004**, *114*, 5–14.
28. Godwin, J.G.; Ge, X.; Stephan, K.; Jurisch, A.; Tullius, S.G.; Iacomini, J. Identification of a microRNA signature of renal ischemia reperfusion injury. *Proc. Natl. Acad. Sci. USA* **2010**, *107*, 14339–14344.

29. Bhatt, K.; Zhou, L.; Mi, Q.S.; Huang, S.; She, J.X.; Dong, Z. MicroRNA-34a is induced via p53 during cisplatin nephrotoxicity and contributes to cell survival. *Mol. Med.* **2010**, *16*, 409–416.
30. Krupa, A.; Jenkins, R.; Luo, D.D.; Lewis, A.; Phillips, A.; Fraser, D. Loss of microRNA-192 promotes fibrogenesis in diabetic nephropathy. *J. Am. Soc. Nephrol.* **2010**, *21*, 438–447.
31. Pollak, M.R.; Brown, E.M.; Chou, Y.H.; Hebert, S.C.; Marx, S.J.; Steinmann, B.; Levi, T.; Seidman, C.E.; Seidman, J.G. Mutations in the human Ca(2+)-sensing receptor gene cause familial hypocalciuric hypercalcemia and neonatal severe hyperparathyroidism. *Cell* **1993**, *75*, 1297–1303.
32. Simon, D.B.; Lu, Y.; Choate, K.A.; Velazquez, H.; Al-Sabban, E.; Praga, M.; Casari, G.; Bettinelli, A.; Colussi, G.; Rodriguez-Soriano, J.; *et al.* Paracellin-1, a renal tight junction protein required for paracellular Mg^{2+} resorption. *Science* **1999**, *285*, 103–106.
33. Konrad, M.; Schaller, A.; Seelow, D.; Pandey, A.V.; Waldegger, S.; Lesslauer, A.; Vitzthum, H.; Suzuki, Y.; Luk, J.M.; Becker, C.; *et al.* Mutations in the tight-junction gene claudin 19 (CLDN19) are associated with renal magnesium wasting, renal failure, and severe ocular involvement. *Am. J. Hum. Genet.* **2006**, *79*, 949–957.
34. Thorleifsson, G.; Holm, H.; Edvardsson, V.; Walters, G.B.; Styrkarsdottir, U.; Gudbjartsson, D.F.; Sulem, P.; Halldorsson, B.V.; de Vegt, F.; d'Ancona, F.C.; *et al.* Sequence variants in the CLDN14 gene associate with kidney stones and bone mineral density. *Nat. Genet.* **2009**, *41*, 926–930.
35. Hou, J.; Renigunta, A.; Konrad, M.; Gomes, A.S.; Schneeberger, E.E.; Paul, D.L.; Waldegger, S.; Goodenough, D.A. Claudin-16 and claudin-19 interact and form a cation-selective tight junction complex. *J. Clin. Invest.* **2008**, *118*, 619–628.
36. Gong, Y.; Renigunta, V.; Himmerkus, N.; Zhang, J.; Renigunta, A.; Bleich, M.; Hou, J. Claudin-14 regulates renal Ca^{++} transport in response to CaSR signalling via a novel microRNA pathway. *EMBO J.* **2012**, *31*, 1999–2012.
37. Weber, S.; Schneider, L.; Peters, M.; Misselwitz, J.; Rönnefarth, G.; Böswald, M.; Bonzel, K.E.; Seeman, T.; Suláková, T.; Kuwertz-Bröking, E.; *et al.* Novel paracellin-1 mutations in 25 families with familial hypomagnesemia with hypercalciuria and nephrocalcinosis. *J. Am. Soc. Nephrol.* **2001**, *12*, 1872–1881.
38. Hou, J.; Shan, Q.; Wang, T.; Gomes, A.S.; Yan, Q.; Paul, D.L.; Bleich, M.; Goodenough, D.A. Transgenic RNAi depletion of claudin-16 and the renal handling of magnesium. *J. Biol. Chem.* **2007**, *282*, 17114–17122.
39. Hou, J.; Renigunta, A.; Gomes, A.S.; Hou, M.; Paul, D.L.; Waldegger, S.; Goodenough, D.A. Claudin-16 and claudin-19 interaction is required for their assembly into tight junctions and for renal reabsorption of magnesium. *Proc. Natl. Acad. Sci. USA* **2009**, *106*, 15350–15355.
40. Ma, L.; Young, J.; Prabhala, H.; Pan, E.; Mestdagh, P.; Muth, D.; Teruya-Feldstein, J.; Reinhardt, F.; Onder, T.T.; Valastyan, S.; *et al.* miR-9, a MYC/MYCN-activated microRNA, regulates E-cadherin and cancer metastasis. *Nat. Cell Biol.* **2010**, *12*, 247–256.
41. Hu, H.; Du, L.; Nagabayashi, G.; Seeger, R.C.; Gatti, R.A. ATM is down-regulated by *N*-Myc-regulated microRNA-421. *Proc. Natl. Acad. Sci. USA* **2010**, *107*, 1506–1511.

42. Brennan, S.C.; Thiem, U.; Roth, S.; Aggarwal, A.; Fetahu, I.S.; Tennakoon, S.; Gomes, A.R.; Brandi, M.L.; Bruggeman, F.; Mentaverri, R.; *et al.* Calcium sensing receptor signalling in physiology and cancer. *Biochim. Biophys. Acta* **2012**, *1833*, 1732–1744.
43. Buchert, M.; Papin, M.; Bonnans, C.; Darido, C.; Raye, W.S.; Garambois, V.; Pélegrin, A.; Bourgaux, J.F.; Pannequin, J.; Joubert, D.; *et al.* Symplekin promotes tumorigenicity by up-regulating claudin-2 expression. *Proc. Natl. Acad. Sci. USA* **2010**, *107*, 2628–2633.
44. Nübel, T.; Preobraschenski, J.; Tuncay, H.; Weiss, T.; Kuhn, S.; Ladwein, M.; Langbein, L.; Zöller, M. Claudin-7 regulates EpCAM-mediated functions in tumor progression. *Mol. Cancer Res.* **2009**, *7*, 285–299.
45. Mladinov, D.; Liu, Y.; Mattson, D.L.; Liang, M. MicroRNAs contribute to the maintenance of cell-type-specific physiological characteristics: miR-192 targets Na^+/K^+-ATPase β1. *Nucleic Acids Res.* **2013**, *41*, 1273–1283.
46. Guyton, A.C. Blood pressure control—Special role of the kidneys and body fluids. *Science* **1991**, *252*, 1813–1816.
47. Elvira-Matelot, E.; Zhou, X.O.; Farman, N.; Beaurain, G.; Henrion-Caude, A.; Hadchouel, J.; Jeunemaitre, X. Regulation of WNK1 expression by miR-192 and aldosterone. *J. Am. Soc. Nephrol.* **2010**, *21*, 1724–1731.
48. McCormick, J.A.; Yang, C.L.; Ellison, D.H. WNK kinases and renal sodium transport in health and disease: An integrated view. *Hypertension* **2008**, *51*, 588–596.
49. Wilson, F.H.; Disse-Nicodeme, S.; Choate, K.A.; Ishikawa, K.; Nelson-Williams, C.; Desitter, I.; Gunel, M.; Milford, D.V.; Lipkin, G.W.; Achard, J.M.; *et al.* Human hypertension caused by mutations in WNK kinases. *Science* **2001**, *293*, 1107–1112.
50. Lin, D.H.; Yue, P.; Pan, C.; Sun, P.; Wang, W.H. MicroRNA 802 stimulates ROMK channels by suppressing caveolin-1. *J. Am. Soc. Nephrol.* **2011**, *22*, 1087–1098.
51. Huang, W.; Liu, H.; Wang, T.; Zhang, T.; Kuang, J.; Luo, Y.; Chung, S.S.; Yuan, L.; Yang, J.Y. Tonicity-responsive microRNAs contribute to the maximal induction of osmoregulatory transcription factor OREBP in response to high-NaCl hypertonicity. *Nucleic Acids Res.* **2011**, *39*, 475–485.
52. Burg, M.B.; Ferraris, J.D.; Dimitrieva, N.I. Cellular response to hyperosmotic stresses. *Physiol. Rev.* **2007**, *87*, 1441–1474.
53. Krützfeldt, J.; Rajewsky, N.; Braich, R.; Rajeev, K.G.; Tuschl, T.; Manoharan, M.; Stoffel, M. Silencing of microRNAs *in vivo* with "antagomirs". *Nature* **2005**, *438*, 685–689.
54. Obad, S.; dos Santos, C.O.; Petri, A.; Heidenblad, M.; Broom, O.; Ruse, C.; Fu, C.; Lindow, M.; Stenvang, J.; Straarup, E.M.; *et al.* Silencing of microRNA families by seed-targeting tiny LNAs. *Nat. Genet.* **2011**, *43*, 371–378.
55. Castoldi, M.; Vujic Spasic, M.; Altamura, S.; Elmén, J.; Lindow, M.; Kiss, J.; Stolte, J.; Sparla, R.; D'Alessandro, L.A.; Klingmüller, U.; *et al.* The liver-specific microRNA miR-122 controls systemic iron homeostasis in mice. *J. Clin. Invest.* **2011**, *121*, 1386–1396.
56. Yi, R.; Poy, M.N.; Stoffel, M.; Fuchs, E. A skin microRNA promotes differentiation by repressing "stemness". *Nature* **2008**, *452*, 225–229.

57. Thum, T.; Gross, C.; Fiedler, J.; Fischer, T.; Kissler, S.; Bussen, M.; Galuppo, P.; Just, S.; Rottbauer, W.; Frantz, S.; *et al.* MicroRNA-21 contributes to myocardial disease by stimulating MAP kinase signalling in fibroblasts. *Nature* **2008**, *456*, 980–984.
58. Liu, G.; Friggeri, A.; Yang, Y.; Milosevic, J.; Ding, Q.; Thannickal, V.J.; Kaminski, N.; Abraham, E. miR-21 mediates fibrogenic activation of pulmonary fibroblasts and lung fibrosis. *J. Exp. Med.* **2010**, *207*, 1589–1597.
59. Jopling, C.L.; Yi, M.; Lancaster, A.M.; Lemon, S.M.; Sarnow, P. Modulation of hepatitis C virus RNA abundance by a liver-specific MicroRNA. *Science* **2005**, *309*, 1577–1581.
60. Lanford, R.E.; Hildebrandt-Eriksen, E.S.; Petri, A.; Persson, R.; Lindow, M.; Munk, M.E.; Kauppinen, S.; Ørum, H. Therapeutic silencing of microRNA-122 in primates with chronic hepatitis C virus infection. *Science* **2010**, *327*, 198–201.
61. Putta, S.; Lanting, L.; Sun, G.; Lawson, G.; Kato, M.; Natarajan, R. Inhibiting microRNA-192 ameliorates renal fibrosis in diabetic nephropathy. *J. Am. Soc. Nephrol.* **2012**, *23*, 458–469.
62. Ebert, M.S.; Neilson, J.R.; Sharp, P.A. MicroRNA sponges: Competitive inhibitors of small RNAs in mammalian cells. *Nat. Methods* **2007**, *4*, 721–726.
63. Krützfeldt, J.; Kuwajima, S.; Braich, R.; Rajeev, K.G.; Pena, J.; Tuschl, T.; Manoharan, M.; Stoffel, M. Specificity, duplex degradation and subcellular localization of antagomirs. *Nucleic Acids Res.* **2007**, *35*, 2885–2892.
64. Van Rooij, E.; Sutherland, L.B.; Thatcher, J.E.; DiMaio, J.M.; Naseem, R.H.; Marshall, W.S.; Hill, J.A.; Olson, E.N. Dysregulation of microRNAs after myocardial infarction reveals a role of miR-29 in cardiac fibrosis. *Proc. Natl. Acad. Sci. USA* **2008**, *105*, 13027–13032.
65. Lorenzen, J.M.; Thum, T. Circulating and urinary microRNAs in kidney disease. *Clin. J. Am. Soc. Nephrol.* **2012**, *7*, 1528–1533.

5

Detecting and Comparing Non-Coding RNAs in the High-Throughput Era

Giovanni Bussotti [1,*], Cedric Notredame [2] and Anton J. Enright [1]

[1] European Bioinformatics Institute, Wellcome Trust Genome Campus, Hinxton, Cambridge CB10 1SD, UK; E-Mail: aje@ebi.ac.uk
[2] Bioinformatics and Genomics Program, Centre for Genomic Regulation (CRG), Aiguader, 88, 08003 Barcelona, Spain; E-Mail: Cedric.Notredame@crg.eu

* Author to whom correspondence should be addressed; E-Mail: giovanni.bussotti@ebi.ac.uk;

Abstract: In recent years there has been a growing interest in the field of non-coding RNA. This surge is a direct consequence of the discovery of a huge number of new non-coding genes and of the finding that many of these transcripts are involved in key cellular functions. In this context, accurately detecting and comparing RNA sequences has become important. Aligning nucleotide sequences is a key requisite when searching for homologous genes. Accurate alignments reveal evolutionary relationships, conserved regions and more generally any biologically relevant pattern. Comparing RNA molecules is, however, a challenging task. The nucleotide alphabet is simpler and therefore less informative than that of amino-acids. Moreover for many non-coding RNAs, evolution is likely to be mostly constrained at the structural level and not at the sequence level. This results in very poor sequence conservation impeding comparison of these molecules. These difficulties define a context where new methods are urgently needed in order to exploit experimental results to their full potential. This review focuses on the comparative genomics of non-coding RNAs in the context of new sequencing technologies and especially dealing with two extremely important and timely research aspects: the development of new methods to align RNAs and the analysis of high-throughput data.

Keywords: lncRNAs; ncRNAs; high-throughput; RNA-seq; comparative biology; sequencing

1. Introduction

1.1. The Non-Coding RNA (New)-World

In recent years, the non-coding RNA (ncRNA) field has rapidly expanded (Figure 1) with a rapid increase in the number of newly identified and biologically relevant ncRNAs. Just a decade ago, the number of known ncRNAs was restricted to a small amount of housekeeping genes (including ribosomal RNAs, transfer RNAs and small nucleolar RNAs) and an even more limited collection of regulatory RNAs, such as lin-4 in *Caenorhabditis elegans* [1] and Xist in mammals [2]. Since then, the number of novel ncRNAs has increased dramatically and far more is known about their function, biogenesis, length, structural and sequence features. New and ever more sophisticated high-throughput technologies, such as tiling arrays and next generation sequencing (NGS) have been applied to comprehensively profile the transcriptome of various organisms.

Figure 1. Number of publications in PubMed found using the keyword "ncRNA" (dark grey) and "regulatory RNA" (pale gray). The x-axis represents the timeline, the y-axis the number of times the words "ncRNA" and "regulatory RNA" match a publication in PubMed normalized by the total number of publications in that year (expressed as one part per ten thousand).

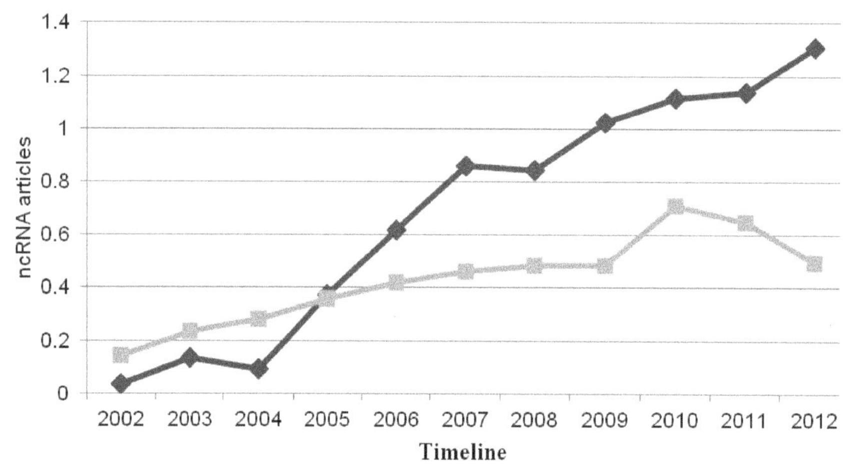

This wealth of data has allowed the identification of thousands of novel short ncRNAs, including PIWI interacting RNAs [3] and small nucleolar RNAs [4] and has resulted in the compilation or the update of many publicly available databases [5–10]. Furthermore, high-throughput approaches have revealed extensive and pervasive transcription of long ncRNAs (lncRNAs) [11–13], operationally defined as functional RNA longer than 200 base pairs that does not template protein synthesis. In the human genome, for instance, the GENCODE consortium annotated 9,640 lncRNA loci representing 15,512 transcripts [3,14] and in [15] the authors estimated that total number of human lncRNAs genes to be about 50,000, more than two-fold greater than the number of protein-coding genes. These discoveries were very timely in the context of growing concern for the lack of a significant correlation between the number of protein-coding genes and the commonly accepted concept of "organism complexity" [4,16,17]. It was proposed that alternative splicing and ncRNAs could be accountable for complex gene regulation architectures, meaning that the "Central Dogma" of genetic programming enunciated by Francis Crick in 1958 (RNA is transcribed from DNA and translated into protein) [18] had to be slightly altered and at least in higher eukaryotes may be inadequate [16,17]. The biological role of most of these novel long untranslated molecules is still a controversial issue. Some authors have even raised doubts about whether these transcripts are functional at all [19]. The lack of shared discernible features hampers our ability to define lncRNA classes, thus impeding function prediction [20]. However mounting experimental evidence illustrates that lncRNAs are implicated in a variety of biological processes [21] and are linked to various diseases including cancer [22]. Additionally, the functional roles of lncRNA transcripts have been uncovered in signaling sensors [23], embryonic stem cell differentiation [11], brain function [24,25], subcellular compartmentalization and chromatin remodeling [26]. Some examples include X chromosome inactivation by Xist, the silencing of autosomal imprinted genes accomplished by Air, nuclear trafficking regulated by NRON and muscle differentiation controlled by linc-MD1 [2,27–29]. In [30] the authors identified a class of lncRNAs named ncRNA-a (ncRNA-activator) able to stimulate the expression of proximal protein-coding genes, and a recent update on ncRNA-a [31] showed that the co-activator complex Mediator plays a central role in the activation process. See [21] and [32] for more examples and lncRNAdb [33] for the central repository of known lncRNAs in eukaryotes. lncRNAs are expressed, some are spliced, they are often conserved across vertebrates, and their expression is frequently tissue- and/or cell-specific and localized to specific subcellular compartments [11,25,34]. It has been shown that lncRNAs can act both in *cis* [30,35] and in *trans* [36], some acting as precursors for short ncRNAs [37–39], while others act independently as long transcripts. As in [40] lncRNAs can be classified as "intergenic" or "genic" depending on their position/orientation with respect to protein-coding genes. lncRNAs not overlapping any protein-coding gene are tagged as intergenic and then further classified according to their transcription orientation with the closest protein-coding loci (same sense, convergent, or divergent). The genic lncRNA set are catalogued as "exonic" if overlapping a protein-coding exon. Otherwise, lncRNAs are labeled as "intronic", when positioned within protein-coding introns or as "overlapping", in presence of a protein-coding transcript located within the intron of the lncRNA [40].

1.2. lncRNA Challenges

Although the conservation level of different lncRNAs may be not always directly comparable (e.g., the evolutionary conservation of genic lncRNAs may be biased by the presence of the protein-coding genes), overall approximately half of reported human lncRNA exhibit significant conservation across mammals [40]. These levels suggest some key cellular function, even though only a small fraction of these transcripts have so far been functionally characterized. Such functional analyses remain however, very superficial and lack precise molecular mechanisms explaining the activity of these novel transcripts. Our low level of understanding can be in part attributed to the difficulty with working experimentally with lncRNAs: detection is difficult for a combination of biological and technical aspects. The first relates to the low levels of non-coding genetic expression. After ribosomal RNA (rRNA), protein-coding mRNA represents the highest population of RNA species [41]. In previous studies [34,42,43] it has been reported that lncRNAs are on average 3 to 10 fold less expressed than mRNAs. Besides the complicated task of capturing weaker expression signals, many lncRNAs have pronounced tissue/stage specificity [43,44]. In other words, lncRNA genes can easily be left undetected unless the correct cell type and condition are considered. One more complication for ncRNA discovery has been the difficulty of sequencing deep enough, a hurdle only recently overcome by NGS. Additionally, our ability to assemble and annotate genomes was less advanced than currently and we had simplified notions of transcriptome complexity. Most of the classical low-throughput approaches, such as RT-PCR and northern blotting, have been successfully used to analyze the expression of small numbers of genes, but they were not adequate to address the "pervasive transcription" aspect of genomes [45,46]. Furthermore, there are specific classes of ncRNAs, such as circular RNAs (circRNAs), that have been extremely hard to identify. circRNAs are a class of non-coding RNA family that were discovered more than 20 years ago [47–50]. These RNAs form circles that arise from non-canonical splicing events (also known as exon shuffling) that join a splice donor to an upstream splice acceptor to produce a circular RNA molecule. Recent studies [51,52] show that the human circRNA CDR1as, antisense to the Cerebellar Degeneration-Related protein 1 (CDR1), hosts around 70 binding sites for the miR-7 microRNA and is highly associated with the Argonaute protein Ago2 as demonstrated by PAR-CLIP and HITS-CLIP experiments [51,52]. Mainly because of their non-canonical splicing behavior, circRNAs have eluded detection by next generation sequencing until recently. These latest studies adopted a novel computational approach to identify circRNAs from high-throughput RNA-seq data and demonstrated their widespread abundance within transcriptomes [51,53].

In general, a major obstacle for ncRNA detection is the difficulty to perform informative sequence comparisons. Standard primary sequence alignment is hampered by the low complexity of the nucleic alphabet, making it difficult to produce statistically meaningful RNA alignments. Ribonucleic acid chemistry relies on just four primary residues: two purines and two pyrimidines. Consequently, RNA gene sequences do not have strong statistical signals, unlike protein-coding genes. For instance two RNA sequences must share an identity of at least ~60% to be considered significant in homology relationships prediction [54]. Below this level, common ancestry is hard to infer with certainty. By comparison, this threshold is around ~20%–35% for proteins [55]. Furthermore, ncRNA appears to be evolving rapidly [56] or are under the influence of very specific evolutionary constraints [56]. It was proposed that most ncRNAs evolve at higher mutation rates, with the maintenance of secondary structures being the main source of selection [57,58]. This assumption makes sense from an evolutionary standpoint. As ncRNAs will be left untranslated, the nucleotide sequence itself is not constrained to keep the codon reading frame. Of course many exceptions exist. Specific ncRNAs types can hold functional sequences and act via their primary sequence (*i.e.*, miRNAs). Previous reports have shown that at least some miRNA genes are well conserved across species [59–61], reinforcing the idea that sequences encoding a function evolve under purifying selection. Aside from these specific and relatively rare examples, it seems that for most known ncRNAs, evolution is limited by structural constraints [62,63]. This induces a characteristic pattern of covariance that occurs when a mutation is affecting a nucleotide pairing to another in a structured domain (Figure 2). If the mutation breaks the base pairing so that the functionality of such a domain is compromised, the matching nucleotide is favored to mutate in turn, *i.e.*, is co-varying to restore the base pairing and keep the structure unchanged.

Figure 2. RNA mutations are tightly linked to the RNA structure conservation. (**a**) Example where the mutation of a C into an A is compensated by the change G-U. The two positions are not independent, but communicating one with the other to maintain the structure unvaried; (**b**) Same hairpin as shown in (a). The presence of the compensatory mutation is highlighted by the multiple sequence comparison.

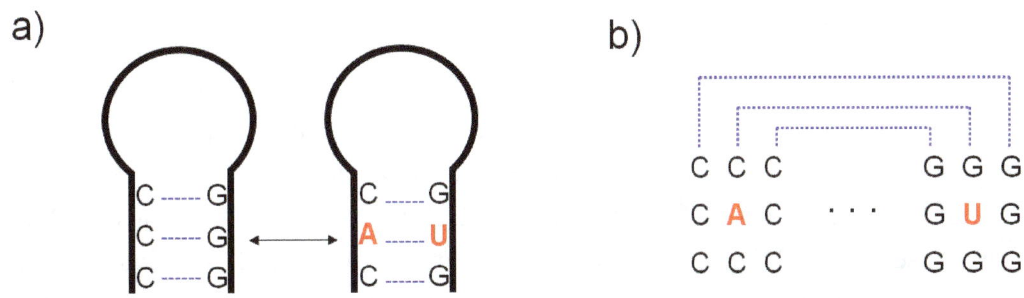

For most aligners these features of RNA are hard to account for when using standard alignment procedures that postulate positional independence and seek only to maximize identity. Furthermore, RNA can hold functional pseudo-knots. These are structural configurations where at least two RNA stem-loops are interposed one into the other. Although some comparative approaches including pseudo-knots exist [64,65], these are disregarded by most software due to reasons of computational complexity [66]. As a consequence ncRNA sequences are harder to align than proteins, a limitation that affects our ability to accurately detect and classify them. The difficulty in comparing ncRNAs calls for other information sources that alignment algorithms can use. More than ever, the issue of accurately comparing and aligning ncRNAs is of critical importance. This is precisely the problem discussed in the following section, where we review established and more recent methodologies able to make the best of available RNA information (Section 2). Next we discuss different homology based strategies for ncRNA detection (Section 3) and the analysis of high-throughput expression data (Section 4). See Table 1 for a summary of the resources described in the text.

2. Comparing Non-Coding RNAs

As mentioned, generating meaningful ncRNA alignments is a challenging task and at least in some cases, the best accuracy could be achieved by exploiting RNA structural information. However, in many situations using such information is complicated. In spite of the development of aligners that take into account the RNA secondary structure information, one major issue is the poor availability of high quality structures. The problem is at least in part due to the difficulties encountered at experimental level in crystallization. Getting crystals from RNA molecules is complicated because of their chemical specificity. The accumulation of crystals is prevented by the high RNA flexibility. RNAs have flexible structures adopting inter-domain movements and with respect to proteins have weaker tertiary interactions [67]. The polyanionic charge of the phosphate backbone makes the nucleotide sequence move much more than in proteins and this makes the packaging of crystals much harder to achieve. As a consequence, the crystals are either hard to grow or uninformative. Even when trying to resolve RNA molecules in solution using NMR, the resonance assignment is more difficult for RNA than for proteins [68]. RNAs have only 4 primary nucleosides instead of the 20 different amino-acid side chains found in proteins [69]. Thus, the chemical shift dispersion is narrower in RNA than in proteins, resulting in less informative spectra [69].

2.1. RNA Structure Prediction

Because of these limitations, RNA structure is usually computationally predicted without any experimental support [70,71]. RNA secondary structure inference amounts to the computation of base-pairings that shape the *in vivo* molecule structure. The prediction is performed using primary RNA sequence data. Another possibility is including other sources of statistical information to constrain structure prediction, for instance an alignment of structurally homologous RNA sequences. Regarding single sequence RNA secondary structure predictions, there are two main groups of approaches: empirical free-energy parameters [72] and knowledge based [73–75]. The first considers a biophysical model to calculate the structure whose folding has the minimum Gibbs free energy ΔG). In this approach, [76–80] the nearest stable folding is employed to compute the conformational stability of the Minimum Free Energy (MFE) structure. The energy parameters needed in this approach were assessed on a set of optical melting experiments on model systems [77–79]. The two most popular implementations of the MFE structure prediction algorithm are Mfold [70] and RNAfold [81] packages. The latter implements McCaskill's algorithm [82], an approach to calculate the probability of a certain secondary structure in the whole thermodynamic ensemble. This approach is based on the partition function, which sums all Boltzmann weighted free energies of each secondary structure that is possible given an RNA sequence. In this model, the confidence estimate in a particular base pair *i,j* is given by the sum of the probabilities of all structures containing that base pair *i,j* divided by the sum over all structures [83]. Knowledge based approaches rely on probabilistic models, where statistical learning procedures are used instead of empirical measurement of thermodynamic parameters. The Stochastic Context Free Grammar (SCFG) model [73] represents one popular example of such probabilistic models. The parameters used by the SCFG models are estimated on the set of RNAs with known structures (e.g., rRNA).

Prediction consistency is the main limit of both MFE and knowledge based methods [84]. (See the example in Figure 3). The percentage of known base pairs predicted correctly by the secondary structure prediction methods ranges from 40% to 75% [73–75,85]. This low figure may be the result of three confounding factors. Firstly, folding *in vivo* can be influenced by RNA chaperones [86], RNA editing [87], and even by the transcriptional process itself [88]. At present, there is no software able to account for these effects. Secondly, looking for a single structure can sometimes be inadequate. There are cases, such as the ribo-switches [89,90], where multiple functional structures can be derived from the same sequence depending on conditions such as temperature or other external factors. Standard predictors are not well suited to deal with such situations and require dedicated tools able to identify potential conformational switches [91,92]. Thirdly, RNAs might contain pseudo-knots, which are ignored by most tools due to reasons of computational complexity [66].

Figure 3. Consistency of RNA secondary structure predictions. In this example the human mir-3180 (Rfam accession id RF02010; AJ323057.1/363-249) was folded using different approaches yielding different output structures. (**a**) Secondary structure of the family as estimated by Rfam release 10.1; (**b**) RNAfold web server prediction based on Vienna RNA package version 2.0.0. [93]; (**c**) Mfold web server prediction, running Mfold version 4.6 [71].

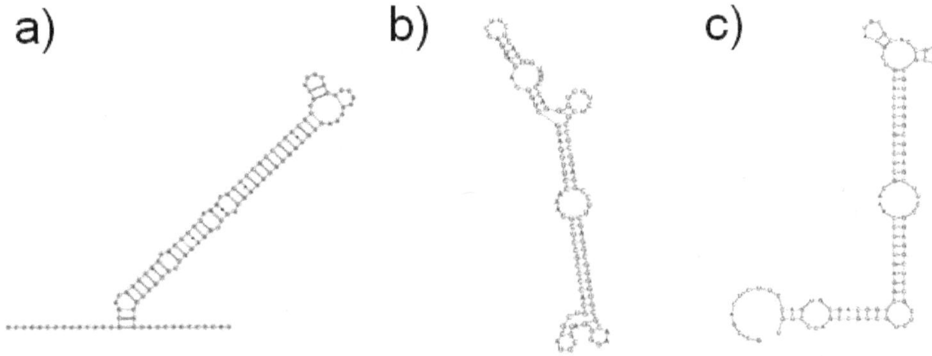

The best secondary structure prediction accuracy can be achieved using comparative methods [66]. These apply to a set of structurally homologous RNA sequences being aligned. For some of these computation tools, the output will be the prediction of each individual homologous structure, while in other situations the result will be a unique consensus structure. The consensus structure does not exist *in vivo*, but rather it is a model that captures the common, relevant structural aspects conserved within the family.

2.2. Structure Prediction and Alignment Strategies

Due to the close relationship between sequence and structure, structure prediction and sequence alignment can be described as interdependent problems [63]. As theorized by Sankoff [94], the most suitable approach should involve the simultaneous alignment and folding of the considered sequences. Unfortunately, a strict application of this approach would be computationally prohibitive and the lack of an appropriate heuristic solution is reflected by the wealth of available alternative solutions. The web server WAR [95] is a good example. This tool allows the execution of a total of 14 different strategies to align and predict the common secondary structure of multiple RNA sequences. Over the years, so many methods have been described that some kind of classification is needed to catalogue them. Paul Gardner proposed three categories he refers to as "plans" [66,96]. In plan A, one starts with the estimation of a multiple sequence alignment and then the aligned sequences are folded jointly (as a kind of consensus). The initial alignment can be done by any standard MSA aligner (e.g., ClustalW [97], T-Coffee [98]), and the folding of the aligned sequences can be performed by a plethora of tools (e.g., RNAalifold [99], PFOLD [100], ILM [101], ConStruct [102]) optimizing a score based on compensated mutations and thermodynamics. However this strategy is effective just in a determined sequence similarity range. On one hand, sequences that are too similar do not carry any covariance or purifying selection information and are less informative from an evolutionary standpoint. On the other hand, sequences need to be similar enough to be meaningfully aligned as below the "twilight zone" of similarity sequence alignment tends to obscure the covariance signal [96]. Plan B makes it possible to use evolutionary signals to help improve the reliability of structure predictions. This approach accounts for an underlying RNA substitution model where mutation probabilities are affected by structural dependencies. The maintenance of a 3D fold is a major evolutionary constraint influencing the acceptance of point mutations. From this perspective, a nucleotide located in the stem is not as free to mutate as a nucleotide located in a loop. Substitutions taking place in structured functional domains of RNAs can disrupt the wild-type conformation and seriously affect the molecular function. As a consequence, a nucleotide whose pairing has been disrupted by the mutation of its mate, is more likely to mutate itself so as to recover the original structure and rescue the function. Back in 1985 Sankoff developed a dynamic programming algorithm able to take into account sequence and structure of an RNA molecule simultaneously [94]. Unfortunately this algorithm is computationally expensive, with a running time equal to $O(N^{3m})$, where m is the number of sequences and N their length. This means that a pairwise comparison has the tremendous CPU cost of $O(N^6)$ which makes this algorithm inapplicable most of the times and calls for simplified heuristics. Several banded modifications of the Sankoff algorithm impose limits on the size or shape of substructures, like Dynalign [103,104], Foldalign [105,106], Stemloc [107], Consan [108]. Another example is pmmulti [109], a Sankoff algorithm variant able to perform multiple secondary structure alignments by aligning consensus base pair probability matrices. Plan C is used by programs such as R-Coffee [110] or RNAcast [111]. In these methods each individual sequence is folded separately before running the alignment. Equivalent secondary structures between two RNAs can be used to enhance the alignment accuracy. For instance, let seq1 and seq2 be two RNA sequences, x and y be two nucleotides matching in a secondary structure in seq1, and x' and y' be two nucleotides matching in a secondary structure in seq2. If x aligns to x' then implicitly y should be driven to align to y'. For example, R-Coffee uses RNAplfold [112] to compute the secondary structure of the provided sequences. After that, R-Coffee computes the multiple sequence alignment having the best agreement between sequences and structures. This pre-folding approach is especially valuable when RNAs are too different to be meaningfully aligned merely by using an off-the-shelf multiple alignment tools (*i.e.*, ClustalW [97]). Plan C is particularly well suited to situations where experimental secondary structures are available.

The situation is radically different when experimental 3D structure information is available. In this case the RNA alignment problem becomes similar to the protein structural alignment problem. This problem is nondeterministic polynomial-time complete (NP-complete) and it involves the alignment of two distance matrices. In most cases the problem can be solved in a rather satisfying way by using heuristics making the best of the geometric information contained in the PDB models. Examples of pairwise structural alignment methods for RNA are SARA [113], DIAL [114], iPARTS [115], ARTS [116] and SARSA [117]. Besides this, recently several 3D RNA structure database search programs were developed, such as LaJolla [118] and FRASS [119].

Giving an exhaustive overview of the methods available for folding and aligning structured RNA sequences is well beyond the scope of this review. Over the last twenty years, more than 30 methods have been described that deal with these issues which is an indication of the complexity of this problem, despite 25 years of research following its formal description by Sankoff.

3. Detecting ncRNA Homologues

In the ncRNA field another critical step is the collection of homologues to genes of interest. Homologues can be used in several situations, such as the detection of functional motifs, inference of possible evolutionary steps or the design of laboratory experiments. For instance, the conservation across species of a certain ncRNA can be used to estimate how likely a gene is to be functionally important. Such information can be used to prioritize experiments, e.g., knockdown experiments of the orthologous gene in a model organism. Over the last few years many different methods have been developed to approach the problem of RNA homology detection. As previously shown [120], homology search methods can be grouped in three sets: sequence-based, profiles and structure-based methods. The first and most straightforward approach to look for homologues is by comparing the nucleotide sequences. Already in 1981 Smith and Waterman developed a dynamic programming algorithm that allows for pairwise local alignment [121]. Nevertheless, this approach is CPU time demanding and implementations of this method have been until recently unpractical for large-scale database and genome wide screenings [122]. For this reason, alternative approaches such as FASTA [123] or BLAST [124] have been frequently preferred. These methods apply heuristics that boost computational speed at the cost of reduced accuracy. In both BLAST and FASTA, the underlying idea is to skip the time consuming comparison of entire query and target sequences, but rather to start identifying short conserved words in a first step called seeding. After that, more accurate time-consuming local alignments are performed. The second class of approaches are based on profiles, including HMMs. Profile HMMs are probabilistic models that are generated out of an input multiple sequence alignment where each position of the alignment is used to estimate nucleotide frequency. These models can be used to screen databases and look for homologs. Profile HMMs are heuristics having usually superior accuracy over methods based on single sequences [125,126]. However, such models have a linear architecture best suitable for modeling linear protein sequences (as opposed to secondary structure relationships). A more appropriate modeling of an RNA alignment should also consider RNA base pair interactions. The best sensitivity can be attained when applying approaches taking into account at the same time sequence similarity and secondary structures, as the Sankoff algorithm does. Unfortunately, the Sankoff algorithm is too computationally demanding, hence the need for approximate heuristic implementations of this exact algorithm. Such approximations include banded Sankoff tools [104,106,108,127], genetic algorithm implementations such as RAGA [128] and covariance models (CMs). CMs are the most commonly used method to carry out efficient database screening and can be described as special form of stochastic context free grammar (profile SCFGs). CMs were first introduced by Sean Eddy in [129] and implemented in Infernal [130]. This and other related applications such as RSEARCH [131] belong to a class of broadly used homology search tools based on the automatic learning of statistical models (the CMs) estimated from an input multiple RNA alignment decorated with the consensus secondary structure. CMs are probabilistic models that can be derived unambiguously out of a structure-annotated sequence alignment and can be used in turn to query a target sequence database to find homologs. Conceptually CMs are similar to profile HMM but able to include RNA base-pairs interactions information. The modeling of such information results in a higher complexity and CMs are represented by a tree-like model architecture, where the tree shape directly mirrors the consensus RNA structure. Unlike HMM states that only allow the emission of matches and indels, CMs embed new states to handle paired/not-paired and directionality information. For instance, in the paired sites, deletions can involve either a single 5' or 3' nucleotide, or the complete base pair. The direction also matters for the insertions that can now concern either the 5' or 3' ends of a base pair. In order to permit multi-loops, the bifurcation states are incorporated as well. In spite of their superior accuracy, CM cannot be used in all situations and are restricted to the identification of unsplit genes. The mapping of queries composed by multiple exons is impossible due to the impossibility of aligning secondary structures to a target interrupted by introns whose position is unknown. Moreover CMs need to "learn" from a set of homologous transcripts, but the set of sequences required to train the model are not always available. There is some circularity in this problem where the CM is used to search homologs that are themselves needed to estimate the model. Another layer of complexity results from the need to assemble a multiple sequence alignment of homologous sequences, needed to train the CM. In the CM the alignment will be used for a probabilistic description of matches, mismatches, insertions and deletions. However, generating accurate RNA alignments is difficult. In Rfam [132] CMs parameters are trained on high quality alignments (seed alignment) obtained from literature with manual curation. This input is used to estimate CMs, which are then passed to Infernal for homology searches. This CM/Infernal strategy is analogous to HMM/HMMER used for Pfam [133]. An option for spotting promising sequence segments and accelerate the detection procedure is to include a pre-filtering step as done for the Rfam setup [134]. This can be accomplished by means of *ad hoc* algorithms [135], profile HMMs like ML-heuristic [126] or BLAST with relaxed expectation values (*E*-values) to avoid losing sensitivity as achieved in Rfam [136]. A number of studies have been dedicated to the optimization of BLASTn parameters for seeking RNA homologs. For instance, in one study [120,137] the effectiveness of BLAST and other popular homology search methods tuned for ncRNA screenings were benchmarked. In [138], BlastR is introduced, a method that both takes advantage of di-nucleotide conservation and BLASTp as search engine to discover distantly related homologs. BlastR can be mounted on the top of computationally demanding algorithms to serve as a pre-filtering tool. One merit of this approach is that it neither require profiles nor secondary structure information, but relies solely on information encoded in primary nucleotide sequences.

Together with sequence-based, profiles and structure-based methods, another possibility for detecting inter-species homologs involves the use of multiple genome alignments [43]. Once established reciprocity between blocks of genomes belonging to different organisms (*i.e.*, syntenic regions), coordinate transfer from one gene to its homolog is straightforward and implies the projection of corresponding positions. This has been made possible thanks to the availability of genome sequences [139–142] and the development of alignment tools able to detect orthologous genomic regions, *i.e.*, loci that proceeded from the same genomic position in the ancestral genome [143]. Although comparing ncRNAs is currently still a complicated task, there exist several bioinformatics options to workaround the poor sequence conservation and effectively perform homology based prediction of novel ncRNAs.

4. High-Throughput Technologies and Genome-Wide Annotation of ncRNAs

4.1. Approaches for the High-Throughput Expression Detection

Recent technological advances have allowed the collection of an unprecedented amount of RNA sequence data coming from a wide range of organisms and conditions. For many years the main strategy for transcript discovery had been the sequencing of cloned complementary DNA (cDNA) of expressed sequence tags (ESTs) [144–146]. EST sequencing was then successfully used for the generation of large-scale expression datasets [147], and already by 1991 this approach had been utilized for human gene discovery [148]. Although it is widely acknowledged that ESTs represent a valuable resource to detect gene expression, they also came with severe limitations such as cost and sequencing requirements. Their dependence on bacterial cloning is an important source of bias and contamination that can lead to redundancy and under-representation or over-representation of host-selected transcripts [149–151]. More recently, oligonucleotide microarray technologies have made high throughput expression analysis much more practical, while the even more recently developed RNA-seq technologies promising transcriptomic analysis of unprecedented accuracy thanks to the application of NGS methods to transcriptome sequencing. Microarrays rely on a collection of nucleotide probe spots attached to a solid support. RNAs are labeled with fluorescent dyes, hybridized to the arrays, washed, and scanned with a laser [152]. Such arrays have been used for the investigation of known or predicted genes and have been until recently one of the most widespread technology for transcriptome exploration. Standard expression arrays are affected by several limitations including the hybridization and cross-hybridization artefacts [153–155], dye-based identification problems [156–160] and physical manufacturing restrictions, impeding the detection of splicing events and the discovery of unannotated genes [151]. A variant of traditional expression array is represented by tiling arrays. These are chips that use extremely densely spotted and probes representing overlapping contiguous regions of genome. Several works relying on this technology and aiming at transcript discovery have been published [38,161–164]. However tiling arrays require a substantial quantity of RNA and have further limitations affecting their sensitivity, specificity and the detection of splicing [151]. For instance, as shown in [165], microarrays lack sensitivity for genes expressed either at low or very high levels and if compared with RNA-seq have much smaller dynamic range. As a consequence, microarrays are inadequate for the quantification of both the prevailing RNA classes, and the less abundant ones. For genes with medium levels of expression, RNA-seq and microarrays return comparable results [165–167]. Still, each approach presents very specific advantages and disadvantages. A thorough comparison of these two approaches lies outside the purpose of this text (for reference, see [152,166,168]). Additional methods for high-throughput RNA discovery include the serial analysis of gene expression (SAGE) [169,170], several updated variants such as LongSAGE [171], RL-SAGE [172], SuperSAGE [173] and analogous approaches like the massively parallel signature sequencing (MPSS) [174]. In general, SAGE-like methods consist in the cloning and then the sequencing of short tags (17–25 nucleotides) coming from RNA extract. The resulting tag sequences can be compared against the source genome or a reference RNA database to attain the digital count of transcript quantities. Two other protocols that can be used in combination with high-throughput sequencing are the paired-end ditags (PETs) [175] and the rapid amplification of cDNA ends (RACE) [176–178]. Both approaches can be used to demarcate transcript boundaries, *i.e.*, define start and end of a transcript. Such information is extremely valuable *in situ* ations where the first and last exons can be respectively 5' and 3' associated with other transcript isoforms, thus making it difficult to define gene boundaries. Similarly, the cap analysis of gene expression (CAGE) [179,180] is a technique that allows high-throughout profiling of transcriptional starts points. Another promising application for ncRNA discovery, named RNA CaptureSeq, has been recently reported [181]. This approach is able to reach unprecedented sequencing depth. RNA CaptureSeq is inspired from exome sequencing techniques and relies on the use of tiling arrays in order to enrich the population of RNAs one wants to sequence. This enrichment step allows a sequencing depth that would be impossible when dealing with the full transcriptome. Although RNA CaptureSeq is not suited to generate full transcriptome profile, it can be used to target specific genomic sites and detect transcript isoforms expressed at very low abundance. As shown in [181] RNA CaptureSeq can be used to fuel the detection of ncRNAs that are missed by genome-wide standard RNA sequencing.

4.2. Datasets

Undoubtedly, high-throughput technologies enable the tremendous possibility to get both qualitative and quantitative information on whole transcripts mass produced by cells. This has resulted in high-resolution views of RNA expression dynamics throughout different tissues and time points [182–184] and fueled the development of ncRNA specific databases, such as Rfam [132], NRED [20], lncRNAdb [33], RNAdb [185], fRNAdb [186] and NONCODE [187]. Furthermore, various groups and projects, such as RefSeq [188], GENCODE [14,189], HAVANA team [190,191], Ensembl [192] and FANTOM [193] undertook the task to comprehensively annotate functional elements, including ncRNAs, of a number of species using experimental data. The RefSeq repository houses annotations resulting from automated analyses, collaboration and manual curation [188,194]. The GENCODE pipeline combines HAVANA and Ensembl automatic annotations to annotate the human gene features generated in the context of the ENCODE project [14,45,189]. The HAVANA team has the goal to provide manually curated annotations of transcripts aligned to human, mouse and zebrafish genomes. Ensembl runs an automatic *genebuild* process including *ab initio* gene predictions and release 64 supported a total of 61 species [192]. The Ensembl *genebuild* system is adapted to every species in the set according to the data that is available. For instance Ensembl imports and merges high quality HAVANA annotations exclusively for human and mouse. The FANTOM consortium aims to provide functional annotations to the full-length cDNAs [193]. The annotations generated by these consortia are freely available through genome browsers, including UCSC [195], Ensembl [196] and VEGA [197]. As new genomic regions get annotated and new transcript sequences become publicly available, these gene sets continue to growth [14,188,194]. A recent publication [14] indicated that in the last years the number of annotated protein-coding and non-coding transcripts in GENCODE has dramatically increased. For instance, passing from GENCODE version 3c (July 2009, http://www.gencodegenes.org/archive_stats.html) to version 7 (December 2010, http://www.gencodegenes.org/archive_stats.html), the number of protein-coding transcripts increased from 68,880 to 76,052, and the number of lncRNAs jumped from 10,457 to 15,512. In terms of gene annotations, the number of known protein-coding genes has remained almost unchanged, while the ncRNA gene annotations expanded tremendously (see Figure 4).

Figure 4. Number of non-coding and protein-coding genes annotated over the last Ensembl releases. The *x*-axis indicates the number and the date of the release. The vertical axis reports the number of ncRNA (blue line) and protein-coding genes (red line).

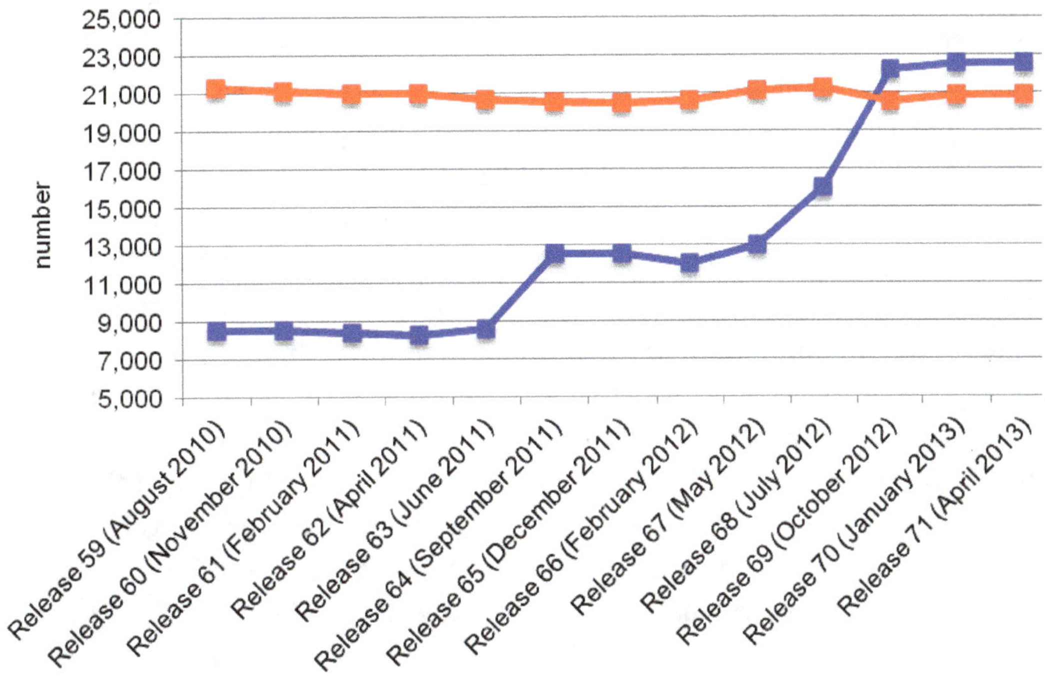

The overall picture, however, remains blurred by inconsistent findings, suggesting that more analyses are still needed. For instance, the recent estimates reported by the ENCODE project indicate that about the 62% of human genomic bases are expressed in long transcripts, while 5.5% only of the whole genome is found within the GENCODE annotated exons [198]. This discrepancy can be in part explained by the fact that GENCODE catalogues transcripts using cDNA/EST alignments [14] rather than RNA-seq short-read data. A classic low-throughput EST sequencing operated by the Sanger technology can identify mostly high abundant transcripts [199], while deep coverage RNA-seq experiments can reveal rare but potentially regulatory transcripts. Nonetheless, ESTs are longer than RNA-seq reads, and can provide more reliable transcriptional evidence [200].

4.3. NGS Challenges

To make the most out of the extraordinary possibilities that NGS offers, it is essential to understand the current limitations. One important point is that the reads returned by standard NGS platforms are usually short (35–500 base pairs [201]) and as a consequence it becomes necessary to reassemble the full-length transcripts. Small non-coding RNAs (*i.e.*, miRNA and piRNAs) represent an exception and there is no need to reassemble them, as they are small enough to be entirely covered by the read length. Unfortunately the process of reassembling transcriptomes starting from short reads is difficult. Normally RNA-seq dataset are big (gigabases to terabases), and thus need to be handled by sufficient large memory and by multi-CPU computers able to execute the algorithms in parallel with sufficient high-performance storage to store primary, temporary and output data. Although various short-read assemblers [202–204] were successfully applied to genome assembly, these packages cannot be easily used to reconstruct transcriptomes. Applying tools normally designed for genome reconstruction to the problem of transcriptome assembly leads to multiple complications. A key issue is that the DNA sequencing depth is supposed to be identical over the entire genome while transcriptome sequencing depth is expected to fluctuate significantly. For this reason, DNA short-read assemblers could erroneously interpret highly abundant transcripts as repetitive genomic regions. Furthermore, when using genome short-read assemblers the read strand is not taken into account. On the contrary, when available, a transcriptome assembler should exploit the strand information to unravel possible antisense expressions on different strands. Finally, the transcript modeling is involved as transcript variants coming from the same gene can share exons and are difficult to resolve unambiguously [199].

It is possible to work out the transcriptome assembly following a reference-based approach, a *de novo* assembly or combinations of each [199]. The first considers the initial mapping of the reads on a reference genome, and then the usage of transcript assemblers. To the end of labeling each read with the genomic location they come from, a new class of software, generally referred as read mapper, has recently shown up. In this context, the availability and the quality of the underlying reference genome are critical. Besides that, when dealing with massive amount of short-read data the CPU and the memory costs can be challenging, and several algorithms are being tailored to achieve best mapping efficiency [205–211]. Other important issues relate to the mapping of reads crossing exon-junction boundaries [212,213] and the uncertainty or lack of accuracy in read alignments. For most downstream applications, the accurate positioning of the reads back to the source genome is crucial. To improve the mapping accuracy, the process can take into account the read quality information [214,215]. The quality scores, introduced by the Phred algorithm [216,217], indicate the reliability of each base call in each read in a log-likelihood scale. Since bases with reduced quality scores have an increased probability to be sequencing errors, a read mapper should either use less severe penalization for mismatches at positions with low base-call quality, or not align such positions at all. The information about the quality score is particularly relevant when mapping reads of large size. This is a result of the fact that 3' ends of longer reads are affected by sequencing errors at higher rates [215]. Besides choosing a threshold for accepted mismatches, other important and sometimes arbitrary decisions regard the split mapping and multiple mapping reads. The first refers to reads that could not be aligned to the reference genome unless split in subparts. Such reads could either highlight the presence of an unreported exon-junction boundary, or be sequencing artefacts. The second indicates reads that align multiple times across the reference genome. This mapping uncertainty is caused by repeated elements and may results in flawed expression establishments. On one hand, removing multiple mapping reads from the analysis would imply an underestimation of the expression of genes embedding repeats. On the other hand, considering multiple mapping reads would lead to artefactual expression measurements. Once mapped the reads, additional issues concern the application of transcript assemblers. Several computational tools have been developed with the purpose of reconstructing transcripts in their entire length, *i.e.*, annotating exon-intron transcript structures. These methods include Cufflinks [218], Isolazo [219] and Scripture [42]. In [220] the authors have shown that variations across transcript assemblers can be source of confusion, with low consistency across methods and a high number of false positives [200,219]. Transcript assemblers seem to have a better agreement when reconstructing protein-coding transcripts [43] with the agreement dropping dramatically when modeling large intervening ncRNAs (lincRNAs). For instance, Cufflinks and

Scripture share only 46% agreement for lincRNA transcript models [43]. Such discrepancies are caused by the differences in how each assembler reconstruct lowly expressed transcripts [43]. In other words, about half of the isoforms estimated by a method in areas with low read density do not correspond to isoforms called by the other method. This poor agreement between transcript assemblers highlights the need for further improvements, calling for the development of new algorithms to accurately represent low abundant transcripts.

Another possibility to assemble a transcriptome from-short reads is *de novo* assembly of transcripts. This strategy does not require any reference genome and is therefore independent on the correct alignment of the reads to the splice sites. Advantages of this approach are that it is less reliant on accurate genome annotation and can be applied to organisms without sequenced or fully annotated genomes. Examples of applications adopting this strategy are described in [221–223]. Nevertheless, the application of *de novo* assembly to complex transcriptomes (e.g., higher eukaryotes) is complicated by the dataset sizes and the dense network of alternatively spliced variants. Furthermore, *de novo* transcriptome assemblers need much deeper sequencing than reference-based assemblers and are largely affected by sequencing errors [199].

Once transcriptome dataset is generated, there are additional complications in the downstream analysis if trying to distinguish genuine ncRNAs from mRNAs. Currently, this issue is becoming increasingly important as many researchers are only interested in one or the other. The most straightforward procedure would be to compare a newly generated transcriptome against existing gene annotations. However in most cases annotations are far from complete and the great majority of genes they include are protein-coding. As a consequence, in a normal RNA-seq experiment a substantial fraction of read contigs map outside of annotated exons [198]. Previously unreported transcripts can be either classified as ncRNA or mRNA according to the protein-coding potential they have. The *in-silico* assignment of a transcript to one of these two groups is not always trivial and may require dedicated expert curation [190]. Some transcript isoforms might insert coding exons and therefore could be partially translated, *i.e.*, generating small peptides. There are further ambiguities for coding transcripts whose untranslated structured molecules are also functional as ncRNAs [224] and for genes having both coding and non-coding isoforms [225]. A commonly used approach to predict the coding potential involves the codon substitution frequency (CSF) estimation [226,227]. This measure is based on an input multiple alignment of orthologous sequences. The CSF score deems a region to be coding depending on how the sequences of the multiple alignment evolved, *i.e.*, showing distinctive mutation patterns, as are expected in coding and non-coding loci. A coding region is expected to embed prevalently conservative amino acid substitutions and synonymous codon substitutions, while showing low occurrence of nonsense and missense mutations. Although CSF has been successfully applied in various research projects [226,228,229], the score is not always easy to estimate with the availability of trustworthy orthologues being the main limiting factor when dealing with new transcriptome datasets. Issues include scarcity or even the absence of orthologs, erroneous insert of pseudogenes in the set and absence of informative variations. For instance, as shown in [40] many putative human lncRNAs are not found in other species, and cannot be analyzed using CSF. Besides this, primate specific lncRNAs rarely show sufficient changes to highlight sense/nonsense mutations patterns. In addition to CSF, other strategies not relying on evolutionary signatures can be effectively used to predict if a transcript is going to be translated into protein or not. For example, there are dedicated

BLAST flavors including BLASTx and RPS-Blast [124,230] that can be used to identify transcripts whose translational product possesses a match in protein databases such as Pfam [133] and UniProt [231]. Other algorithms include CPC (Coding Potential Calculator), a support vector machine (SVM) classifier including both Open Reading Frame (ORF) and homology predictions features [232], PORTRAIT (Prediction of transcriptomic ncRNA by ab initio methods) a SVM classifier not using homology information [233] and CPAT (Coding Potential Assessment Tool), a logistic regression model built with four sequence features including ORF predictions [234]. Unfortunately, bioinformatics predictions can easily return mistaken assignments when dealing with ncRNAs closely related to coding mRNAs, and result in some confusion when transferring annotation across species, or within a genome. Such observations may wrongly suggest pseudogenization events or a turnover between proteins and ncRNAs.

4.4. Other Approaches

Over the last few years, other approaches alternative or complementary to RNA-seq have been attempted to generate high-throughput ncRNA annotations. In 2009, Mitchell Guttman and co-workers published the first of a series of analysis that recently came out linking lncRNA detection to histone modifications [13]. In this work, the authors pioneered a chromatin-state based method to identify well-defined transcriptional units occurring between known protein-coding genes. Their analysis relied on the observation by [235] that promoters of genes expressed by the RNA polymerase II (Pol II) are signed by trimethylation of lysine 4 of histone H3 (H3K4me3) while the transcribed area is marked by trimethylation of lysine 36 of histone H3 (H3K36me3). Following this observation, the authors did chromatin immunoprecipitation followed by high-throughput sequencing (ChIP-seq) [235] to generate profiles of chromatin states. This approach revealed 1600 mouse lincRNAs, corresponding to H3K4me3-H3K36me3 chromatin domains and lying outside of protein-coding regions. The prediction reliability has been estimated by additional analysis showing that lincRNAs are more conserved than neutrally evolving sequences and that most of experimentally tested loci were found to be expressed [13]. An alternative strategy used for ncRNA detection involves a combination of different high-throughput data sources and their integration using bioinformatics [236]. This approach, named incRNA, relies on a machine learning method and has been applied to the genome-wide identification of *Caenorhabditis elegans* ncRNAs. incRNA combines predicted and experimental data for a total of nine different information sources. These include the expression data coming from various developmental stages and conditions, as well as the GC content, the predictions of RNA secondary structure folding energy, the prediction of evolutionary conserved DNA sequence and secondary structure. These results illustrate how the integration of multiple information sources ends in highly accurate predictions of novel ncRNA genes.

Recently, a number of works reporting a massive quantity of novel ncRNA genes in various species have been published [40,43,44,237,238]. Such rapid growth has been possible thanks to the parallel development of new and ever more sophisticated bioinformatics approaches. Nevertheless, such analyses remain superficial with uncertainties of different type and degree affecting most predictions. For example, the homology search pipeline described in [40] is not sensitive enough to map rapidly evolving lncRNAs, hence the limit to play comprehensive evolutionary study. Moreover such lncRNA predictions should be taken with care, not just because they are not experimentally verified, but also because they are far from representing the complete genome-wide lncRNA figure. For validation purpose, some works provide the number of predicted lncRNA supported by expression evidences. For instance in [13] the authors confirmed by tiling array the expression of ~70% lncRNA predictions. In other cases as in [44], RT-PCR has been used to validate 15 newly identified lncRNAs. On the short run available transcription data is expected to increase very rapidly, and the necessity to accurately and quickly validate ncRNAs is becoming more pressing than ever.

Table 1. A summary of methods, datasets and browsers for non-coding RNA analysis. The first column indicates the resource type. The second column the resource name. The third column reports the PubMed ID when available, if not the web address. The fourth column provides a brief description of the resource.

	Resource	Pubmed ID	Description
Comparing ncRNAs (Section 2)	Mfold	6163133	Single sequence RNA secondary structure prediction.
	RNAfold	12824340	
	WAR	18492721	WEB server allowing the execution of different alignment methods
	RNAalifold	12079347	Folding previously aligned RNAs (Plan A)
	PFOLD	12824339	
	ILM	14693809	
	Construct	10518612	
	Dynalign	11902836	Sankoff derived algorithm for the simultaneous alignment and secondary structure prediction (Plan B)
	Foldalign	9278497	
	Stemloc	15790387	
	Consan	16952317	
	pmmulti	15073017	
	R-Coffee	18420654	Aligners taking into account previously estimated secondary structure (Plan C)
	RNAcast	16020472	
	SARA	18689811	3D structure alignment method
	DIAL	17567620	
	iPARTS	20507908	
	ARTS	16204124	
	SARSA	18502774	
	LaJolla	http://www.mdpi.com/1999-4893/2/2/692	
	FRASS	20553602	

Table 1. *Cont.*

	Resource	Pubmed ID	Description
Detecting ncRNAs (section 3)	ML-heuristic	16267089	Profile HMM
	RAGA	9358168	Genetic algorithm
	RSEARCH	14499004	Covariance model
	Infernal	12095421	
	BlastR	21624887	BLAST-based dinucleotide homology search
Datasets and browsers (section 4)	ENCODE	22955616	Consortium
	Ensembl	22086963	
	FANTOM	11217851	
	HAVANA	http://www.sanger.ac.uk/research/projects/vertebrategenome/havana/	Annotation team
	GENCODE	22955987	Project for the annotation of all human gene features
	UCSC	12045153	Genome browser
	VEGA	18003653	
	RefSeq	18927115	Collection of DNA, transcripts, and proteins
	Rfam	12520045	ncRNA database
	NRED	18829717	
	lncRNAdb	21112873	
	RNAdb	17145715	
	fRNAdb	17099231	
	NONCODE	15608158	

5. Discussion and Conclusions

ncRNA functional characterization is a rapidly expanding research area. In the past few years, it has become clear that the majority of the transcripts in cells are more than mere intermediates between the hereditary information encoded in DNA and the mechanical operative component represented by proteins. Indeed, it appears that numerous transcripts may not be translated at all while still being involved in critical biological functions such as cell differentiation and chromatin remodeling. Taking together 15 human cell lines, the cumulative coverage of transcribed regions is ~62% and ~75% of the whole human genome for processed and primary transcripts, respectively [239]. This "pervasive transcription" is strikingly high, especially when considering that a mere 3% of the human genome codes for protein-coding exons. [198]. Numerous novel, previously uncharacterized RNA species have been recently detected. A sizeable fraction of them are defined as lncRNA, *i.e.*, functional molecules longer than 200 nucleotides that do not show any coding potential. Some of these molecules are spliced, capped, differentially expressed in tissues/cells or developmental stages and tend to be more conserved across species than would result from neutral evolution. For these reasons and because of the increasing number of transcripts whose function has been experimentally validated, it is believed that many of these new ncRNAs belong to an important, relatively unexplored class of regulatory elements. Thanks to ongoing improvements in sequencing technologies it has become possible to collect a significant amount of these uncharacterized transcripts. The latest generation of sequencing technologies makes it possible to perform large scale sequencing of entire transcriptomes. This technique, known as RNA-seq has already had a dramatic impact on our perception of the human transcription landscape [183,239]. Similar studies have been carried out in a number of genetic model organisms including rodents [44,151], plants [240], insects [184], worms [241] and yeasts [242]. In [243] the author argues that RNA-seq represents the most promising technology for transcriptome research. The main strength of RNA-seq approaches are the high degree of dynamic range they offers, returning better sensitivities than microarrays without the need of *a priori* speculation regarding the genomic loci being transcribed [244]. If the pace of scientific progress is maintained and if costs keep decreasing, one can reasonably expect this technology to rapidly become a key component of personalized medicine, especially when considering the new venues of development that are currently being considered [152,245].

From a functional perspective, much remains to be done for accurate characterization and functional analysis of ncRNAs. To infer the function of novel ncRNAs one possibility is looking for functional motifs. This can be done by running motif finders algorithms to predict structurally conserved and potentially functional motifs [246–252]. Furthermore, the functional characterization of a novel ncRNA can be aided by the detection of protein-RNA binding motifs and the identification of protein interaction partners. Experimental approaches suited for this include RIP (Rna Immunoprecipitation) and CLIP (Cross-Linking and ImmunoPrecipitation) [253]. Comparative studies also offer a very efficient way to have functional insights and prioritizing analysis. They can be used to predict function by homology, assess phylogenetic relationships, detect functional motifs or classify related molecules in order to identify families. A major challenge when tackling ncRNA comparisons results from the remarkable variability of traits and functions. Considering sizes only, ncRNA molecules can be as short as a miRNA (~22 nt) and up to ~17 kb long in the case of Xist [2]. Another difficulty when comparing ncRNAs is that most of these genes have poorly conserved sequences. Such diversity challenges our ability to compare, classify and search with conventional alignment tools. In addition ncRNA genes have no equivalent of codon bias and ORFs that help powering the statistical component of machine learning approaches when doing protein prediction [254]. The strongest signal contained by RNA sequences is usually evolutionarily conserved secondary structures. Many efficient algorithms exist that are able to predict potential structures using MFE or SCFG computations. Unfortunately, these predictions ignore the contribution of the environment and are not always accurate enough to significantly improve alignment accuracy and homology modeling. Emerging technologies allowing the high-throughput generation of experimentally derived secondary structures [255] will hopefully help addressing this problem. Unfortunately, taking into account secondary structures while comparing sequences is a challenging procedure, too intensive from a computational point of view to be practical in most circumstances [108]. This makes it is difficult to compare mono-exonic genes while taking the secondary structure into account, and totally impossible when the transcripts are multi-exonic (*i.e.,* the secondary structures are interrupted by introns). It has been shown [40,237] that BLAST can be effectively used for lncRNA homolog prediction, in combination with splicing informed heuristics such as exonerate [256] or GeneWise [257]. This strategy is not new, and similar approaches have already been used for the discovery of protein-coding homologs [258–260]. As one would expect, homology based RNA searches are severely limited by the capacity to align distant homologues. For instance, when searching the human lncRNA complement against mammalian genomes [40] or when using an estimated pig complement [237], the authors only managed to find, beyond primates, less than 50% of the query genes across cow, mouse or dog. This result may reflect a high turnover, but the conservation/disappearance patterns, poorly correlated to phylogenetic history, are most likely indicative of a limited detection capacity. Other confounding factors include misassembled or partially sequenced genomes. Additional analysis would be needed to validate the Blast/exonerate mapping approach. At this stage, it is therefore impossible, without further experiments, to establish whether lncRNA queries that failed to map are really absent in the target species or undetected. In this context, high quality templates, such as the GENCODE queries used in [40], offer better likelihood to return precise annotation. In the same publication it is also predicted that sizeable fraction of the human lncRNAs is primate specific [40]. This result is in agreement with a recently published study [44] where the authors identified lncRNAs expressed in rodents' adult liver, and then compared the expression of the orthologous genomic regions. This work illustrates that loss of lncRNA transcription among rodents is associated with loss of sequence constraints and that many lncRNA genes seems to be species or lineage specific. Another application of homology based approaches is the possibility to identify novel human lncRNA genes candidates by using non-human templates as query [237]. As shown in the paper [237], there are 131 pig lncRNAs mapping to unannotated regions of the human genome. This result suggests that although human is probably one of the most extensively annotated higher-eukaryote, extra improvements might be achieved using data gathered in other non-model organisms. In [40] the authors also extend the lncRNA conservation study to a multiple genome alignments strategy based on PhastCons conservation scores. The analysis is in agreement with previous reports [13,30] and confirms that lncRNAs sequences are less constrained than those of protein-coding genes. Remarkably, it was shown that the distribution of lncRNA exons conservation is bimodal, with a fraction substantially approximate to ancestral repeats, and another group appreciably shifted toward the protein-coding set. This indicates that some lncRNA are under a selection as strong as that seen for proteins and suggests that a sizeable fraction of lncRNA genes are probably functional. The fraction of lncRNAs having a mutation rate almost indistinguishable from repeats suggests that at least some lncRNAs (close to a third) may be transcriptional noise. However, despite this abundance of lncRNA sequences that do not appear to be under selection, the transcript product itself might still have a biological role and as shown in [261,262] the transcription process itself of some ncRNA can bear regulative functions.

Despite the difficulties encountered when comparing ncRNAs, homology search of ncRNAs can be successfully used to detect new genes. New and ever more sophisticated algorithms will help addressing the challenges brought by NGS technologies. The ultimate goal is the creation of thorough transcriptome annotations and unbiased expression profiling of each individual transcript. It is still too early to tell. However, if they live up to their promises and expectation, the discovery of this new large class of RNAs may well define one of the turning points of modern biology.

Acknowledgments

The authors wish to acknowledge Roderic Guigó for the helpful comments and suggestions.

References

1. Lee, R.C.; Feinbaum, R.L.; Ambros, V. The *C. elegans* heterochronic gene lin-4 encodes small RNAs with antisense complementarity to lin-14. *Cell* **1993**, *75*, 843–854.
2. Brown, C.J.; Hendrich, B.D.; Rupert, J.L.; Lafreniere, R.G.; Xing, Y.; Lawrence, J.; Willard, H.F. The human XIST gene: Analysis of a 17 kb inactive X-specific RNA that contains conserved repeats and is highly localized within the nucleus. *Cell* **1992**, *71*, 527–542.
3. Farazi, T.A.; Juranek, S.A.; Tuschl, T. The growing catalog of small RNAs and their association with distinct Argonaute/Piwi family members. *Development* **2008**, *135*, 1201–1214.
4. Bachellerie, J.P.; Cavaille, J.; Huttenhofer, A. The expanding snoRNA world. *Biochimie* **2002**, *84*, 775–790.
5. Barrett, T.; Suzek, T.O.; Troup, D.B.; Wilhite, S.E.; Ngau, W.C.; Ledoux, P.; Rudnev, D.; Lash, A.E.; Fujibuchi, W.; Edgar, R. NCBI GEO: Mining millions of expression profiles—database and tools. *Nucleic Acids Res.* **2005**, *33*, D562–D566.
6. Parkinson, H.; Sarkans, U.; Shojatalab, M.; Abeygunawardena, N.; Contrino, S.; Coulson, R.; Farne, A.; Lara, G.G.; Holloway, E.; *et al.* ArrayExpress—A public repository for microarray gene expression data at the EBI. *Nucleic Acids Res.* **2005**, *33*, D553–D555.
7. Griffiths-Jones, S.; Grocock, R.J.; van Dongen, S.; Bateman, A.; Enright, A.J. miRBase: MicroRNA sequences, targets and gene nomenclature. *Nucleic Acids Res.* **2006**, *34*, D140–D144.
8. Fraser, B.A.; Weadick, C.J.; Janowitz, I.; Rodd, F.H.; Hughes, K.A. Sequencing and characterization of the guppy (Poecilia reticulata) transcriptome. *BMC Genomics* **2011**, *12*, 202.
9. Tuda, J.; Mongan, A.E.; Tolba, M.E.; Imada, M.; Yamagishi, J.; Xuan, X.; Wakaguri, H.; Sugano, S.; Sugimoto, C.; Suzuki, Y. Full-parasites: database of full-length cDNAs of apicomplexa parasites, 2010 update. *Nucleic Acids Res.* **2011**, *39*, D625–D631.
10. Mamidala, P.; Wijeratne, A.J.; Wijeratne, S.; Kornacker, K.; Sudhamalla, B.; Rivera-Vega, L.J.; Hoelmer, A.; Meulia, T.; Jones, S.C.; Mittapalli, O. RNA-Seq and molecular docking reveal multi-level pesticide resistance in the bed bug. *BMC Genomics* **2012**, *13*, 6.
11. Dinger, M.E.; Amaral, P.P.; Mercer, T.R.; Pang, K.C.; Bruce, S.J.; Gardiner, B.B.; Askarian-Amiri, M.E.; Ru, K.; Solda, G.; Simons, C.; *et al.* Long noncoding RNAs in mouse embryonic stem cell pluripotency and differentiation. *Genome Res.* **2008**, *18*, 1433–1445.
12. Okazaki, Y.; Furuno, M.; Kasukawa, T.; Adachi, J.; Bono, H.; Kondo, S.; Nikaido, I.; Osato, N.; Saito, R.; Suzuki, H.; *et al.* Analysis of the mouse transcriptome based on functional annotation of 60,770 full-length cDNAs. *Nature* **2002**, *420*, 563–573.
13. Guttman, M.; Amit, I.; Garber, M.; French, C.; Lin, M.F.; Feldser, D.; Huarte, M.; Zuk, O.; Carey, B.W.; Cassady, J.P.; *et al.* Chromatin signature reveals over a thousand highly conserved large non-coding RNAs in mammals. *Nature* **2009**, *458*, 223–227.
14. Harrow, J.; Frankish, A.; Gonzalez, J.M.; Tapanari, E.; Diekhans, M.; Kokocinski, F.; Aken, B.L.; Barrell, D.; Zadissa, A.; Searle, S.; *et al.* GENCODE: The reference human genome annotation for The ENCODE Project. *Genome Res.* **2012**, *22*, 1760–1774.
15. Managadze, D.; Lobkovsky, A.E.; Wolf, Y.I.; Shabalina, S.A.; Rogozin, I.B.; Koonin, E.V. The vast, conserved mammalian lincRNome. *PLoS Comput. Biol.* **2013**, *9*, e1002917.

16. Mattick, J.S.; Gagen, M.J. The evolution of controlled multitasked gene networks: The role of introns and other noncoding RNAs in the development of complex organisms. *Mol. Biol. Evol.* **2001**, *18*, 1611–1630.
17. Mattick, J.S. Non-coding RNAs: The architects of eukaryotic complexity. *EMBO Rep.* **2001**, *2*, 986–991.
18. Crick, F.H. On protein synthesis. *Symp. Soc. Exp. Biol.* **1958**, *12*, 138–163.
19. Wang, J.; Zhang, J.; Zheng, H.; Li, J.; Liu, D.; Li, H.; Samudrala, R.; Yu, J.; Wong, G.K. Mouse transcriptome: Neutral evolution of "non-coding" complementary DNAs. *Nature* **2004**, *431*, doi:10.1038/nature03016.
20. Dinger, M.E.; Pang, K.C.; Mercer, T.R.; Crowe, M.L.; Grimmond, S.M.; Mattick, J.S. NRED: A database of long noncoding RNA expression. *Nucleic Acids Res.* **2009**, *37*, D122–D126.
21. Mattick, J.S. The genetic signatures of noncoding RNAs. *PLoS Genet.* **2009**, *5*, e1000459.
22. Wapinski, O.; Chang, H.Y. Long noncoding RNAs and human disease. *Trends Cell. Biol.* **2011**, *21*, 354–361.
23. Wang, X.; Song, X.; Glass, C.K.; Rosenfeld, M.G. The long arm of long noncoding RNAs: roles as sensors regulating gene transcriptional programs. *Cold Spring Harb. Perspect. Biol.* **2011**, *3*, a003756.
24. Satterlee, J.S.; Barbee, S.; Jin, P.; Krichevsky, A.; Salama, S.; Schratt, G.; Wu, D.Y. Noncoding RNAs in the brain. *J. Neurosci.* **2007**, *27*, 11856–11859.
25. Mercer, T.R.; Dinger, M.E.; Sunkin, S.M.; Mehler, M.F.; Mattick, J.S. Specific expression of long noncoding RNAs in the mouse brain. *Proc. Natl. Acad. Sci. USA* **2008**, *105*, 716–721.
26. Kaikkonen, M.U.; Lam, M.T.; Glass, C.K. Non-coding RNAs as regulators of gene expression and epigenetics. *Cardiovasc. Res.* **2011**, *90*, 430–440.
27. Braidotti, G.; Baubec, T.; Pauler, F.; Seidl, C.; Smrzka, O.; Stricker, S.; Yotova, I.; Barlow, D.P. The Air noncoding RNA: An imprinted *cis*-silencing transcript. *Cold Spring Harb. Symp. Quant. Biol.* **2004**, *69*, 55–66.
28. Willingham, A.T.; Orth, A.P.; Batalov, S.; Peters, E.C.; Wen, B.G.; Aza-Blanc, P.; Hogenesch, J.B.; Schultz, P.G. A strategy for probing the function of noncoding RNAs finds a repressor of NFAT. *Science* **2005**, *309*, 1570–1573.
29. Cesana, M.; Cacchiarelli, D.; Legnini, I.; Santini, T.; Sthandier, O.; Chinappi, M.; Tramontano, A.; Bozzoni, I. A long noncoding RNA controls muscle differentiation by functioning as a competing endogenous RNA. *Cell* **2011**, *147*, 358–369.
30. Ørom, U.A.; Derrien, T.; Beringer, M.; Gumireddy, K.; Gardini, A.; Bussotti, G.; Lai, F.; Zytnicki, M.; Notredame, C.; Huang, Q.; *et al.* Long noncoding RNAs with enhancer-like function in human cells. *Cell* **2010**, *143*, 46–58.
31. Lai, F.; Orom, U.A.; Cesaroni, M.; Beringer, M.; Taatjes, D.J.; Blobel, G.A.; Shiekhattar, R. Activating RNAs associate with Mediator to enhance chromatin architecture and transcription. *Nature* **2013**, *494*, 497–501.
32. Rinn, J.L.; Chang, H.Y. Genome regulation by long noncoding RNAs. *Annu. Rev. Biochem.* **2012**, *81*, 145–166.

33. Amaral, P.P.; Clark, M.B.; Gascoigne, D.K.; Dinger, M.E.; Mattick, J.S. lncRNAdb: A reference database for long noncoding RNAs. *Nucleic Acids Res.* **2011**, *39*, D146–D151.
34. Ravasi, T.; Suzuki, H.; Pang, K.C.; Katayama, S.; Furuno, M.; Okunishi, R.; Fukuda, S.; Ru, K.; Frith, M.C.; Gongora, M.M.; *et al.* Experimental validation of the regulated expression of large numbers of non-coding RNAs from the mouse genome. *Genome Res.* **2006**, *16*, 11–19.
35. Wang, X.; Arai, S.; Song, X.; Reichart, D.; Du, K.; Pascual, G.; Tempst, P.; Rosenfeld, M.G.; Glass, C.K.; Kurokawa, R. Induced ncRNAs allosterically modify RNA-binding proteins in *cis* to inhibit transcription. *Nature* **2008**, *454*, 126–130.
36. Rinn, J.L.; Kertesz, M.; Wang, J.K.; Squazzo, S.L.; Xu, X.; Brugmann, S.A.; Goodnough, L.H.; Helms, J.A.; Farnham, P.J.; Segal, E.; *et al.* Functional demarcation of active and silent chromatin domains in human HOX loci by noncoding RNAs. *Cell* **2007**, *129*, 1311–1323.
37. Rodriguez, A.; Griffiths-Jones, S.; Ashurst, J.L.; Bradley, A. Identification of mammalian microRNA host genes and transcription units. *Genome Res.* **2004**, *14*, 1902–1910.
38. Kapranov, P.; Cheng, J.; Dike, S.; Nix, D.A.; Duttagupta, R.; Willingham, A.T.; Stadler, P.F.; Hertel, J.; Hackermuller, J.; Hofacker, I.L.; *et al.* RNA maps reveal new RNA classes and a possible function for pervasive transcription. *Science* **2007**, *316*, 1484–1488.
39. Ogawa, Y.; Sun, B.K.; Lee, J.T. Intersection of the RNA interference and X-inactivation pathways. *Science* **2008**, *320*, 1336–1341.
40. Derrien, T.; Johnson, R.; Bussotti, G.; Tanzer, A.; Djebali, S.; Tilgner, H.; Guernec, G.; Martin, D.; Merkel, A.; Knowles, D.G.; *et al.* The GENCODE v7 catalog of human long noncoding RNAs: Analysis of their gene structure, evolution, and expression. *Genome Res.* **2012**, *22*, 1775–1789.
41. Chen, Z.; Duan, X. Ribosomal RNA depletion for massively parallel bacterial RNA-sequencing applications. *Methods Mol. Biol.* **2011**, *733*, 93–103.
42. Guttman, M.; Garber, M.; Levin, J.Z.; Donaghey, J.; Robinson, J.; Adiconis, X.; Fan, L.; Koziol, M.J.; Gnirke, A.; Nusbaum, C.; *et al. Ab. initio* reconstruction of cell type-specific transcriptomes in mouse reveals the conserved multi-exonic structure of lincRNAs. *Nat. Biotechnol.* **2010**, *28*, 503–510.
43. Cabili, M.N.; Trapnell, C.; Goff, L.; Koziol, M.; Tazon-Vega, B.; Regev, A.; Rinn, J.L. Integrative annotation of human large intergenic noncoding RNAs reveals global properties and specific subclasses. *Genes Dev.* **2011**, *25*, 1915–1927.
44. Kutter, C.; Watt, S.; Stefflova, K.; Wilson, M.D.; Goncalves, A.; Ponting, C.P.; Odom, D.T.; Marques, A.C. Rapid turnover of long noncoding rnas and the evolution of gene expression. *PLoS Genet.* **2012**, *8*, e1002841.
45. ENCODE Project Consortium. Identification and analysis of functional elements in 1% of the human genome by the ENCODE pilot project. *Nature* **2007**, *447*, 799–816.
46. Clark, M.B.; Amaral, P.P.; Schlesinger, F.J.; Dinger, M.E.; Taft, R.J.; Rinn, J.L.; Ponting, C.P.; Stadler, P.F.; Morris, K.V.; Morillon, A.; *et al.* The reality of pervasive transcription. *PLoS Biol.* **2011**, *9*, doi:10.1371/journal.pbio.1000625.
47. Capel, B.; Swain, A.; Nicolis, S.; Hacker, A.; Walter, M.; Koopman, P.; Goodfellow, P.; Lovell-Badge, R. Circular transcripts of the testis-determining gene Sry in adult mouse testis. *Cell* **1993**, *73*, 1019–1030.

48. Cocquerelle, C.; Mascrez, B.; Hétuin, D.; Bailleul, B. Mis-splicing yields circular RNA molecules. *FASEB J.* **1993**, *7*, 155–160.
49. Nigro, J.M.; Cho, K.R.; Fearon, E.R.; Kern, S.E.; Ruppert, J.M.; Oliner, J.D.; Kinzler, K.W.; Vogelstein, B. Scrambled exons. *Cell* **1991**, *64*, 607–613.
50. Zaphiropoulos, P.G. Exon skipping and circular RNA formation in transcripts of the human cytochrome P-450 2C18 gene in epidermis and of the rat androgen binding protein gene in testis. *Mol. Cell. Biol.* **1997**, *17*, 2985–2993.
51. Memczak, S.; Jens, M.; Elefsinioti, A.; Torti, F.; Krueger, J.; Rybak, A.; Maier, L.; Mackowiak, S.D.; Gregersen, L.H.; Munschauer, M.; *et al.* Circular RNAs are a large class of animal RNAs with regulatory potency. *Nature* **2013**, *495*, 333–338.
52. Hansen, T.B.; Jensen, T.I.; Clausen, B.H.; Bramsen, J.B.; Finsen, B.; Damgaard, C.K.; Kjems, J. Natural RNA circles function as efficient microRNA sponges. *Nature* **2013**, *495*, 384–388.
53. Jeck, W.R.; Sorrentino, J.A.; Wang, K.; Slevin, M.K.; Burd, C.E.; Liu, J.; Marzluff, W.F.; Sharpless, N.E. Circular RNAs are abundant, conserved, and associated with ALU repeats. *RNA* **2013**, *19*, 141–157.
54. Capriotti, E.; Marti-Renom, M.A. Quantifying the relationship between sequence and three-dimensional structure conservation in RNA. *BMC Bioinforma.* **2010**, *11*, 322.
55. Rost, B. Twilight zone of protein sequence alignments. *Protein Eng.* **1999**, *12*, 85–94.
56. Pang, K.C.; Frith, M.C.; Mattick, J.S. Rapid evolution of noncoding RNAs: lack of conservation does not mean lack of function. *Trends Genet.* **2006**, *22*, 1–5.
57. Bernhart, S.H.; Hofacker, I.L. From consensus structure prediction to RNA gene finding. *Brief Funct. Genomic Proteomic* **2009**, *8*, 461–471.
58. Sun, Y.; Aljawad, O.; Lei, J.; Liu, A. Genome-scale NCRNA homology search using a Hamming distance-based filtration strategy. *BMC Bioinforma.* **2012**, *13*, S12.
59. Bentwich, I.; Avniel, A.; Karov, Y.; Aharonov, R.; Gilad, S.; Barad, O.; Barzilai, A.; Einat, P.; Einav, U.; Meiri, E.; *et al.* Identification of hundreds of conserved and nonconserved human microRNAs. *Nat. Genet.* **2005**, *37*, 766–770.
60. Berezikov, E.; van Tetering, G.; Verheul, M.; van de Belt, J.; van Laake, L.; Vos, J.; Verloop, R.; van de Wetering, M.; Guryev, V.; Takada, S.; *et al.* Many novel mammalian microRNA candidates identified by extensive cloning and RAKE analysis. *Genome Res.* **2006**, *16*, 1289–1298.
61. Guerra-Assuncao, J.A.; Enright, A.J. Large-scale analysis of microRNA evolution. *BMC Genomics* **2012**, *13*, 218.
62. Missal, K.; Rose, D.; Stadler, P.F. Non-coding RNAs in *Ciona intestinalis*. *Bioinformatics* **2005**, *21*, ii77–ii78.
63. Lindgreen, S.; Gardner, P.P.; Krogh, A. MASTR: Multiple alignment and structure prediction of non-coding RNAs using simulated annealing. *Bioinformatics* **2007**, *23*, 3304–3311.
64. Sperschneider, J.; Datta, A.; Wise, M.J. Predicting pseudoknotted structures across two RNA sequences. *Bioinformatics* **2012**, *28*, 3058–3065.
65. Wong, T.K.F.; Wan, K.-L.; Hsu, B.-Y.; Cheung, B.W.Y.; Hon, W.-K.; Lam, T.-W.; Yiu, S.-M. RNASAlign: RNA structural alignment system. *Bioinformatics* **2011**, *27*, 2151–2152.

66. Gardner, P.P.; Giegerich, R. A comprehensive comparison of comparative RNA structure prediction approaches. *BMC Bioinforma.* **2004**, *5*, 140.
67. Ravindran, P.P.; Heroux, A.; Ye, J.D. Improvement of the crystallizability and expression of an RNA crystallization chaperone. *J. Biochem.* **2011**, *150*, 535–543.
68. Furtig, B.; Richter, C.; Wohnert, J.; Schwalbe, H. NMR spectroscopy of RNA. *Chembiochem* **2003**, *4*, 936–962.
69. Tzakos, A.G.; Grace, C.R.; Lukavsky, P.J.; Riek, R. NMR techniques for very large proteins and rnas in solution. *Annu. Rev. Biophys. Biomol. Struct.* **2006**, *35*, 319–342.
70. Zuker, M.; Stiegler, P. Optimal computer folding of large RNA sequences using thermodynamics and auxiliary information. *Nucleic Acids Res.* **1981**, *9*, 133–148.
71. Zuker, M. Mfold web server for nucleic acid folding and hybridization prediction. *Nucleic Acids Res.* **2003**, *31*, 3406–3415.
72. Mathews, D.H.; Turner, D.H.; Zuker, M. RNA secondary structure prediction. *Curr. Protoc. Nucleic Acid Chem.* **2007**, doi:10.1002/0471142700.nc1102s28.
73. Dowell, R.D.; Eddy, S.R. Evaluation of several lightweight stochastic context-free grammars for RNA secondary structure prediction. *BMC Bioinforma.* **2004**, *5*, 71.
74. Dima, R.I.; Hyeon, C.; Thirumalai, D. Extracting stacking interaction parameters for RNA from the data set of native structures. *J. Mol. Biol.* **2005**, *347*, 53–69.
75. Do, C.B.; Woods, D.A.; Batzoglou, S. CONTRAfold: RNA secondary structure prediction without physics-based models. *Bioinformatics* **2006**, *22*, e90–e98.
76. Xia, T.; SantaLucia, J.J.; Burkard, M.E.; Kierzek, R.; Schroeder, S.J.; Jiao, X.; Cox, C.; Turner, D.H. Thermodynamic parameters for an expanded nearest-neighbor model for formation of RNA duplexes with Watson-Crick base pairs. *Biochemistry* **1998**, *37*, 14719–14735.
77. Mathews, D.H.; Sabina, J.; Zuker, M.; Turner, D.H. Expanded sequence dependence of thermodynamic parameters improves prediction of RNA secondary structure. *J. Mol. Biol.* **1999**, *288*, 911–940.
78. Mathews, D.H.; Disney, M.D.; Childs, J.L.; Schroeder, S.J.; Zuker, M.; Turner, D.H. Incorporating chemical modification constraints into a dynamic programming algorithm for prediction of RNA secondary structure. *Proc. Natl. Acad. Sci. USA* **2004**, *101*, 7287–7292.
79. Lu, Z.J.; Turner, D.H.; Mathews, D.H. A set of nearest neighbor parameters for predicting the enthalpy change of RNA secondary structure formation. *Nucleic Acids Res.* **2006**, *34*, 4912–4924.
80. Lu, Z.J.; Gloor, J.W.; Mathews, D.H. Improved RNA secondary structure prediction by maximizing expected pair accuracy. *RNA* **2009**, *15*, 1805–1813.
81. Hofacker, I.L.; Fontana W.; Stadler, P.F.; Bonhoeffer S.; Tacker M.; Schuster, P. Fast folding and comparison of rna secondary structures. *Monatshefte f. Chem.* **1994**, *125*, 167–188.
82. McCaskill, J.S. The equilibrium partition function and base pair binding probabilities for RNA secondary structure. *Biopolymers* **1990**, *29*, 1105–1119.
83. Durbin, R.; Eddy, S.R.; Krogh, A.; Mitchison, G. *Biological Sequence Analysis. Probabilistic Models of Proteins and Nucleic Acids*; Cambridge University Press: Cambridge, UK, 1998.
84. Deigan, K.E.; Li, T.W.; Mathews, D.H.; Weeks, K.M. Accurate SHAPE-directed RNA structure determination. *Proc. Natl. Acad. Sci. USA* **2009**, *106*, 97–102.

85. Doshi, K.J.; Cannone, J.J.; Cobaugh, C.W.; Gutell, R.R. Evaluation of the suitability of free-energy minimization using nearest-neighbor energy parameters for RNA secondary structure prediction. *BMC Bioinforma.* **2004**, *5*, 105.
86. Herschlag, D. RNA chaperones and the RNA folding problem. *J. Biol. Chem.* **1995**, *270*, 20871–20874.
87. Brennicke, A.; Marchfelder, A.; Binder, S. RNA editing. *FEMS Microbiol. Rev.* **1999**, *23*, 297–316.
88. Pan, T.; Sosnick, T. RNA folding during transcription. *Annu. Rev. Biophys. Biomol. Struct.* **2006**, *35*, 161–175.
89. Mandal, M.; Breaker, R.R. Gene regulation by riboswitches. *Nat. Rev. Mol. Cell. Biol.* **2004**, *5*, 451–463.
90. Soukup, J.K.; Soukup, G.A. Riboswitches exert genetic control through metabolite-induced conformational change. *Curr. Opin. Struct. Biol.* **2004**, *14*, 344–349.
91. Bengert, P.; Dandekar, T. Riboswitch finder—A tool for identification of riboswitch RNAs. *Nucleic Acids Res.* **2004**, *32*, W154–W159.
92. Voss, B.; Meyer, C.; Giegerich, R. Evaluating the predictability of conformational switching in RNA. *Bioinformatics* **2004**, *20*, 1573–1582.
93. Hofacker, I.L. Vienna RNA secondary structure server. *Nucleic Acids Res.* **2003**, *31*, 3429–3431.
94. Sankoff, D. Simultaneous solution of the RNA folding, alignment and protosequence problems. *SIAM J. Appl. Math.* **1985**, *45*, 810–825.
95. Torarinsson, E.; Lindgreen, S. WAR: Webserver for aligning structural RNAs. *Nucleic Acids Res.* **2008**, *36*, W79–W84.
96. Bremges, A.; Schirmer, S.; Giegerich, R. Fine-tuning structural RNA alignments in the twilight zone. *BMC Bioinforma.* **2010**, *11*, 222.
97. Thompson, J.D.; Higgins, D.G.; Gibson, T.J. CLUSTAL W: Improving the sensitivity of progressive multiple sequence alignment through sequence weighting, position-specific gap penalties and weight matrix choice. *Nucleic Acids Res.* **1994**, *22*, 4673–4680.
98. Notredame, C.; Higgins, D.G.; Heringa, J. T-Coffee: A novel method for fast and accurate multiple sequence alignment. *J. Mol. Biol.* **2000**, *302*, 205–217.
99. Hofacker, I.L.; Fekete, M.; Stadler, P.F. Secondary structure prediction for aligned RNA sequences. *J. Mol. Biol.* **2002**, *319*, 1059–1066.
100. Knudsen, B.; Hein, J. Pfold: RNA secondary structure prediction using stochastic context-free grammars. *Nucleic Acids Res.* **2003**, *31*, 3423–3428.
101. Ruan, J.; Stormo, G.D.; Zhang, W. An iterated loop matching approach to the prediction of RNA secondary structures with pseudoknots. *Bioinformatics* **2004**, *20*, 58–66.
102. Luck, R.; Graf, S.; Steger, G. ConStruct: A tool for thermodynamic controlled prediction of conserved secondary structure. *Nucleic Acids Res.* **1999**, *27*, 4208–4217.
103. Mathews, D.H.; Turner, D.H. Dynalign: An algorithm for finding the secondary structure common to two RNA sequences. *J. Mol. Biol.* **2002**, *317*, 191–203.
104. Mathews, D.H. Predicting a set of minimal free energy RNA secondary structures common to two sequences. *Bioinformatics* **2005**, *21*, 2246–2253.

105. Gorodkin, J.; Heyer, L.J.; Stormo, G.D. Finding the most significant common sequence and structure motifs in a set of RNA sequences. *Nucleic Acids Res.* **1997**, *25*, 3724–3732.
106. Havgaard, J.H.; Lyngso, R.B.; Stormo, G.D.; Gorodkin, J. Pairwise local structural alignment of RNA sequences with sequence similarity less than 40%. *Bioinformatics* **2005**, *21*, 1815–1824.
107. Holmes, I. Accelerated probabilistic inference of RNA structure evolution. *BMC Bioinforma.* **2005**, *6*, 73.
108. Dowell, R.D.; Eddy, S.R. Efficient pairwise RNA structure prediction and alignment using sequence alignment constraints. *BMC Bioinforma.* **2006**, *7*, 400.
109. Hofacker, I.L.; Bernhart, S.H.; Stadler, P.F. Alignment of RNA base pairing probability matrices. *Bioinformatics* **2004**, *20*, 2222–2227.
110. Wilm, A.; Higgins, D.G.; Notredame, C. R-Coffee: A method for multiple alignment of non-coding RNA. *Nucleic Acids Res.* **2008**, *36*, e52.
111. Reeder, J.; Giegerich, R. Consensus shapes: An alternative to the Sankoff algorithm for RNA consensus structure prediction. *Bioinformatics* **2005**, *21*, 3516–3523.
112. Bernhart, S.H.; Hofacker, I.L.; Stadler, P.F. Local RNA base pairing probabilities in large sequences. *Bioinformatics* **2006**, *22*, 614–615.
113. Capriotti, E.; Marti-Renom, M.A. RNA structure alignment by a unit-vector approach. *Bioinformatics* **2008**, *24*, i112–i118.
114. Ferre, F.; Ponty, Y.; Lorenz, W.A.; Clote, P. DIAL: A web server for the pairwise alignment of two RNA three-dimensional structures using nucleotide, dihedral angle and base-pairing similarities. *Nucleic Acids Res.* **2007**, *35*, W659–W668.
115. Wang, C.W.; Chen, K.T.; Lu, C.L. iPARTS: An improved tool of pairwise alignment of RNA tertiary structures. *Nucleic Acids Res.* **2010**, *38*, W340–W347.
116. Dror, O.; Nussinov, R.; Wolfson, H. ARTS: Alignment of RNA tertiary structures. *Bioinformatics* **2005**, *21*, ii47–ii53.
117. Chang, Y.F.; Huang, Y.L.; Lu, C.L. SARSA: A web tool for structural alignment of RNA using a structural alphabet. *Nucleic Acids Res.* **2008**, *36*, W19–W24.
118. Bauer, R.P.; Rother, K.P.; Moor, P.P.; Reinert, K.P.; Steinke, T.P.; Bujnicki, J.P.; Preissner, R.P. Fast structural alignment of biomolecules using a hash table, N-grams and string descriptors. *Algorithms* **2009**, *2*, 692–709.
119. Kirillova, S.; Tosatto, S.C.; Carugo, O. FRASS: The web-server for RNA structural comparison. *BMC Bioinforma.* **2010**, *11*, 327.
120. Freyhult, E.K.; Bollback, J.P.; Gardner, P.P. Exploring genomic dark matter: A critical assessment of the performance of homology search methods on noncoding RNA. *Genome Res.* **2007**, *17*, 117–125.
121. Smith, T.F.; Waterman, M.S. Identification of common molecular subsequences. *J. Mol. Biol.* **1981**, *147*, 195–197.
122. Rognes, T. Faster Smith-Waterman database searches with inter-sequence SIMD parallelisation. *BMC Bioinforma.* **2011**, *12*, 221.
123. Lipman, D.J.; Pearson, W.R. Rapid and sensitive protein similarity searches. *Science* **1985**, *227*, 1435–1441.

124. Altschul, S.F.; Gish, W.; Miller, W.; Myers, E.W.; Lipman, D.J. Basic local alignment search tool. *J. Mol. Biol.* **1990**, *215*, 403–410.
125. Eddy, S.R. Profile hidden Markov models. *Bioinformatics* **1998**, *14*, 755–763.
126. Weinberg, Z.; Ruzzo, W.L. Sequence-based heuristics for faster annotation of non-coding RNA families. *Bioinformatics* **2006**, *22*, 35–39.
127. Holmes, I. A probabilistic model for the evolution of RNA structure. *BMC Bioinforma.* **2004**, *5*, 166.
128. Notredame, C.; O'Brien, E.A.; Higgins, D.G. RAGA: RNA sequence alignment by genetic algorithm. *Nucleic Acids Res.* **1997**, *25*, 4570–4580.
129. Eddy, S.R.; Durbin, R. RNA sequence analysis using covariance models. *Nucleic Acids Res.* **1994**, *22*, 2079–2088.
130. Eddy, S.R. A memory-efficient dynamic programming algorithm for optimal alignment of a sequence to an RNA secondary structure. *BMC Bioinforma.* **2002**, *3*, 18.
131. Klein, R.J.; Eddy, S.R. RSEARCH: Finding homologs of single structured RNA sequences. *BMC Bioinforma.* **2003**, *4*, 44.
132. Griffiths-Jones, S.; Bateman, A.; Marshall, M.; Khanna, A.; Eddy, S.R. Rfam: An RNA family database. *Nucleic Acids Res.* **2003**, *31*, 439–441.
133. Finn, R.D.; Tate, J.; Mistry, J.; Coggill, P.C.; Sammut, S.J.; Hotz, H.R.; Ceric, G.; Forslund, K.; Eddy, S.R.; Sonnhammer, E.L.; *et al.* The Pfam protein families database. *Nucleic Acids Res.* **2008**, *36*, D281–D288.
134. Griffiths-Jones, S.; Moxon, S.; Marshall, M.; Khanna, A.; Eddy, S.R.; Bateman, A. Rfam: Annotating non-coding RNAs in complete genomes. *Nucleic Acids Res.* **2005**, *33*, D121–D124.
135. Zhang, S.; Borovok, I.; Aharonowitz, Y.; Sharan, R.; Bafna, V. A sequence-based filtering method for ncRNA identification and its application to searching for riboswitch elements. *Bioinformatics* **2006**, *22*, e557–e565.
136. Gardner, P.P.; Daub, J.; Tate, J.G.; Nawrocki, E.P.; Kolbe, D.L.; Lindgreen, S.; Wilkinson, A.C.; Finn, R.D.; Griffiths-Jones, S.; Eddy, S.R.; *et al.* Rfam: Updates to the RNA families database. *Nucleic Acids Res.* **2009**, *37*, D136–D140.
137. Roshan, U.; Chikkagoudar, S.; Livesay, D.R. Searching for evolutionary distant RNA homologs within genomic sequences using partition function posterior probabilities. *BMC Bioinforma.* **2008**, *9*, 61.
138. Bussotti, G.; Raineri, E.; Erb, I.; Zytnicki, M.; Wilm, A.; Beaudoing, E.; Bucher, P.; Notredame, C. BlastR—Fast and accurate database searches for non-coding RNAs. *Nucleic Acids Res.* **2011**, *39*, 6886–6895.
139. Lander, E.S.; Linton, L.M.; Birren, B.; Nusbaum, C.; Zody, M.C.; Baldwin, J.; Devon, K.; Dewar, K.; Doyle, M.; FitzHugh, W.; *et al.* Initial sequencing and analysis of the human genome. *Nature* **2001**, *409*, 860–921.
140. Venter, J.C.; Adams, M.D.; Myers, E.W.; Li, P.W.; Mural, R.J.; Sutton, G.G.; Smith, H.O.; Yandell, M.; Evans, C.A.; Holt, R.A.; *et al.* The sequence of the human genome. *Science* **2001**, *291*, 1304–1351.

141. Aparicio, S.; Chapman, J.; Stupka, E.; Putnam, N.; Chia, J.M.; Dehal, P.; Christoffels, A.; Rash, S.; Hoon, S.; Smit, A.; *et al.* Whole-genome shotgun assembly and analysis of the genome of Fugu rubripes. *Science* **2002**, *297*, 1301–1310.
142. Waterston, R.H.; Lindblad-Toh, K.; Birney, E.; Rogers, J.; Abril, J.F.; Agarwal, P.; Agarwala, R.; Ainscough, R.; Alexandersson, M.; An, P.; *et al.* Initial sequencing and comparative analysis of the mouse genome. *Nature* **2002**, *420*, 520–562.
143. Schwartz, S.; Kent, W.J.; Smit, A.; Zhang, Z.; Baertsch, R.; Hardison, R.C.; Haussler, D.; Miller, W. Human-mouse alignments with BLASTZ. *Genome Res.* **2003**, *13*, 103–107.
144. Boguski, M.S.; Tolstoshev, C.M.; Bassett, D.E.J. Gene discovery in dbEST. *Science* **1994**, *265*, 1993–1994.
145. Dias Neto, E.; Correa, R.G.; Verjovski-Almeida, S.; Briones, M.R.; Nagai, M.A.; da Silva, W.J.; Zago, M.A.; Bordin, S.; Costa, F.F.; Goldman, G.H.; *et al.* Shotgun sequencing of the human transcriptome with ORF expressed sequence tags. *Proc. Natl. Acad. Sci. USA* **2000**, *97*, 3491–3496.
146. Gerhard, D.S.; Wagner, L.; Feingold, E.A.; Shenmen, C.M.; Grouse, L.H.; Schuler, G.; Klein, S.L.; Old, S.; Rasooly, R.; Good, P.; *et al.* The status, quality, and expansion of the NIH full-length cDNA project: The Mammalian Gene Collection (MGC). *Genome Res.* **2004**, *14*, 2121–2127.
147. Boguski, M.S.; Lowe, T.M.; Tolstoshev, C.M. dbEST—Database for "expressed sequence tags." *Nat. Genet.* **1993**, *4*, 332–333.
148. Adams, M.D.; Kelley, J.M.; Gocayne, J.D.; Dubnick, M.; Polymeropoulos, M.H.; Xiao, H.; Merril, C.R.; Wu, A.; Olde, B.; Moreno, R.F.; *et al.* Complementary DNA sequencing: Expressed sequence tags and human genome project. *Science* **1991**, *252*, 1651–1656.
149. Bonaldo, M.F.; Lennon, G.; Soares, M.B. Normalization and subtraction: Two approaches to facilitate gene discovery. *Genome Res.* **1996**, *6*, 791–806.
150. Nagaraj, S.H.; Gasser, R.B.; Ranganathan, S. A hitchhiker's guide to expressed sequence tag (EST) analysis. *Brief. Bioinforma.* **2007**, *8*, 6–21.
151. Mortazavi, A.; Williams, B.A.; McCue, K.; Schaeffer, L.; Wold, B. Mapping and quantifying mammalian transcriptomes by RNA-Seq. *Nat. Methods* **2008**, *5*, 621–628.
152. Malone, J.H.; Oliver, B. Microarrays, deep sequencing and the true measure of the transcriptome. *BMC Biol.* **2011**, *9*, 34.
153. Eklund, A.C.; Turner, L.R.; Chen, P.; Jensen, R.V.; deFeo, G.; Kopf-Sill, A.R.; Szallasi, Z. Replacing cRNA targets with cDNA reduces microarray cross-hybridization. *Nat. Biotechnol.* **2006**, *24*, 1071–1073.
154. Okoniewski, M.J.; Miller, C.J. Hybridization interactions between probesets in short oligo microarrays lead to spurious correlations. *BMC Bioinforma.* **2006**, *7*, 276.
155. Casneuf, T.; van de Peer, Y.; Huber, W. *In situ* analysis of cross-hybridisation on microarrays and the inference of expression correlation. *BMC Bioinforma.* **2007**, *8*, 461.
156. Cox, W.G.; Beaudet, M.P.; Agnew, J.Y.; Ruth, J.L. Possible sources of dye-related signal correlation bias in two-color DNA microarray assays. *Anal. Biochem.* **2004**, *331*, 243–254.
157. Dombkowski, A.A.; Thibodeau, B.J.; Starcevic, S.L.; Novak, R.F. Gene-specific dye bias in microarray reference designs. *FEBS Lett.* **2004**, *560*, 120–124.

158. Rosenzweig, B.A.; Pine, P.S.; Domon, O.E.; Morris, S.M.; Chen, J.J.; Sistare, F.D. Dye bias correction in dual-labeled cDNA microarray gene expression measurements. *Environ. Health Perspect.* **2004**, *112*, 480–487.
159. Dobbin, K.K.; Kawasaki, E.S.; Petersen, D.W.; Simon, R.M. Characterizing dye bias in microarray experiments. *Bioinformatics* **2005**, *21*, 2430–2437.
160. Martin-Magniette, M.L.; Aubert, J.; Cabannes, E.; Daudin, J.J. Evaluation of the gene-specific dye bias in cDNA microarray experiments. *Bioinformatics* **2005**, *21*, 1995–2000.
161. Bertone, P.; Stolc, V.; Royce, T.E.; Rozowsky, J.S.; Urban, A.E.; Zhu, X.; Rinn, J.L.; Tongprasit, W.; Samanta, M.; Weissman, S.; *et al.* Global identification of human transcribed sequences with genome tiling arrays. *Science* **2004**, *306*, 2242–2246.
162. Cheng, J.; Kapranov, P.; Drenkow, J.; Dike, S.; Brubaker, S.; Patel, S.; Long, J.; Stern, D.; Tammana, H.; Helt, G.; *et al.* Transcriptional maps of 10 human chromosomes at 5-nucleotide resolution. *Science* **2005**, *308*, 1149–1154.
163. Royce, T.E.; Rozowsky, J.S.; Bertone, P.; Samanta, M.; Stolc, V.; Weissman, S.; Snyder, M.; Gerstein, M. Issues in the analysis of oligonucleotide tiling microarrays for transcript mapping. *Trends Genet.* **2005**, *21*, 466–475.
164. Kapranov, P.; Willingham, A.T.; Gingeras, T.R. Genome-wide transcription and the implications for genomic organization. *Nat. Rev. Genet.* **2007**, *8*, 413–423.
165. Wang, Z.; Gerstein, M.; Snyder, M. RNA-Seq: A revolutionary tool for transcriptomics. *Nat. Rev. Genet.* **2009**, *10*, 57–63.
166. Marioni, J.C.; Mason, C.E.; Mane, S.M.; Stephens, M.; Gilad, Y. RNA-seq: An assessment of technical reproducibility and comparison with gene expression arrays. *Genome Res.* **2008**, *18*, 1509–1517.
167. Fu, X.; Fu, N.; Guo, S.; Yan, Z.; Xu, Y.; Hu, H.; Menzel, C.; Chen, W.; Li, Y.; Zeng, R.; *et al.* Estimating accuracy of RNA-Seq and microarrays with proteomics. *BMC Genomics* **2009**, *10*, 161.
168. Ozsolak, F.; Milos, P.M. RNA Sequencing: Advances, challenges and opportunities. *Nat. Rev. Genet.* **2011**, *12*, 87–98.
169. Velculescu, V.E.; Zhang, L.; Vogelstein, B.; Kinzler, K.W. Serial analysis of gene expression. *Science* **1995**, *270*, 484–487.
170. Harbers, M.; Carninci, P. Tag-based approaches for transcriptome research and genome annotation. *Nat. Methods* **2005**, *2*, 495–502.
171. Saha, S.; Sparks, A. B.; Rago, C.; Akmaev, V.; Wang, C.J.; Vogelstein, B.; Kinzler, K.W.; Velculescu, V.E. Using the transcriptome to annotate the genome. *Nat. Biotechnol.* **2002**, *20*, 508–512.
172. Gowda, M.; Jantasuriyarat, C.; Dean, R.A.; Wang, G.L. Robust-LongSAGE (RL-SAGE): A substantially improved LongSAGE method for gene discovery and transcriptome analysis. *Plant. Physiol.* **2004**, *134*, 890–897.
173. Matsumura, H.; Ito, A.; Saitoh, H.; Winter, P.; Kahl, G.; Reuter, M.; Kruger, D.H.; Terauchi, R. SuperSAGE. *Cell. Microbiol.* **2005**, *7*, 11–18.
174. Brenner, S.; Johnson, M.; Bridgham, J.; Golda, G.; Lloyd, D.H.; Johnson, D.; Luo, S.; McCurdy, S.; Foy, M.; Ewan, M.; *et al.* Gene expression analysis by massively parallel signature sequencing (MPSS) on microbead arrays. *Nat. Biotechnol.* **2000**, *18*, 630–634.

175. Ng, P.; Wei, C.L.; Sung, W.K.; Chiu, K.P.; Lipovich, L.; Ang, C.C.; Gupta, S.; Shahab, A.; Ridwan, A.; Wong, C.H.; *et al.* Gene identification signature (GIS) analysis for transcriptome characterization and genome annotation. *Nat. Methods* **2005**, *2*, 105–111.
176. Schaefer, B.C. Revolutions in rapid amplification of cDNA ends: New strategies for polymerase chain reaction cloning of full-length cDNA ends. *Anal. Biochem.* **1995**, *227*, 255–273.
177. Kapranov, P.; Drenkow, J.; Cheng, J.; Long, J.; Helt, G.; Dike, S.; Gingeras, T.R. Examples of the complex architecture of the human transcriptome revealed by RACE and high-density tiling arrays. *Genome Res.* **2005**, *15*, 987–997.
178. Olivarius, S.; Plessy, C.; Carninci, P. High-throughput verification of transcriptional starting sites by Deep-RACE. *BioTechniques* **2009**, *46*, 130–132.
179. Kodzius, R.; Kojima, M.; Nishiyori, H.; Nakamura, M.; Fukuda, S.; Tagami, M.; Sasaki, D.; Imamura, K.; Kai, C.; Harbers, M.; *et al.* CAGE: Cap analysis of gene expression. *Nat. Methods* **2006**, *3*, 211–222.
180. Shiraki, T.; Kondo, S.; Katayama, S.; Waki, K.; Kasukawa, T.; Kawaji, H.; Kodzius, R.; Watahiki, A.; Nakamura, M.; Arakawa, T.; *et al.* Cap analysis gene expression for high-throughput analysis of transcriptional starting point and identification of promoter usage. *Proc. Natl. Acad. Sci. USA* **2003**, *100*, 15776–15781.
181. Mercer, T.R.; Gerhardt, D.J.; Dinger, M.E.; Crawford, J.; Trapnell, C.; Jeddeloh, J.A.; Mattick, J.S.; Rinn, J.L. Targeted RNA sequencing reveals the deep complexity of the human transcriptome. *Nat. Biotechnol.* **2011**, *30*, 99–104.
182. Mathavan, S.; Lee, S.G.; Mak, A.; Miller, L.D.; Murthy, K.R.; Govindarajan, K.R.; Tong, Y.; Wu, Y.L.; Lam, S.H.; Yang, H.; *et al.* Transcriptome analysis of zebrafish embryogenesis using microarrays. *PLoS Genet.* **2005**, *1*, 260–276.
183. Wang, E.T.; Sandberg, R.; Luo, S.; Khrebtukova, I.; Zhang, L.; Mayr, C.; Kingsmore, S.F.; Schroth, G.P.; Burge, C.B. Alternative isoform regulation in human tissue transcriptomes. *Nature* **2008**, *456*, 470–476.
184. Graveley, B.R.; Brooks, A.N.; Carlson, J.W.; Duff, M.O.; Landolin, J.M.; Yang, L.; Artieri, C.G.; van Baren, M.J.; Boley, N.; Booth, B.W.; *et al.* The developmental transcriptome of Drosophila melanogaster. *Nature* **2011**, *471*, 473–479.
185. Pang, K.C.; Stephen, S.; Dinger, M.E.; Engström, P.G.; Lenhard, B.; Mattick, J.S. RNAdb 2.0—An expanded database of mammalian non-coding RNAs. *Nucleic Acids Res.* **2007**, *35*, D178–D182.
186. Kin, T.; Yamada, K.; Terai, G.; Okida, H.; Yoshinari, Y.; Ono, Y.; Kojima, A.; Kimura, Y.; Komori, T.; Asai, K. fRNAdb: A platform for mining/annotating functional RNA candidates from non-coding RNA sequences. *Nucleic Acids Res.* **2007**, *35*, D145–D148.
187. Liu, C.; Bai, B.; Skogerbø, G.; Cai, L.; Deng, W.; Zhang, Y.; Bu, D.; Zhao, Y.; Chen, R. NONCODE: An integrated knowledge database of non-coding RNAs. *Nucleic Acids Res.* **2005**, *33*, D112–D125.
188. Pruitt, K.D.; Tatusova, T.; Klimke, W.; Maglott, D.R. NCBI reference Sequences: Current status, policy and new initiatives. *Nucleic Acids Res.* **2009**, *37*, D32–D36.
189. Harrow, J.; Denoeud, F.; Frankish, A.; Reymond, A.; Chen, C.K.; Chrast, J.; Lagarde, J.; Gilbert, J.G.; Storey, R.; Swarbreck, D.; *et al.* GENCODE: Producing a reference annotation for ENCODE. *Genome Biol.* **2006**, *7*, S4 1–S4 9.

190. Havana team. Available online: http://www.sanger.ac.uk/research/projects/vertebrategenome/havana/ (accessed 6 March 2013).
191. Loveland, J.E.; Gilbert, J.G.; Griffiths, E.; Harrow, J.L. Community gene annotation in practice. *Database (Oxford)* **2012**, *2012*, doi:10.1093/database/bas009.
192. Flicek, P.; Amode, M.R.; Barrell, D.; Beal, K.; Brent, S.; Carvalho-Silva, D.; Clapham, P.; Coates, G.; Fairley, S.; Fitzgerald, S.; *et al.* Ensembl 2012. *Nucleic Acids Res.* **2012**, *40*, D84–D90.
193. Kawai, J.; Shinagawa, A.; Shibata, K.; Yoshino, M.; Itoh, M.; Ishii, Y.; Arakawa, T.; Hara, A.; Fukunishi, Y.; Konno, H.; *et al.* Functional annotation of a full-length mouse cDNA collection. *Nature* **2001**, *409*, 685–690.
194. Pruitt, K.D.; Tatusova, T.; Brown, G.R.; Maglott, D.R. NCBI reference sequences (RefSeq): Current status, new features and genome annotation policy. *Nucleic Acids Res.* **2012**, *40*, D130–D135.
195. Kent, W.J.; Sugnet, C.W.; Furey, T.S.; Roskin, K.M.; Pringle, T.H.; Zahler, A.M.; Haussler, D. The human genome browser at UCSC. *Genome Res.* **2002**, *12*, 996–1006.
196. Stalker, J.; Gibbins, B.; Meidl, P.; Smith, J.; Spooner, W.; Hotz, H.R.; Cox, A.V. The Ensembl web site: Mechanics of a genome browser. *Genome Res.* **2004**, *14*, 951–955.
197. Wilming, L.G.; Gilbert, J.G.; Howe, K.; Trevanion, S.; Hubbard, T.; Harrow, J.L. The vertebrate genome annotation (Vega) database. *Nucleic Acids Res.* **2008**, *36*, D753–D760.
198. ENCODE Project Consortium An integrated encyclopedia of DNA elements in the human genome. *Nature* **2012**, *489*, 57–74.
199. Martin, J.A.; Wang, Z. Next-generation transcriptome assembly. *Nat. Rev. Genet.* **2011**, *12*, 671–682.
200. Rogers, M.F.; Thomas, J.; Reddy, A.S.; Ben-Hur, A. SpliceGrapher: Detecting patterns of alternative splicing from RNA-Seq data in the context of gene models and EST data. *Genome Biol.* **2012**, *13*, R4.
201. Metzker, M.L. Sequencing technologies—the next generation. *Nat. Rev. Genet.* **2010**, *11*, 31–46.
202. Butler, J.; MacCallum, I.; Kleber, M.; Shlyakhter, I.A.; Belmonte, M.K.; Lander, E.S.; Nusbaum, C.; Jaffe, D.B. ALLPATHS: *de novo* assembly of whole-genome shotgun microreads. *Genome Res.* **2008**, *18*, 810–820.
203. Zerbino, D.R.; Birney, E. Velvet: Algorithms for de novo short read assembly using de Bruijn graphs. *Genome Res.* **2008**, *18*, 821–829.
204. Simpson, J.T.; Wong, K.; Jackman, S.D.; Schein, J.E.; Jones, S.J.; Birol, I. ABySS: A parallel assembler for short read sequence data. *Genome Res.* **2009**, *19*, 1117–1123.
205. Li, R.; Li, Y.; Kristiansen, K.; Wang, J. SOAP: Short oligonucleotide alignment program. *Bioinformatics* **2008**, *24*, 713–714.
206. Lin, H.; Zhang, Z.; Zhang, M.Q.; Ma, B.; Li, M. ZOOM! Zillions of oligos mapped. *Bioinformatics* **2008**, *24*, 2431–2437.
207. Langmead, B.; Trapnell, C.; Pop, M.; Salzberg, S.L. Ultrafast and memory-efficient alignment of short DNA sequences to the human genome. *Genome Biol.* **2009**, *10*, R25.
208. Li, H.; Durbin, R. Fast and accurate short read alignment with Burrows-Wheeler transform. *Bioinformatics* **2009**, *25*, 1754–1760.
209. Schatz, M.C. CloudBurst: Highly sensitive read mapping with MapReduce. *Bioinformatics* **2009**, *25*, 1363–1369.

210. Ahmadi, A.; Behm, A.; Honnalli, N.; Li, C.; Weng, L.; Xie, X. Hobbes: Optimized gram-based methods for efficient read alignment. *Nucleic Acids Res.* **2012**, *40*, e41.
211. Derrien, T.; Estelle, J.; Marco Sola, S.; Knowles, D.G.; Raineri, E.; Guigo, R.; Ribeca, P. Fast computation and applications of genome mappability. *PLoS One* **2012**, *7*, e30377.
212. Cloonan, N.; Xu, Q.; Faulkner, G.J.; Taylor, D.F.; Tang, D.T.; Kolle, G.; Grimmond, S.M. RNA-MATE: A recursive mapping strategy for high-throughput RNA-sequencing data. *Bioinformatics* **2009**, *25*, 2615–2616.
213. Trapnell, C.; Pachter, L.; Salzberg, S.L. TopHat: Discovering splice junctions with RNA-Seq. *Bioinformatics* **2009**, *25*, 1105–1111.
214. Li, H.; Ruan, J.; Durbin, R. Mapping short DNA sequencing reads and calling variants using mapping quality scores. *Genome Res.* **2008**, *18*, 1851–1858.
215. Smith, A.D.; Xuan, Z.; Zhang, M.Q. Using quality scores and longer reads improves accuracy of Solexa read mapping. *BMC Bioinforma.* **2008**, *9*, 128.
216. Ewing, B.; Green, P. Base-calling of automated sequencer traces using phred. II. Error probabilities. *Genome Res.* **1998**, *8*, 186–194.
217. Ewing, B.; Hillier, L.; Wendl, M.C.; Green, P. Base-calling of automated sequencer traces using phred. I. Accuracy assessment. *Genome Res.* **1998**, *8*, 175–185.
218. Trapnell, C.; Williams, B.A.; Pertea, G.; Mortazavi, A.; Kwan, G.; van Baren, M.J.; Salzberg, S.L.; Wold, B.J.; Pachter, L. Transcript assembly and quantification by RNA-Seq reveals unannotated transcripts and isoform switching during cell differentiation. *Nat. Biotechnol.* **2010**, *28*, 511–515.
219. Li, W.; Feng, J.; Jiang, T. IsoLasso: A LASSO regression approach to RNA-Seq based transcriptome assembly. *J. Comput. Biol.* **2011**, *18*, 1693–1707.
220. Palmieri, N.; Nolte, V.; Suvorov, A.; Kosiol, C.; Schlötterer, C. Evaluation of different reference based annotation strategies using RNA-Seq—A case study in *Drososphila pseudoobscura. PLoS One* **2012**, doi:10.1371/journal.pone.0046415.
221. Garg, R.; Patel, R.K.; Tyagi, A.K.; Jain, M. *De novo* assembly of chickpea transcriptome using short reads for gene discovery and marker identification. *DNA Res.* **2011**, *18*, 53–63.
222. Grabherr, M.G.; Haas, B.J.; Yassour, M.; Levin, J.Z.; Thompson, D.A.; Amit, I.; Adiconis, X.; Fan, L.; Raychowdhury, R.; Zeng, Q.; *et al.* Full-length transcriptome assembly from RNA-Seq data without a reference genome. *Nat. Biotechnol.* **2011**, *29*, 644–652.
223. Jager, M.; Ott, C.E.; Grunhagen, J.; Hecht, J.; Schell, H.; Mundlos, S.; Duda, G.N.; Robinson, P.N.; Lienau, J. Composite transcriptome assembly of RNA-seq data in a sheep model for delayed bone healing. *BMC Genomics* **2011**, *12*, 158.
224. Keiler, K.C. Biology of trans-translation. *Annu. Rev. Microbiol.* **2008**, *62*, 133–151.
225. Novikova, I.V.; Hennelly, S.P.; Sanbonmatsu, K.Y. Structural architecture of the human long non-coding RNA, steroid receptor RNA activator. *Nucleic Acids Res.* **2012**, *40*, 5034–5051.
226. Clamp, M.; Fry, B.; Kamal, M.; Xie, X.; Cuff, J.; Lin, M.F.; Kellis, M.; Lindblad-Toh, K.; Lander, E.S. Distinguishing protein-coding and noncoding genes in the human genome. *Proc. Natl. Acad. Sci. USA* **2007**, *104*, 19428–19433.
227. Lin, M.F.; Jungreis, I.; Kellis, M. PhyloCSF: A comparative genomics method to distinguish protein coding and non-coding regions. *Bioinformatics* **2011**, *27*, i275–i282.

228. Lin, M.F.; Carlson, J.W.; Crosby, M.A.; Matthews, B.B.; Yu, C.; Park, S.; Wan, K.H.; Schroeder, A.J.; Gramates, L.S.; St Pierre, S.E.; *et al.* Revisiting the protein-coding gene catalog of Drosophila melanogaster using 12 fly genomes. *Genome Res.* **2007**, *17*, 1823–1836.
229. Liao, Q.; Liu, C.; Yuan, X.; Kang, S.; Miao, R.; Xiao, H.; Zhao, G.; Luo, H.; Bu, D.; Zhao, H.; *et al.* Large-scale prediction of long non-coding RNA functions in a coding-non-coding gene co-expression network. *Nucleic Acids Res.* **2011**, *39*, 3864–3878.
230. Marchler-Bauer, A.; Panchenko, A.R.; Shoemaker, B.A.; Thiessen, P.A.; Geer, L.Y.; Bryant, S.H. CDD: A database of conserved domain alignments with links to domain three-dimensional structure. *Nucleic Acids Res.* **2002**, *30*, 281–283.
231. UniProt Consortium. Reorganizing the protein space at the Universal Protein Resource (UniProt). *Nucleic Acids Res.* **2012**, *40*, D71–D75.
232. Kong, L.; Zhang, Y.; Ye, Z.-Q.; Liu, X.-Q.; Zhao, S.-Q.; Wei, L.; Gao, G. CPC: Assess the protein-coding potential of transcripts using sequence features and support vector machine. *Nucleic Acids Res.* **2007**, *35*, W345–W349.
233. Arrial, R.T.; Togawa, R.C.; Brigido, M.M. Screening non-coding RNAs in transcriptomes from neglected species using PORTRAIT: Case study of the pathogenic fungus Paracoccidioides brasiliensis. *BMC Bioinforma.* **2009**, *10*, 239.
234. Wang, L.; Park, H.J.; Dasari, S.; Wang, S.; Kocher, J.-P.; Li, W. CPAT: Coding-Potential Assessment Tool using an alignment-free logistic regression model. *Nucleic Acids Res.* **2013**, *41*, e74.
235. Mikkelsen, T.S.; Ku, M.; Jaffe, D.B.; Issac, B.; Lieberman, E.; Giannoukos, G.; Alvarez, P.; Brockman, W.; Kim, T.K.; Koche, R.P.; *et al.* Genome-wide maps of chromatin state in pluripotent and lineage-committed cells. *Nature* **2007**, *448*, 553–560.
236. Lu, Z.J.; Yip, K.Y.; Wang, G.; Shou, C.; Hillier, L.W.; Khurana, E.; Agarwal, A.; Auerbach, R.; Rozowsky, J.; Cheng, C.; *et al.* Prediction and characterization of noncoding RNAs in *C. elegans* by integrating conservation, secondary structure, and high-throughput sequencing and array data. *Genome Res.* **2011**, *21*, 276–285.
237. Esteve-Codina, A.; Kofler, R.; Palmieri, N.; Bussotti, G.; Notredame, C.; Perez-Enciso, M. Exploring the gonad transcriptome of two extreme male pigs with RNA-seq. *BMC Genomics* **2011**, *12*, 552.
238. Nam, J.W.; Bartel, D. Long non-coding RNAs in *C. elegans*. *Genome Res.* **2012**, *22*, 2529–2540.
239. Djebali, S.; Davis, C.A.; Merkel, A.; Dobin, A.; Lassmann, T.; Mortazavi, A.; Tanzer, A.; Lagarde, J.; Lin, W.; Schlesinger, F.; *et al.* Landscape of transcription in human cells. *Nature* **2012**, *489*, 101–108.
240. Eveland, A.L.; McCarty, D.R.; Koch, K.E. Transcript profiling by 3'-untranslated region sequencing resolves expression of gene families. *Plant Physiol.* **2008**, *146*, 32–44.
241. Hillier, L.W.; Reinke, V.; Green, P.; Hirst, M.; Marra, M.A.; Waterston, R.H. Massively parallel sequencing of the polyadenylated transcriptome of *C. elegans*. *Genome Res.* **2009**, *19*, 657–666.
242. Nagalakshmi, U.; Wang, Z.; Waern, K.; Shou, C.; Raha, D.; Gerstein, M.; Snyder, M. The transcriptional landscape of the yeast genome defined by RNA sequencing. *Science* **2008**, *320*, 1344–1349.

243. Shendure, J. The beginning of the end for microarrays? *Nat. Methods* **2008**, *5*, 585–587.
244. Morozova, O.; Hirst, M.; Marra, M.A. Applications of new sequencing technologies for transcriptome analysis. *Annu. Rev. Genomics Hum. Genet.* **2009**, *10*, 135–151.
245. Auer, P.L.; Doerge, R.W. Statistical design and analysis of RNA sequencing data. *Genetics* **2010**, *185*, 405–416.
246. Hiller, M.; Pudimat, R.; Busch, A.; Backofen, R. Using RNA secondary structures to guide sequence motif finding towards single-stranded regions. *Nucleic Acids Res.* **2006**, *34*, e117.
247. Chang, T.-H.; Huang, H.-D.; Chuang, T.-N.; Shien, D.-M.; Horng, J.-T. RNAMST: Efficient and flexible approach for identifying RNA structural homologs. *Nucleic Acids Res.* **2006**, *34*, W423–W428.
248. Yao, Z.; Weinberg, Z.; Ruzzo, W.L. CMfinder—A covariance model based RNA motif finding algorithm. *Bioinformatics* **2006**, *22*, 445–452.
249. Ji, Y.; Xu, X.; Stormo, G.D. A graph theoretical approach for predicting common RNA secondary structure motifs including pseudoknots in unaligned sequences. *Bioinformatics* **2004**, *20*, 1591–1602.
250. Riccitelli, N.J.; Lupták, A. Computational discovery of folded RNA domains in genomes and *in vitro* selected libraries. *Methods* **2010**, *52*, 133–140.
251. Gautheret, D.; Major, F.; Cedergren, R. Pattern searching/alignment with RNA primary and secondary structures: An effective descriptor for tRNA. *Comput. Appl. Biosci.* **1990**, *6*, 325–331.
252. Gautheret, D.; Lambert, A. Direct RNA motif definition and identification from multiple sequence alignments using secondary structure profiles. *J. Mol. Biol.* **2001**, *313*, 1003–1011.
253. König, J.; Zarnack, K.; Luscombe, N.M.; Ule, J. Protein-RNA interactions: New genomic technologies and perspectives. *Nat. Rev. Genet.* **2011**, *13*, 77–83.
254. Rivas, E.; Eddy, S.R. Secondary structure alone is generally not statistically significant for the detection of noncoding RNAs. *Bioinformatics* **2000**, *16*, 583–605.
255. Kertesz, M.; Wan, Y.; Mazor, E.; Rinn, J.L.; Nutter, R.C.; Chang, H.Y.; Segal, E. Genome-wide measurement of RNA secondary structure in yeast. *Nature* **2010**, *467*, 103–107.
256. Slater, G.S.; Birney, E. Automated generation of heuristics for biological sequence comparison. *BMC Bioinforma.* **2005**, *6*, 31.
257. Birney, E.; Clamp, M.; Durbin, R. GeneWise and genomewise. *Genome Res.* **2004**, *14*, 988–995.
258. Eyras, E.; Reymond, A.; Castelo, R.; Bye, J.M.; Camara, F.; Flicek, P.; Huckle, E.J.; Parra, G.; Shteynberg, D.D.; Wyss, C.; *et al.* Gene finding in the chicken genome. *BMC Bioinforma.* **2005**, *6*, 131.
259. Mariotti, M.; Guigo, R. Selenoprofiles: Profile-based scanning of eukaryotic genome sequences for selenoprotein genes. *Bioinformatics* **2010**, *26*, 2656–2663.
260. Vieira, F.G.; Rozas, J. Comparative genomics of the odorant-binding and chemosensory protein gene families across the Arthropoda: Origin and evolutionary history of the chemosensory system. *Genome Biol. Evol.* **2011**, *3*, 476–490.
261. Latos, P.A.; Pauler, F.M.; Koerner, M.V.; Şenergin, H.B.; Hudson, Q.J.; Stocsits, R.R.; Allhoff, W.; Stricker, S.H.; Klement, R.M.; Warczok, K.E.; *et al.* Airn transcriptional overlap, but not its lncRNA products, induces imprinted Igf2r silencing. *Science* **2012**, *338*, 1469–1472.
262. Santoro, F.; Pauler, F.M. Silencing by the imprinted Airn macro lncRNA: Transcription is the answer. *Cell. Cycle* **2013**, *12*, 711–712.

6

Non-Coding RNAs and Cancer

Federica Calore [†], Francesca Lovat [†] and Michela Garofalo *

Department of Molecular Virology, Immunology and Medical Genetics,
Comprehensive Cancer Center, Ohio State University, Columbus, OH 43210, USA;
E-Mails: federica.calore@osumc.edu (F.C.); francesca.lovat@osumc.edu (F.L.)

[†] These authors contributed equally to this work.

* Author to whom correspondence should be addressed; E-Mail: michela.garofalo@osumc.edu;

Abstract: The discovery of the biological relevance of non-coding RNA (ncRNAs) molecules represents one of the most significant advances in contemporary molecular biology. Expression profiling of human tumors, based on the expression of miRNAs and other short or long ncRNAs, has identified signatures associated with diagnosis, staging, progression, prognosis, and response to treatment. In this review we will discuss the recent remarkable advancement in the understanding the biological functions of human ncRNAs in cancer, the mechanisms of expression and the therapeutic potential.

Keywords: small non-coding RNAs; long non-coding RNAs; cancer

---

1. Introduction

The human genome sequencing performed by the International Human Genome Sequencing Consortium revealed that the number of protein-coding genes corresponded only to 20–25,000 [1]. While, at first, it was common belief that the remaining, bigger portion of the human genome was not functional and considered as “junk DNA”, several studies based on advanced technologies such as tiling arrays and RNA deep sequencing have recently pointed out the identification of thousands of RNA transcripts not derived from known genes and not encoding for a protein [2,3]. These molecules have been classified as non-coding RNAs (ncRNAs).

NcRNAs could be grouped into two major classes based on the transcript size: small ncRNAs less than 200 bp, such as piRNAs (Piwi-associated RNAs), miRNAs (microRNAs), and snoRNAs (small nucleolar RNAs), and long ncRNAs (lncRNAs), greater than 200 bp. Each of these classes can be further divided, whereas novel subclasses are still being discovered and characterized. All these ncRNAs form huge molecular networks and play a central role in regulating cellular activities in Eukaryotes. The alteration and dysregulation of several ncRNA has been reported in various human diseases, including cancer, providing evidence for targeting these molecules as anticancer agents. Here, we will summarize the current knowledge about regulatory functions of ncRNAs, with special emphasis on their effects in cancer formation and progression.

2. piRNAs

PIWI-family proteins and their associated small RNAs (piRNAs) provide an essential protection for germ-cell genomes against the activity of transposable elements (TE). They help to maintain genome integrity, silencing TE [4] and this role is highly conserved across animal species. Unlike the other classes of small noncoding RNAs, which are 24–32 nt in length, they are generated from single-stranded RNA precursors through a Dicer-independent mechanism [5–7]. PiRNAs associate with PIWI proteins, which are germline-specific members of the Argonaute protein family, while siRNAs and miRNAs associate with ubiquitously expressed AGO subfamily members. The PIWI protein family is highly conserved across a variety of species and organisms. MIWI, MILI, and MIWI2 (Piwil4) are the three mouse PIWI proteins [8–10], whereas PIWIL1/HIWI, PIWIL2/HILI, PIWIL3, and PIWIL4/HIWI2 are the four PIWI proteins expressed in humans [11]. PIWI mutations in mice, Drosophila, and zebrafish, result in consistent defects in spermatogenic cells, demonstrating the essential role of PIWI proteins in germline development [12–15]. PiRNAs are more expressed in testes than other small noncoding RNAs [16–19] and are involved in spermatogenesis by regulating meiosis and/or suppressing TE. Hundreds of thousands of different piRNA species have been found in mammals [20], with no clear secondary structure motifs but with a common bias 5' uridine. At 3' termini piRNAs present a 2'-*O*-methylation, a process mediated by methyltransferase HEN1, which is associated with PIWI proteins [21,22]. This modification protects piRNA from 3'→5' exonucleases activity, suggesting an increase of their stability [23]. PiRNAs are not distributed across the whole genome but they are clustered in few hundred genomic loci called piRNA clusters [6]. The biogenesis of piRNA could be divided into two main pathways: primary processing and ping-pong amplification cycle (Figure 1). First, piRNA clusters are transcribed in both directions and provide a pool of fragmented primary piRNAs. Primary piRNA transcripts are exported to the cytoplasm where numerous factors (*i.e.*, Zucchini, Armitage and YB) participate in piRNA processing and loading onto PIWI proteins [24,25]. Piwi-piRNA complexes are transported into the nucleus, where they inhibit transcription of TE [26,27]. This first process (primary processing) is similar between germline and somatic cells. The next phase, the ping-pong amplification, is restricted to germline cells and requires slicer activity of PIWI proteins [25,26]. Recently, few works have demonstrated, by deep sequencing, that piRNA population is present in many more cell types than germline cells. For example, Lee *et al.* indentified the presence of a limited set of piRNAs in the mouse hippocampus. The most up-regulated, *DQ541777,* controls spine shape [28]. Moreover, another study described piRNA expression in more than 130 fruit fly, mouse, and rhesus macaque samples. Further, in mouse pancreas and macaque epididymis, piRNA are abundant as much as piRNA abundance in the germline [29]. An emerging number of studies highlights the role of piRNAs or PIWI proteins in the regulation of tumorigenesis. Indeed, piRNAs have been described in HeLa cells [30] and gastric, colon, lung, and breast cancer tissues [31]. These discoveries should not be surprising considering that cancer cells and germ cells share common features such as rapid proliferation and potentially infinite self-renewal. The first evidence of the role of piRNAs in cancer is described by Qiao *et al.* Hiwi, a Piwi family member, is over-expressed in seminomas but not in nonseminomas or in somatic tumors of the adult testis [32]. Moreover, HIWI over-expression has been also shown in cervical, pancreatic, colorectal, endometrial, esophageal, liver cancer, and gliomas [33–39]. Recently, Cheng and colleagues demonstrated that the expression of *piR-651* in gastric, colon, lung, and breast cancer tissues was higher compared to normal adjacent tissues. The levels of *piR-651* were associated with tumor-node-metastasis (TNM) stages. Inhibition of *piR-651* caused the arrest of gastric cancer cells at the G_2/M phase [31]; therefore this pi-RNA shows an oncogenic role and plays a crucial function in carcinogenesis. Another study demonstrated the down-regulation of *piR-823* in gastric cancer tissues compared to normal tissues suggesting its potential tumor suppressive role [40]. *In vivo* studies showed that the over-expression of *piR-823* significantly inhibited tumor growth in a dose-dependent manner. Moreover, *piR-823* was significantly lower in peripheral blood of gastric cancer patients compared to healthy controls. The levels of *piR-823* were positively associated with TNM stages and distant metastasis, suggesting that *piR-823* should be tested as a biomarker for detecting circulating gastric cancer cells in the blood [41]. All these data may suggest an important role of the axis PIWI and PIWI-associated RNAs going beyond the regulation of the genome in germline tissues and more studies are needed in order to investigate their specific role in tumorigenesis.

3. MicroRNAs

In 1993, Victor Ambros and colleagues discovered a gene, *lin-4*, that affected the development of *Caenorhabditis elegans* and found that its product was a small nonprotein-coding RNA [42]. The number of known small RNAs in different organisms such as *Caenorhabditis elegans*, *Drosophila melanogaster*, plants, and mammals, including humans, has since expanded substantially. MicroRNAs (miRNAs) are 19- to 24- nucleotide non-coding RNA molecules that regulate the expression of target mRNAs both at the transcriptional and translational level [43,44]. While in plants such regulation occurs through perfect base-pairing, usually in the 3' untranslated region (UTR) of the targeted mRNA, in mammals the base-pairing is only partial [45,46].

Each member of this large family of non coding RNAs can have hundreds of different targets and nearly 30% of the genes are regulated by, at least, one miRNA [43]. Several studies have demonstrated the involvement of miRNAs in different biological processes such as proliferation, cell cycle regulation, proliferation, apoptosis, differentiation, development, metabolism, neuronal patterning, and aging [43–53].

Figure 1. PiRNAs and microRNAs biogenesis. On the left, piRNAs biogenesis. PiRNAs are processed from single-stranded RNA precursors. The biogenesis of piRNAs could be divided in two main pathways: primary processing and ping-pong amplification cycle. MIWI2, a PIWI protein, translocates processed piRNAs into the nucleus, where they block the transcription of the TE (trasposon elements). On the right, miRNA biogenesis. Primary transcripts (pri-miRs) are transcribed by the RNA polymerase II. In the nucleus pri-miRs are processed by Drosha-DGCR8 into pre-miRs of ~60–70 nt. The produced pre-miRNAs are exported by the Exportin 5 to the cytoplasm where they are processed in ~18–22-nucleotide miRNA duplexes by the cytoplasmic RNase III Dicer. Normally, one strand of this duplex is degraded, whereas the other strand accumulates as a mature miRNA. From the miRNA-miRNA duplex, only the miRNA enters preferentially in the protein effector complex, formed by the RNA-induced silencing complex (RISC) and miRgonaute. Perfect or nearly perfect complementarities between miRNA and its target 3' UTR induce RISC to cleave the target mRNA, whereas imperfect base matching induces mainly translational silencing of the target.

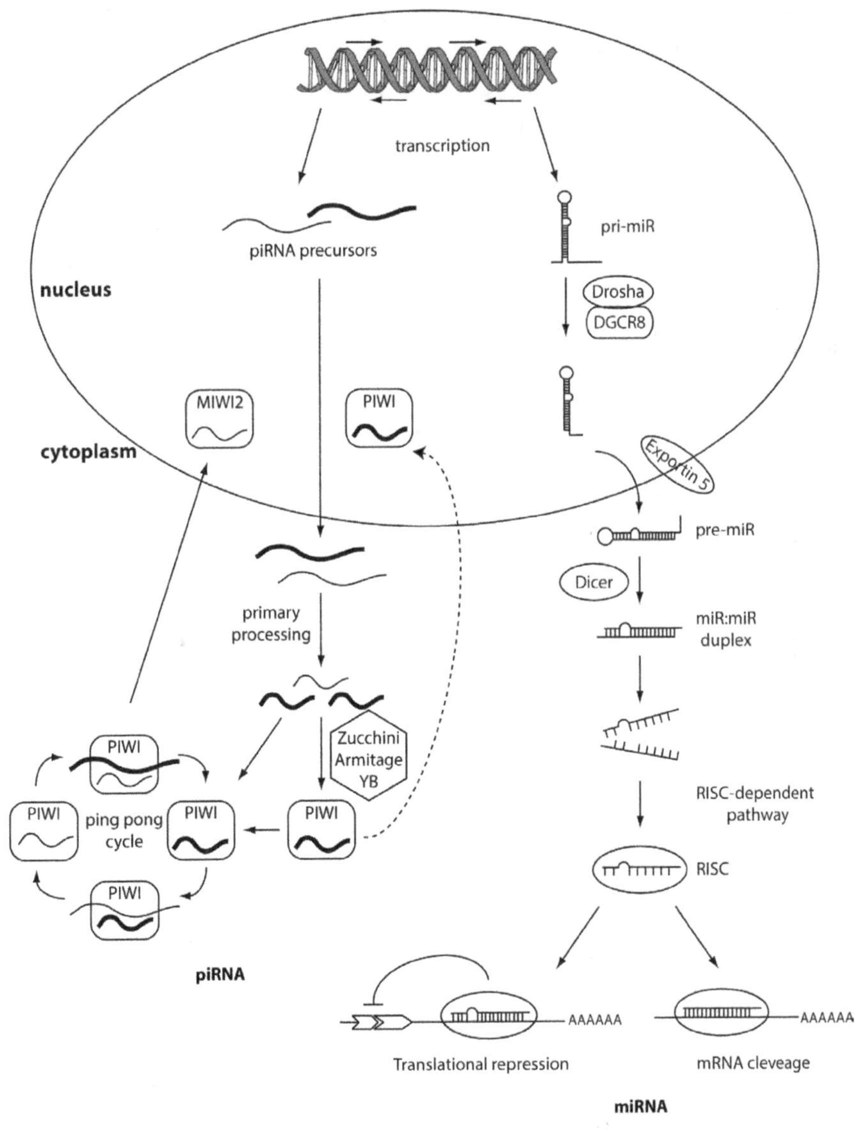

MiRNAs are transcribed by an RNA polymerase II as long, capped and polyadenylated precursors called pri-miRNAs [54,55], which are further cleaved into hairpin-shaped ~70–100 nucleotides precursors (pre-miRNAs) by a ribonuclease III (Drosha) and the double-stranded DNA binding protein DGCR8/Pasha [56] (Figure 1). Exportin 5 then translocates the pre-miRNAs to the cytoplasm [57], where another RNAse III, Dicer [43,58] further processes the precursor in a double strand RNA of about 24 nt. The double-stranded RNA is incorporated into the RISC (RNA-induced silencing complex) but only one strand, the mature microRNA, remains stably associated with the RISC and will drive the complex to the target mRNA. If the base-pairing between miRNA and the 3' UTR of the target mRNA is perfect the messenger is cleaved and degraded, whereas imperfect complementarity will result in translational silencing without mRNA degradation [59,60]. Several studies have demonstrated that miRNAs have a crucial role in cancer formation and spread. These small non-coding RNAs are, in fact, usually located in minimal regions of amplification, loss of heterozygosity, fragile sites, and common breakpoint regions in or in proximity of oncogenes or tumor suppressor genes. Moreover, profiling studies have demonstrated that miRNAs are differentially expressed in tumors *vs.* normal human tissues. These data have allowed the classification of microRNAs into two groups: oncomiRs (which act as oncogenes and are usually overexpressed in cancer, promoting tumor formation and spread) and tumor-suppressor miRs (which impair tumor growth and are usually silenced because of mutations, promoter methylation, or chromosomal rearrangements) [61–64], although some microRNAs can act as both oncogene or tumor-suppressor gene depending on the cellular context [65] (Figure 2 and Table1).

Figure 2. OncomiRs and tumor suppressor miRs. Correct cellular homeostasis is driven by a proper balance between oncomiRs and tumor suppressor miRs. OncomiRs are usually located in the amplified regions of the genome and are frequently over-expressed in cancer, promoting tumor growth and metastasis. Tumor suppressor miRs are often down-regulated in cancer and inhibit tumor growth inducing apoptosis and blocking cell migration.

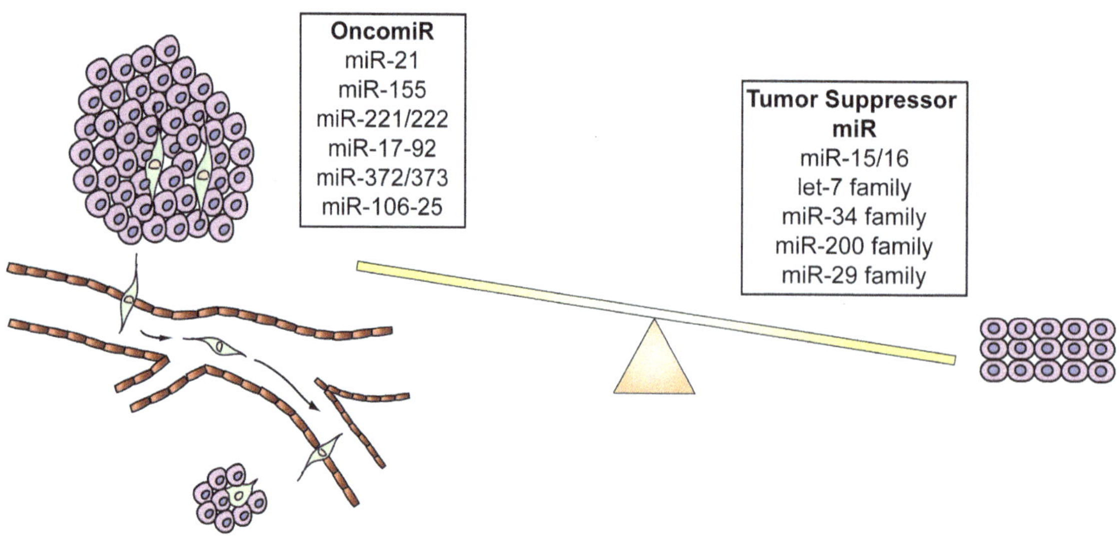

3.1. OncomiRs

One of the most well-known oncomiRs is *miR-21*, overexpressed in different types of cancer such as chronic lymphocytic leukemia (CLL) [62], acute myelogenous leukemia (AML) [66], glioblastoma [67], pancreatic, prostate, colon, gastric, breast, and lung cancer [68]. In 2008, Asangani and coworkers [69] demonstrated that *miR-21* downregulated the tumor suppressor PCDC4 (programmed cell death 4) promoting tumor invasion and metastasis in colorectal cancer. Zhang and colleagues [70] showed that *miR-21* induced growth and invasion in non-small cell lung cancer by repressing PTEN (phosphatase and tensin homolog); moreover, *miR-21* modulate TRAIL sensitivity in glioma cells mainly by modulating caspase-3 and TAp63 expression and TRAIL-induced caspase machinery [71], confirming that *miR-21* acts like an oncogene by blocking the expression of critical apoptosis-related genes.

Another example of oncomiR is represented by *miR-155*. Similarly to *miR-21*, *miR-155* is highly expressed in CLL [72], AML [73], lung, breast and pancreatic cancer [68], Hodgkin disease [72], and primary mediastinal non-Hodgkin's lymphoma [62]. In 2010, Jiang and coworkers demonstrated that *miR-155* targeted the tumor suppressor gene *Socs1* (*suppressor of cytokine signaling 1* gene) in human breast cancer cells, promoting cell proliferation, colony formation, and xenograft tumor growth [74]. *MiR-155* has also been found to be one of the most potent miRNAs suppressing apoptosis in human T cell leukemia (Jurkat cells) and in MDA-MB-453 breast cancer cells [75]. Moreover, in a transgenic mouse model, selective overexpression of *miR-155* in B cells led to early B cells polyclonal proliferation with a high-grade lymphoma-pre-B leukemia, suggesting that *miR-155* promotes the initiation and progression of the disease [76].

MiR-221 and *-222* are also up-regulated in several solid tumors, such as hepatocarcinoma [77], breast estrogen negative cells [78], melanoma cells [79], thyroid cancer [80]. Both these miRNAs induce tumor growth and spread of several cancer cell lines [81–83]. In 2009, our group demonstrated that hepatocyte growth factor receptor (MET) oncogene, through Jun transcriptional activation, upregulated *miR-221* and *-222* expression, which in turn, by targeting *PTEN* and *TIMP3*, conferred resistance to tumor necrosis factor-related apoptosis-inducing ligand (TRAIL)-induced cell death and enhanced tumorigenicity of lung and liver cancer cells. Therefore, the use of microRNAs in therapeutic intervention could sensitize tumor cells to drug-inducing apoptosis and also inhibit their survival, proliferation, and invasive abilities [84].

The oncomiR group is wide, and comprises other microRNAs such as the *miR-17-92* cluster, which is crucial for B-cell proliferation and its absence induces an increase of the proapoptotic protein Bim and inhibits the pro-B to pre-B cell development [85]; *miR-372/373*, which are involved in the development of human testicular germ cell tumors by neutralizing the TP53 pathway [86]; *miR-10b*, which promotes cell migration and invasion in breast cancer [87]; the polycistron *miR-106-25*, which acts as an oncogene by interfering with the synthesis of p21 and Bim [88].

3.2. Tumor Suppressor MicroRNAs

The group of miRNAs able to inhibit cell growth, induce apoptosis, and block cell cycle, are called tumor suppressor miRs. Normally, oncomiRs are located mainly in the amplified regions in human cancers and are frequently over-expressed in neoplastic tissues. Conversely, tumor suppressor miRs are located in the deleted regions and are often down-regulated in cancerous tissues.

The first evidence that miRNAs are involved in cancer comes from the finding that *miR-15* and *miR-16* are down-regulated or deleted in most patients with chronic lymphocytic leukemia [61].

Their expression is inversely related to several oncogenes, such as *Bcl-2* [89], *CCND1*, *WNT3A* [90], *Ccne1*, *Bmi-1* [91], and *VEGF-A* [92], which induce cell proliferation, survival, invasion and angiogenesis. Recently it has been shown that *miR-15* and *-16* are involved in drug resistance. Pouliot *et al.* demonstrated that *miR-15* and *-16* sensitized cisplatin-resistant epidermoid carcinoma cells to apoptosis by targeting *WEE1* and *CHK1* [93].

Another example of tumor suppressor miR is represented by the *let-7* family. Several studies described the down-regulation of *let-7* family in numerous tumors, including lung [94], gastric [95], colon cancer [96], and Burkitt's lymphoma [97]. *Let-7* family targets and inhibits the expression of several oncogenes such as *c-Myc* [97], *Ras* [98], *high-mobility group A* (*HMGA*) [99], *Janus protein tyrosine kinase* (*JAK*) and signal transducer and activator of transcription 3 (*STAT3*) pathway [100]. A recent study also reported that let-7 directly targets *PAK1*, *DIAPH2*, *RDX*, and *ITGB8*, multiple genes involved in the actin cytoskeleton pathway, inhibiting breast cancer cell migration [101].

The tumor suppressor activity of miR 34 family has been demonstrated in cancer cell types of lung [102], liver [103], breast [104], colon [105], brain [106], ovary [107], esophagus [108], and the lymphoid system [109]. In mammals, *miR-34* family comprises three processed miRNAs that are encoded by two different genes: *miR-34a* is encoded by its own transcript, whereas *miR-34b* and *-34c* share a common primary transcript. Their expression is directly induced by p53 in response to DNA damage or oncogenic stress [110]. *MiR-34* family inhibits many different oncogenic pathways involved in the control of cellular proliferation, cell cycle, and senescence by targeting oncogenes such as mitogen-activated protein kinase kinase 1 (*MEK1, MAP2K1*), R-Ras (*RRAS*), platelet-derived growth factor receptors (*PDGFRA and PDGFRB*) [111], and hepatocyte growth factor receptor (*MET*), *BCL2* and survivin.

MiR-200 family is commonly lost in aggressive tumors such as lung, prostate and pancreatic cancer. It has been shown that *miR-200* family members directly target *ZEB1* and *ZEB2*, transcriptional repressors of E-cadherin [112], and *BMI1*, reducing epithelial mesenchimal transition [113].

MiR-29s are also downregulated in multiple cancer types such as CLL [62], breast [114] and cervical cancer [115], and hepatocellular carcinoma [116]. *MiR-29* family targets and inhibits *DNMT3A* and *-3B* (DNA methyltransferases 3A and 3B) [117], *Tcl1* in chronic lymphocytic leukaemia [118] and *Mcl1* in cholangio-carcinoma [119]. Moreover, it has been demonstrated that the down-regulation of *miR-29* by MYC, HDAC, and EZH2 promotes cell survival and growth in MYC-associated lymphomas [120]. In conclusion, the correct cell homeostasis and survival are driven by a proper balance between oncomiRs and tumor suppressor miRs. Up-regulation of oncomiRs or down-regulation of tumor suppressor miRs leads to cancer formation and progression (Figure 2 and Table 1).

Table 1. OncomiRs and tumor suppressor miRs.

miRNA	Tumor type	Status	References
miR-21	CLL, AML, glioblastoma, pancreatic, prostate, colon, gastric, breast and lung cancer	Up-regulated	[62,66–71]
miR-155	CLL, AML, lung, breast and pancreatic cancer, Hodgkin disease, primary mediastinal non-Hodgkin's lymphoma	Up-regulated	[62,66,68,72,74–76]
miR-221/222	hepatocarcinoma, breast cancer, melanoma, thyroid cancer and glioma	Up-regulated	[77–84]
miR-17-92	AML	Up-regulated	[85]
miR-372/373	testicular germ cell tumor	Up-regulated	[86]
miR-10b	breast cancer	Up-regulated	[87]
miR-106-25	gastric cancer	Up-regulated	[88]
miR-15-16	CLL, prostate and ovarian cancer and multiple myeloma	Down-regulated	[61,89–93]
let-7 family	lung, gastric, colon, breast cancer and Burkitt's lymphoma	Down-regulated	[94–101]
miR-34	lung, liver, breast, colon, brain, ovary, esophageal cancer and non-small cell lung cancer (NSCLC)	Down-regulated	[102–111]
miR-200	lung, prostate and pancreatic cancer	Down-regulated	[112,113]
miR-29	CLL, breast and cervical cancer hepatocellular and cholangio-carcinoma	Down-regulated	[62,114–120]

4. snoRNAs

Small nucleolar RNAs (snoRNAs) are small non-coding RNAs whose length ranges from 60 to 300 nucleotides. SnoRNAs are normally located within introns of protein-coding genes and are transcribed by RNA polymerase II, but in some cases they can be found within introns of lncRNAs [121,122]. Within the cell, snoRNAs specifically accumulate in the nucleolar compartment, where they are responsible of the 2'-*O*-ribose methylation and pseudouridylation of specific ribosomal RNA nucleotides, essential modifications for the efficient and accurate production of the ribosome [123].

SnoRNAs can be classified into two groups: H/ACA box and C/D box. In both cases, snoRNAs hybridize specifically to the complementary sequence in the rRNAs and the associated protein complexes (C/D or H/ACA snoRNP) carry out the appropriate modification on the nucleotide that is identified by snoRNAs [124–126].

The H/ACA box snoRNAs family is involved in pseudouridylation of rRNAs. These ncRNAs have two major hairpin elements, connected by a hinge, and followed by a short tail region containing the conserved H and ACA box motifs that are located at the bases of the 5' and 3', respectively. The sequence specificity for the pseudouridylation is carried by two short antisense elements located in an internal loop of the 5' and/or 3' hairpins [127].

The C/D box snoRNAs, instead, are mainly involved in the 2'-*O*-methylation of rRNAs. This group of ncRNAs contains two short sequence motifs, box C (5'-PuUGAUGA-3') and box D (5'-CUGA-3') located near the 5' and the 3' ends, respectively. These elements form a terminal stem-box structure, involving not only elements required for snoRNAs nuclear localization, but also another copy of the box C, named box C', in their central portion, and another box D, named box D'. 2'-*O*-methylation is carried out through one or two antisense elements located upstream box D and/or box D' and complementary to a site of rRNA 2'-*O*-ribose methylation [128]. The process of snoRNAs maturation has not been entirely unveiled, however it has been demonstrated that the maturation of box C/D snoRNAs in yeast can occur through two pathways (Figure 3) [129]. In the first pathway, splicing of a pre-mRNA leads to a snoRNA-containing lariat, which is then linearized by the enzyme Dbr1p. Thanks to the activity of endonucleases and exonucleases the mature snoRNA is finally released. The second pathway, instead, is splicing-independent: the snoRNA is excided from the intron of the pre-mRNA directly, leading to the destruction of the mRNA. However, this latter pathway is still not well defined and the enzymes involved in this process have not been totally identified.

Although the main function of snoRNAs seems to be related to rRNA folding and stabilization, recent discoveries have pointed out a wider regulatory function for these small ncRNAs. For example, snoRNAs seem to be involved in miRNA synthesis. In 2010, Breimer and coworkers identified several box C/D sno-miRNAs, originating from relatively short snoRNAs (such as *U27* and *HBII-336*) displaying miRNA features in mRNAs silencing in different cell types, therefore controlling several biological processes normally regulated by miRNAs [130].

SnoRNAs are also involved in the onset of the Prader-Willy syndrome (PWS), induced by the genetic loss of the 15q11–q13 locus, normally active only on the paternal allele. This site is characterized by several copies of the *HBII-85* snoRNA, whose loss seems to be correlated with the PWS phenotype, both in human and in mice [131,132]. Moreover, recent studies reported the involvement of snoRNAs in cancer formation and progression, although the exact molecular mechanisms by which snoRNAs regulate cancer are still unknown.

Similarly to miRNAs, snoRNA expression has been found deregulated in cancer patient samples. In fact, the expression of *GAS5* (growth arrest specific 5), a gene which encodes an lncRNA but also harbors ten intronic snoRNAs, is downregulated in breast cancer compared to normal adjacent epithelial breast tissue. *GAS5* transcript sensitizes mammalian cells to apoptosis inducers, therefore displaying a tumor-suppressor role [133]. Moreover, Nakamura and coworkers demonstrated that *GAS5* was a partner of BCL6 in a patient with diffuse large B-cell lymphoma, carrying the chromosomal translocation t (1; 3) (q25; q27) [134], while Gee showed that *GAS5* low expression correlated with poor prognosis in breast cancer and head and neck squamous carcinoma [135]. The same authors also reported that snoRNA U50 is frequently transcriptionally downregulated in breast and prostate cancer [136] and that its 2-nucleotides somatic and germline deletion led to increased incidence of homozygosity for the deletion in cancer cells.

Other snoRNAs, such as *snoRNA42*, overexpressed in NSCLC cells, are located at frequently amplified genomic regions in tumors, therefore acting like oncogenes and promoting tumor growth. In 2011 Mei and coworkers found that *snoRNA42* knockdown in NSCLC cells impaired tumorigenicity *in vitro* and *in vivo* promoting apoptosis in a p53-dependent manner; conversely its enforced expression in bronchial epitheliums promoted cell growth [137].

Moreover, Liao *et al.* performed a profiling study on 22 NSCLC tissues. They found an overexpression of six snoRNAs compared to normal specimens. In addition to *snoRNA42*, they identified *SNORD33*, *SNORD66*, *SNORD73B*, *SNORD76* and *SNORD78*. Of these, *SNORD33*, *SNORD66*, and *SNORD76* expression in the plasma of NSCLC patients was higher compared to cancer-free individuals [138]. It has been demonstrated that, in addition to deregulated snoRNAs, also mutations of genes encoding for snoRNPs (snoRNA-associated proteins) can promote tumorigenesis. One of these genes is the human dyskerin, a putative pseudouridine synthase involved in the rRNA pseudouridylation and in the stabilization of the telomerase RNA elements. Mutations of its gene, *DKC1*, cause the X-linked genetic disease dyskeratosis congenita and promote tumor formation in mice [139]. The same effects have been described when point mutations occur in the genes encoding NOP10 and NHP2, both components of the H/ACA snoRNPs.

Figure 3. Intronic snoRNA processing. SnoRNA maturation occurs through two distinct pathways: splicing-dependent and splicing-independent. In the first pathway, the splicing of a pre-mRNA leads to a snoRNA-containing lariat, which is linearized by the enzyme Dbr1p and then endonucleases and exonucleases release the mature snoRNA. In the splicing-independent pathway the snoRNA is directly excided from the intron of the pre-mRNA by endonucleolytic cleavage.

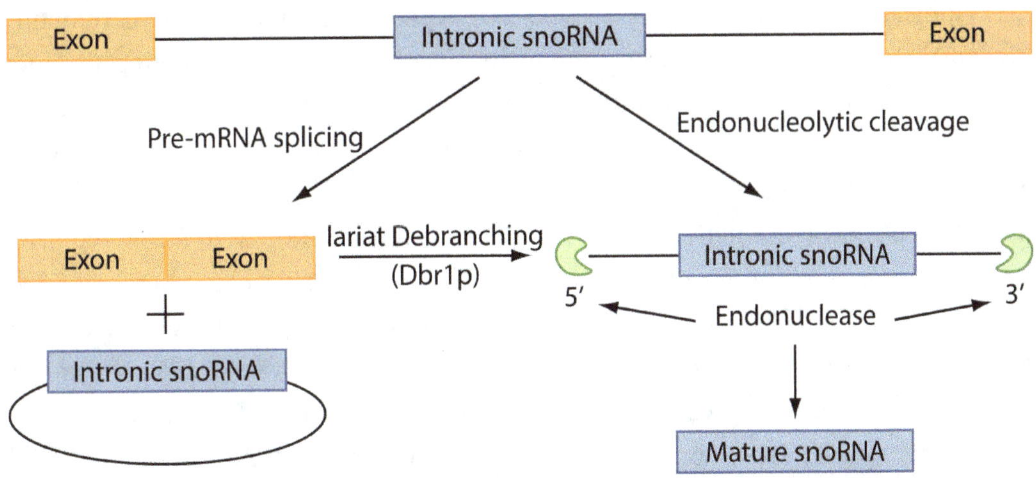

5. Long Noncoding RNAs

Several studies based on RNA deep sequencing and genome-wide analysis have recently pointed out that the genome of mammals, as well of other organisms, contains thousands of long transcripts whose length ranges from 200 nt to 100 kilobases, called long non-coding RNAs (lncRNAs or lincRNA, for long intergenic ncRNA) [3,140–144]. LncRNAs are located within nuclear or cytosolic fractions [145]. They are usually transcribed by RNA polymerase II but have no open reading frame [146], and map to intronic and intergenic regions [147]. Moreover, they display epigenetic features common to protein-coding genes, such as trimethylation of histone 3 lysine 4 (H3K4me3) at the transcriptional start site (TSS), and trimethylation of histone 3 lysine 36 (H3K36me3) throughout the gene body [148,149]. It has been estimated that nearly 15,000 lncRNAs are present in the human genome, but only a small fraction is expressed in a given cell type. All the information regarding identified lncRNAs has been catalogued and is available at the website http://www.lncrnadb.org [150–152].

Although they were initially thought to be the product of a "noisy" inconsequential transcription resulting from low RNA polymerase fidelity [142], recent studies have demonstrated that lncRNAs regulate several biological processes such as transcription [153–156], translation [157], cellular differentiation [158], regulation of gene expression [159], cell cycle regulation [160,161], chromatin modification [143,162,163], and nuclear-cytoplasmic trafficking [159,164–167].

LncRNAs have been also found to guide protein complexes which regulate chromatin modification or transcription to their targets [143,162,168,169]. Finally, it has been demonstrated that lncRNAs are dysregulated in several human diseases, including cancer.

Dysregulated expression of lncRNAs in cancer marks the spectrum of disease progression [170] and may serve as an independent predictor for patient outcomes [171].

Long non-coding RNAs can mediate epigenetic changes by recruiting chromatin-remodeling complexes to specific genomic loci. A recent study found that 20% of 3300 human long non coding RNAs are bound by Polycomb Repressive Complex 2 (PRC2) [162]. Although the specific molecular mechanisms are not defined, there are several examples that illustrate the silencing potential of lncRNAs. The first, most known example is represented by *Xist* (X-inactive-specific transcript) gene, which encodes an lncRNA crucial for the inactivation of the X-chromosome in mammals [172]. Basically, *Xist* physically coats one of the two X-chromosomes and recruits the chromatin regulator PRC2 (Polycomb chromatin remodeling complex) to this chromosome, promoting the formation of heterochromatin through histone modifications [173]. Another important example is represented by the hundreds of long ncRNAs which are sequentially expressed in the human homeobox (Hox) loci, where they define chromatin domains of differential histone methylation and RNA polymerase accessibility [174]. One of these ncRNAs, Hox transcript antisense RNA (*HOTAIR*) regulates *in trans* human *HOXD* genes expression through the induction of a repressive chromatin state. This occurs through the association of *HOTAIR* with the chromatin-modifying complexes PRC2, LSD1, and coREST/REST [143,162,171]. As modulator of epigenetic landmark, it has been shown that *HOTAIR* has a profound effect on tumorigenesis. Indeed, it is upregulated in breast and colon cancers and it is associated with metastasis and poor prognosis [171]. Another important effect of lncRNAs on chromatin modification with important consequences in cancer is represented by the lncRNA *ANRIL*, which controls the epigenetic status of the locus *INK4b/ARF/INK4a* by interacting with subunits of PRC1 and PRC2. High expression of *ANRIL* has been found in some cancer tissues such as melanoma and prostate cancers ([175,176]. The long noncoding RNA *MALAT1* (metastasis-associated lung adenocarcinoma transcript 1), also known as *NEAT2* (nuclear-enriched abundant transcript 2), is a highly conserved nuclear noncoding RNA (ncRNA) which acts as molecular decoy serving as a structural link in ribonucleoprotein (RNPs) complexes. Gutschner and colleagues developed a *MALAT1* knockout model in human lung tumor cells. In lung cancer, *MALAT1* does not alter alternative splicing but actively regulates gene expression including a set of metastasis-associated genes. Consequently, *MALAT1*-deficient cells are impaired in migration and form fewer tumor nodules in a mouse xenograft. Antisense oligonucleotides (ASO), blocking *MALAT1*, prevent metastasis formation after tumor implantation [177]. In addition to these active lncRNAs acting as oncogenes, there are also lncRNAs with tumor suppressor function. One very famous example is the ncRNA *GAS5* (Growth Arrest-Specific 5). It was originally identified based on its increased levels in growth-arrested mouse NIH3T3 fibroblasts [132]. *GAS5* binds to the DNA-binding domain of the glucocorticoid receptor (GR) by acting as a decoy glucocorticoid response element (GRE), thus competing with DNA GREs for binding to the GR [156]. *GAS5* negatively regulates the survival of lymphoid and breast cells, and is aberrantly expressed in several cancers. Pickard *et al.* showed that *GAS5* promotes apoptosis of prostate cells after irradiation with UV-C light and low levels of *GAS5* expression may therefore reduce the effectiveness of chemotherapeutic agents [178].

Recently, lncRNAs have also shown their tumorigenic potential by modulating the transcriptional program of p53 [179].

A 3kb lncRNA, *linc-RNA-p21*, transcriptionally activated by p53, has been shown to collaborate with p53 in order to control gene expression in response to DNA damage. Silencing of *lincRNA-p21* derepresses the expression of hundreds of genes through the interaction with hnRNP-K (Heterogeneous nuclear ribonucleoprotein K), thus, promoting apoptosis of abnormal cells or restraining tumors [179].

LncRNA *PANDA* is induced in response to external stimuli in a p53-dependent manner. After DNA damage p53 directly binds to the *CDKN1A* locus and activates *PANDA*, which enables cell-cycle arrest and impairs the expression of pro-apoptotic genes thanks to its interaction with the transcription factor NF-YA [LNC8] [180].

In addition to the features described above, recent studies have unveiled other properties of lncRNAs. For instance, it has been demonstrated that pseudogene transcripts are biologically active as they can regulate mRNA stability. One example is given by the tumor suppressor pseudogene *PTENP1*, whose 3' UTR region is very similar to the untraslated region of *PTEN* transcript. Both these regions bind the same set of miRNAs, and *PTENP1* pseudogene may act as "decoy" by protecting *PTEN* mRNA from common miRNA binding and allowing the expression of the tumor suppressor protein. *PTENP1* pseudogene therefore belongs to the group of competing endogenous RNAs (ceRNAs). Similarly, *KRAS* and *KRAS1P* transcript levels have been found positively correlated, corroborating that pseudogene functions mirror the role of their cognate genes as explained by a miRNA decoy mechanism. In cancer, specific mutations at the binding site of these pseudogenes impair their activity, therefore promoting tumor progression [181].

Enhancer-like lncRNAs (eRNAs) were discovered by Ørom and colleagues in 2010 [141]. The authors used a GENCODE annotation of the human genome to characterize over a thousand lncRNAs in several cell lines, finding that some of these RNAs displayed an enhancer-like function. Depletion of these ncRNAs led to a decreased expression of their neighboring protein-codon genes, such as the regulator of hematopoiesis *SCL*, *Snai1*, and *Snai2*, indicating that eRNAs play a pivotal role in development and differentiation. Moreover, Melo *et al.* showed through genome-wide chromatin-binding profiles that p53 protein binds also to regions located at distant sites from any known p53 target gene. Such regions were characterized not only by conserved p53-binding sites but displayed also enhancer activity and interacted with multiple neighboring genes allowing long-distance p53-dependent transcription regulation [182].

Finally, Natural Antisense Transcripts (NATs) are a large class of lncRNA transcribed from the opposite DNA strand to other transcripts and overlap in part with sense RNA. NATs play an important role in antisense regulation in gene expression. NATs have been implicated in several processes such as RNA translation [183] and transcriptional interference [184], and they have a pivotal role also in cancer. *aHIF*, a NAT derived from the 3' UTR of *HIF1*, represents the first case of overexpression of a

NAT associated with a specific human malignant disease: non-papillary clear-cell renal carcinomas, but not in papillary renal carcinomas [185]. Moreover, it has been demonstrated that *aHIF* expression is a poor prognosis marker in breast cancer [186].

The already mentioned *ANRIL* is an antisense lncRNA originates from the *INK4B-ARF-INK4A* locus, which contains three tumour suppressor genes, and it is overexpressed in prostate cancer tissues. Repression of *ANRIL* expression was associated with a reduction in cellular proliferation and increased the expression of both $p16^{Ink4A}$ and $p15^{INK4B}$, which are encoded by *CDKN2A* and *CDKN2B*, respectively [176]. *BOKAS* is a natural antisense transcript of Bok, a proapoptotic member in the Bcl-2 family. The expression of *BOKAS* was only detected in testis and different cancer tissues but not in other normal adult tissues. Overexpression of *BOKAS* was able to inhibit Bok-induced apoptosis in HeLa cells [187].

Another example of NAT is represented by *Zeb2/Sip1* NAT. This NAT regulates E-cadherin expression by increasing the levels of Zeb2 protein, a transcriptional repressor of E-cadherin, suggesting a role for noncoding RNAs in the control of epithelial morphology [188].

6. Conclusions

The recent discoveries regarding the biogenesis and function of ncRNAs have definitely improved our undestanding of the complexity of the human genome and the regulation of several processes. In particular, the involvement of microRNAs in the regulation of cell cycle, proliferation, differentiation and, most of all, in cancer formation and progression has certainly opened new fields of research aimed to better elucidate their mechanisms of action. Recently, our group reported that miRNAs secreted through exosomes bind to Toll-like receptor 8 (TLR8) in human and TLR7 in mouse inducing a pro-inflammatory response [189]. Therefore, in addition to their post-transcriptional regulatory function, miRNAs act like hormones and are involved into cell-to-cell communication. Other studies have shown the presence of tumor-derived microRNAs in serum or plasma as an approach for blood-based detection of human cancers, indicating that microRNAs could be used as circulating biomarkers [190]. Several groups are currently investigating the possibility to use microRNAs as therapeutic tools alone or in combination with chemotherapy.

Successfull systemic delivery of miRNAs as anti-cancer approaches in preclinical models using liposomes [191], viral vectors [192], and nanoparticles [193] has been reported. There is no doubt that miRNAs and other ncRNAs play a very important role in the regulation of pathways involved in tumor development and progression. Although there are still several obstacles to overcome before clinical testing of miRNA therapeutics, such as delivery and chemical modification of miRNA modulators, the fact that ncRNAs are natural antisense interactors and regulate many genes involved in survival and proliferation makes them excellent candidates to become powerful therapeutic tools in the near future.

Acknowledgements

We thank Justin Middleton for editorial assistance. Michela Garofalo is recipient of the Kimmel Award.

References

1. International Human Genome Sequencing Consortium. Finishing the euchromatic sequence of the human genome. *Nature* **2004**, *431*, 931–945.
2. Kapranov, P.; Cheng, J.; Dike, S.; Nix, D.A.; Duttagupta, R.; Willingham, A.T.; Stadler, P.F.; Hertel, J.; Hackermuller, J.; Hofacker, I.L.; *et al.* RNA maps reveal new RNA classes and a possible function for pervasive transcription. *Science* **2007**, *316*, 1484–1488.
3. Carninci, P.; Kasukawa, T.; Katayama, S.; Gough, J.; Frith, M.C.; Maeda, N.; Oyama, R.; Ravasi, T.; Lenhard, B.; Wells, C.; *et al.* The transcriptional landscape of the mammalian genome. *Science* **2005**, *309*, 1559–1563.
4. Senti, K.A.; Brennecke, J. The piRNA pathway: A fly's perspective on the guardian of the genome. *Trends Genet.* **2010**, *26*, 499–509.
5. Vagin, V.V.; Sigova, A.; Li, C.; Seitz, H.; Gvozdev, V.; Zamore, P.D. A distinct small RNA pathway silences selfish genetic elements in the germline. *Science* **2006**, *313*, 320–324.
6. Brennecke, J.; Aravin, A.A.; Stark, A.; Dus, M.; Kellis, M.; Sachidanandam, R.; Hannon, G.J. Discrete small RNA-generating loci as master regulators of transposon activity in drosophila. *Cell* **2007**, *128*, 1089–1103.
7. Houwing, S.; Kamminga, L.M.; Berezikov, E.; Cronembold, D.; Girard, A.; van den Elst, H.; Filippov, D.V.; Blaser, H.; Raz, E.; Moens, C.B.; *et al.* A role for piwi and piRNAs in germ cell maintenance and transposon silencing in zebrafish. *Cell* **2007**, *129*, 69–82.
8. Kuramochi-Miyagawa, S.; Kimura, T.; Ijiri, T.W.; Isobe, T.; Asada, N.; Fujita, Y.; Ikawa, M.; Iwai, N.; Okabe, M.; Deng, W.; *et al.* Mili, a mammalian member of *piwi* family gene, is essential for spermatogenesis. *Development* **2004**, *131*, 839–849.
9. Deng, W.; Lin, H. *Miwi*, a murine homolog of *piwi*, encodes a cytoplasmic protein essential for spermatogenesis. *Dev. Cell* **2002**, *2*, 819–830.
10. Carmell, M.A.; Girard, A.; van de Kant, H.J.; Bourc'his, D.; Bestor, T.H.; de Rooij, D.G.; Hannon, G.J. Miwi2 is essential for spermatogenesis and repression of transposons in the mouse male germline. *Dev. Cell* **2007**, *12*, 503–514.
11. Sasaki, T.; Shiohama, A.; Minoshima, S.; Shimizu, N. Identification of eight members of the argonaute family in the human genome small star, filled. *Genomics* **2003**, *82*, 323–330.
12. Chen, Y.; Pane, A.; Schupbach, T. Cutoff and aubergine mutations result in retrotransposon upregulation and checkpoint activation in drosophila. *Curr. Biol.* **2007**, *17*, 637–642.
13. Cox, D.N.; Chao, A.; Baker, J.; Chang, L.; Qiao, D.; Lin, H. A novel class of evolutionarily conserved genes defined by piwi are essential for stem cell self-renewal. *Genes Dev.* **1998**, *12*, 3715–3727.
14. Houwing, S.; Berezikov, E.; Ketting, R.F. Zili is required for germ cell differentiation and meiosis in zebrafish. *EMBO* **2008**, *27*, 2702–2711.
15. Schupbach, T.; Wieschaus, E. Female sterile mutations on the second chromosome of drosophila melanogaster. II. mutations blocking oogenesis or altering egg morphology. *Genetics* **1991**, *129*, 1119–1136.
16. Aravin, A.; Gaidatzis, D.; Pfeffer, S.; Lagos-Quintana, M.; Landgraf, P.; Iovino, N.; Morris, P.; Brownstein, M.J.; Kuramochi-Miyagawa, S.; Nakano, T.; *et al.* A novel class of small RNAs bind to MILI protein in mouse testes. *Nature* **2006**, *442*, 203–207.

17. Girard, A.; Sachidanandam, R.; Hannon, G.J.; Carmell, M.A. A germline-specific class of small RNAs binds mammalian Piwi proteins. *Nature* **2006**, *442*, 199–202.
18. Grivna, S.T.; Beyret, E.; Wang, Z.; Lin, H. A novel class of small RNAs in mouse spermatogenic cells. *Genes Dev.* **2006**, *20*, 1709–1714.
19. Watanabe, T.; Takeda, A.; Tsukiyama, T.; Mise, K.; Okuno, T.; Sasaki, H.; Minami, N.; Imai, H. Identification and characterization of two novel classes of small RNAs in the mouse germline: Retrotransposon-derived siRNAs in oocytes and germline small RNAs in testes. *Genes Dev.* **2006**, *20*, 1732–1743.
20. Das, P.P.; Bagijn, M.P.; Goldstein, L.D.; Woolford, J.R.; Lehrbach, N.J.; Sapetschnig, A.; Buhecha, H.R.; Gilchrist, M.J.; Howe, K.L.; Stark, R.; *et al.* Piwi and piRNAs act upstream of an endogenous siRNA pathway to suppress tc3 transposon mobility in the *Caenorhabditis elegans* germline. *Mol. Cell* **2008**, *31*, 79–90.
21. Horwich, M.D.; Li, C.; Matranga, C.; Vagin, V.; Farley, G.; Wang, P.; Zamore, P.D. The drosophila RNA methyltransferase, dmhen1, modifies germline piRNAs and single-stranded siRNAs in risc. *Curr. Biol.* **2007**, *17*, 1265–1272.
22. Saito, K.; Sakaguchi, Y.; Suzuki, T.; Suzuki, T.; Siomi, H.; Siomi, M.C. Pimet, the drosophila homolog of hen1, mediates 2'-*O*-methylation of piwi-interacting RNAs at their 3' ends. *Genes Dev.* **2007**, *21*, 1603–1608.
23. Kirino, Y.; Mourelatos, Z. Mouse piwi-interacting RNAs are 2'-*O*-methylated at their 3' termini. *Nat. Struct. Mol. Biol.* **2007**, *14*, 347–348.
24. Nishimasu, H.; Ishizu, H.; Saito, K.; Fukuhara, S.; Kamatani, M.K.; Bonnefond, L.; Matsumoto, N.; Nishizawa, T.; Nakanaga, K.; Aoki, J.; *et al.* Structure and function of zucchini endoribonuclease in piRNA biogenesis. *Nature* **2012**, *491*, 284–287.
25. Saito, K.; Ishizu, H.; Komai, M.; Kotani, H.; Kawamura, Y.; Nishida, K.M.; Siomi, H.; Siomi, M.C. Roles for the Yb body components armitage and Yb in primary piRNA biogenesis in drosophila. *Genes Dev.* **2010**, *24*, 2493–2498.
26. Le Thomas, A.; Rogers, A.K.; Webster, A.; Marinov, G.K.; Liao, S.E.; Perkins, E.M.; Hur, J.K.; Aravin, A.A.; Toth, K.F. Piwi induces piRNA-guided transcriptional silencing and establishment of a repressive chromatin state. *Genes Dev.* **2013**, *27*, 390–399.
27. Rozhkov, N.V.; Hammell, M.; Hannon, G.J. Multiple roles for piwi in silencing drosophila transposons. *Genes Dev.* **2013**, *27*, 400–412.
28. Lee, E.J.; Banerjee, S.; Zhou, H.; Jammalamadaka, A.; Arcila, M.; Manjunath, B.S.; Kosik, K.S. Identification of piRNAs in the central nervous system. *RNA* (*New York, N.Y.*) **2011**, *17*, 1090–1099.
29. Yan, Z.; Hu, H.Y.; Jiang, X.; Maierhofer, V.; Neb, E.; He, L.; Hu, Y.; Hu, H.; Li, N.; Chen, W.; *et al.* Widespread expression of piRNA-like molecules in somatic tissues. *Nucleic Acids Res.* **2011**, *39*, 6596–6607.
30. Lu, Y.; Li, C.; Zhang, K.; Sun, H.; Tao, D.; Liu, Y.; Zhang, S.; Ma, Y. Identification of piRNAs in hela cells by massive parallel sequencing. *BMB Rep.***2010**, *43*, 635–641.
31. Cheng, J.; Guo, J.M.; Xiao, B.X.; Miao, Y.; Jiang, Z.; Zhou, H.; Li, Q.N. PiRNA, the new non-coding RNA, is aberrantly expressed in human cancer cells. *Clin. Chim. Acta* **2011**, *412*, 1621–1625.

32. Qiao, D.; Zeeman, A.M.; Deng, W.; Looijenga, L.H.; Lin, H. Molecular characterization of hiwi, a human member of the *piwi* gene family whose overexpression is correlated to seminomas. *Oncogene* **2002**, *21*, 3988–3999.
33. Grochola, L.F.; Greither, T.; Taubert, H.; Moller, P.; Knippschild, U.; Udelnow, A.; Henne-Bruns, D.; Wurl, P. The stem cell-associated *hiwi* gene in human adenocarcinoma of the pancreas: Expression and risk of tumour-related death. *Br. J. Cancer* **2008**, *99*, 1083–1088.
34. He, W.; Wang, Z.; Wang, Q.; Fan, Q.; Shou, C.; Wang, J.; Giercksky, K.E.; Nesland, J.M.; Suo, Z. Expression of *hiwi* in human esophageal squamous cell carcinoma is significantly associated with poorer prognosis. *BMC Cancer* **2009**, *9*, 426.
35. Liu, W.K.; Jiang, X.Y.; Zhang, Z.X. Expression of *psca*, *piwil1*, and *tbx2* in endometrial adenocarcinoma. *Onkologie* **2010**, *33*, 241–245.
36. Liu, W.K.; Jiang, X.Y.; Zhang, Z.X. Expression of *psca*, *piwil1* and *tbx2* and its correlation with *hpv16* infection in formalin-fixed, paraffin-embedded cervical squamous cell carcinoma specimens. *Arch. Virol.* **2010**, *155*, 657–663.
37. Sun, G.; Wang, Y.; Sun, L.; Luo, H.; Liu, N.; Fu, Z.; You, Y. Clinical significance of *hiwi* gene expression in gliomas. *Brain Res.* **2011**, *1373*, 183–188.
38. Zeng, Y.; Qu, L.K.; Meng, L.; Liu, C.Y.; Dong, B.; Xing, X.F.; Wu, J.; Shou, C.C. *Hiwi* expression profile in cancer cells and its prognostic value for patients with colorectal cancer. *Chin. Med. J.* **2011**, *124*, 2144–2149.
39. Zhao, Y.M.; Zhou, J.M.; Wang, L.R.; He, H.W.; Wang, X.L.; Tao, Z.H.; Sun, H.C.; Wu, W.Z.; Fan, J.; Tang, Z.Y.; *et al. Hiwi* is associated with prognosis in patients with hepatocellular carcinoma after curative resection. *Cancer* **2012**, *118*, 2708–2717.
40. Cheng, J.; Deng, H.; Xiao, B.; Zhou, H.; Zhou, F.; Shen, Z.; Guo, J. *PiR-823*, a novel non-coding small RNA, demonstrates *in vitro* and *in vivo* tumor suppressive activity in human gastric cancer cells. *Cancer Lett.* **2012**, *315*, 12–17.
41. Cui, L.; Lou, Y.; Zhang, X.; Zhou, H.; Deng, H.; Song, H.; Yu, X.; Xiao, B.; Wang, W.; Guo, J. Detection of circulating tumor cells in peripheral blood from patients with gastric cancer using piRNAs as markers. *Clin. Biochem.* **2011**, *44*, 1050–1057.
42. Lee, R.C.; Feinbaum, R.L.; Ambros, V. The *C. elegans* heterochronic gene *lin-4* encodes small RNAs with antisense complementarity to *lin-14*. *Cell* **1993**, *75*, 843–854.
43. Bartel, D.P. MicroRNAs: Genomics, biogenesis, mechanism, and function. *Cell* **2004**, *116*, 281–297.
44. Denli, A.M.; Tops, B.B.; Plasterk, R.H.; Ketting, R.F.; Hannon, G.J. Processing of primary microRNAs by the microprocessor complex. *Nature* **2004**, *432*, 231–235.
45. Lagos-Quintana, M.; Rauhut, R.; Lendeckel, W.; Tuschl, T. Identification of novel genes coding for small expressed RNAs. *Science* **2001**, *294*, 853–858.
46. Lee, R.C.; Ambros, V. An extensive class of small RNAs in *Caenorhabditis elegans*. *Science* **2001**, *294*, 862–864.
47. Hu, W.; Chan, C.S.; Wu, R.; Zhang, C.; Sun, Y.; Song, J.S.; Tang, L.H.; Levine, A.J.; Feng, Z. Negative regulation of tumor suppressor p53 by microRNA *miR-504*. *Mol. Cell* **2010**, *38*, 689–699.

48. Bagga, S.; Bracht, J.; Hunter, S.; Massirer, K.; Holtz, J.; Eachus, R.; Pasquinelli, A.E. Regulation by *let-7* and *lin-4* miRNAs results in target mRNA degradation. *Cell* **2005**, *122*, 553–563.
49. Harfe, B.D. MicroRNAs in vertebrate development. *Curr. Opin. Genet. Dev.* **2005**, *15*, 410–415.
50. Boehm, M.; Slack, F.J. MicroRNA control of lifespan and metabolism. *Cell Cycle* **2006**, *5*, 837–840.
51. Calin, G.A.; Garzon, R.; Cimmino, A.; Fabbri, M.; Croce, C.M. MicroRNAs and leukemias: How strong is the connection? *Leuk. Res.* **2006**, *30*, 653–655.
52. Arisawa, T.; Tahara, T.; Shibata, T.; Nagasaka, M.; Nakamura, M.; Kamiya, Y.; Fujita, H.; Hasegawa, S.; Takagi, T.; Wang, F.Y.; *et al.* A polymorphism of *microRNA 27a* genome region is associated with the development of gastric mucosal atrophy in japanese male subjects. *Dig. Dis. Sci.* **2007**, *52*, 1691–1697.
53. Carleton, M.; Cleary, M.A.; Linsley, P.S. MicroRNAs and cell cycle regulation. *Cell Cycle* **2007**, *6*, 2127–2132.
54. Lee, Y.; Kim, M.; Han, J.; Yeom, K.H.; Lee, S.; Baek, S.H.; Kim, V.N. MicroRNA genes are transcribed by RNA polymerase ii. *EMBO* **2004**, *23*, 4051–4060.
55. Cai, X.; Hagedorn, C.H.; Cullen, B.R. Human microRNAs are processed from capped, polyadenylated transcripts that can also function as mRNAs. *RNA* (*New York, N.Y.*) **2004**, *10*, 1957–1966.
56. Ambros, V. The functions of animal microRNAs. *Nature* **2004**, *431*, 350–355.
57. Bohnsack, M.T.; Czaplinski, K.; Gorlich, D. Exportin 5 is a rangtp-dependent dsRNA-binding protein that mediates nuclear export of pre-miRNAs. *RNA* (*New York, N.Y.*) **2004**, *10*, 185–191.
58. Esquela-Kerscher, A.; Johnson, S.M.; Bai, L.; Saito, K.; Partridge, J.; Reinert, K.L.; Slack, F.J. Post-embryonic expression of *C. elegans* microRNAs belonging to the *lin-4* and *let-7* families in the hypodermis and the reproductive system. *Dev. Dyn.* **2005**, *234*, 868–877.
59. Achard, P.; Herr, A.; Baulcombe, D.C.; Harberd, N.P. Modulation of floral development by a gibberellin-regulated microRNA. *Development* **2004**, *131*, 3357–3365.
60. Gregory, R.I.; Chendrimada, T.P.; Shiekhattar, R. MicroRNA biogenesis: Isolation and characterization of the microprocessor complex. *Methods Mol. Biol.* **2006**, *342*, 33–47.
61. Calin, G.A.; Dumitru, C.D.; Shimizu, M.; Bichi, R.; Zupo, S.; Noch, E.; Aldler, H.; Rattan, S.; Keating, M.; Rai, K.; *et al.* Frequent deletions and down-regulation of micro-RNA genes *miR15* and *miR16* at 13q14 in chronic lymphocytic leukemia. *Proc. Natl. Acad. Sci. USA* **2002**, *99*, 15524–15529.
62. Calin, G.A.; Ferracin, M.; Cimmino, A.; Di Leva, G.; Shimizu, M.; Wojcik, S.E.; Iorio, M.V.; Visone, R.; Sever, N.I.; Fabbri, M.; *et al.* A microRNA signature associated with prognosis and progression in chronic lymphocytic leukemia. *N. Engl. J. Med.* **2005**, *353*, 1793–1801.
63. Nakamura, T.; Canaani, E.; Croce, C.M. Oncogenic all1 fusion proteins target drosha-mediated microRNA processing. *Proc. Natl. Acad. Sci. USA* **2007**, *104*, 10980–10985.
64. Saito, Y.; Liang, G.; Egger, G.; Friedman, J.M.; Chuang, J.C.; Coetzee, G.A.; Jones, P.A. Specific activation of *microRNA-127* with downregulation of the proto-oncogene *bcl6* by chromatin-modifying drugs in human cancer cells. *Cancer Cell* **2006**, *9*, 435–443.
65. Fabbri, M.; Croce, C.M.; Calin, G.A. MicroRNAs. *Cancer J.* **2008**, *14*, 1–6.

66. Garzon, R.; Volinia, S.; Liu, C.G.; Fernandez-Cymering, C.; Palumbo, T.; Pichiorri, F.; Fabbri, M.; Coombes, K.; Alder, H.; Nakamura, T.; *et al.* MicroRNA signatures associated with cytogenetics and prognosis in acute myeloid leukemia. *Blood* **2008**, *111*, 3183–3189.
67. Ciafre, S.A.; Galardi, S.; Mangiola, A.; Ferracin, M.; Liu, C.G.; Sabatino, G.; Negrini, M.; Maira, G.; Croce, C.M.; Farace, M.G. Extensive modulation of a set of microRNAs in primary glioblastoma. *Biochem. Biophys. Res. Commun.* **2005**, *334*, 1351–1358.
68. Volinia, S.; Calin, G.A.; Liu, C.G.; Ambs, S.; Cimmino, A.; Petrocca, F.; Visone, R.; Iorio, M.; Roldo, C.; Ferracin, M.; *et al.* A microRNA expression signature of human solid tumors defines cancer gene targets. *Proc. Natl. Acad. Sci. USA* **2006**, *103*, 2257–2261.
69. Asangani, I.A.; Rasheed, S.A.; Nikolova, D.A.; Leupold, J.H.; Colburn, N.H.; Post, S.; Allgayer, H. *MicroRNA-21* (*miR-21*) post-transcriptionally downregulates tumor suppressor pdcd4 and stimulates invasion, intravasation and metastasis in colorectal cancer. *Oncogene* **2008**, *27*, 2128–2136.
70. Zhang, J.G.; Wang, J.J.; Zhao, F.; Liu, Q.; Jiang, K.; Yang, G.H. *MicroRNA-21* (*miR-21*) represses tumor suppressor pten and promotes growth and invasion in non-small cell lung cancer (nsclc). *Clin. Chim. Acta* **2010**, *411*, 846–852.
71. Quintavalle, C.; Donnarumma, E.; Iaboni, M.; Roscigno, G.; Garofalo, M.; Romano, G.; Fiore, D.; De Marinis, P.; Croce, C.M.; Condorelli, G. Effect of *miR-21* and *miR-30b/c* on trail-induced apoptosis in glioma cells. *Oncogene* **2012**, doi:10.1038/onc.2012.410.
72. Kluiver, J.; Poppema, S.; de Jong, D.; Blokzijl, T.; Harms, G.; Jacobs, S.; Kroesen, B.J.; van den Berg, A. *Bic* and *miR-155* are highly expressed in hodgkin, primary mediastinal and diffuse large B cell lymphomas. *J. Pathol.* **2005**, *207*, 243–249.
73. Calin, G.A.; Cimmino, A.; Fabbri, M.; Ferracin, M.; Wojcik, S.E.; Shimizu, M.; Taccioli, C.; Zanesi, N.; Garzon, R.; Aqeilan, R.I.; *et al. Mir-15a* and *miR-16-1* cluster functions in human leukemia. *Proc. Natl. Acad. Sci. USA* **2008**, *105*, 5166–5171.
74. Jiang, S.; Zhang, H.W.; Lu, M.H.; He, X.H.; Li, Y.; Gu, H.; Liu, M.F.; Wang, E.D. *MicroRNA-155* functions as an oncomir in breast cancer by targeting the suppressor of cytokine signaling 1 gene. *Cancer Res.* **2010**, *70*, 3119–3127.
75. Ovcharenko, D.; Kelnar, K.; Johnson, C.; Leng, N.; Brown, D. Genome-scale microRNA and small interfering RNA screens identify small RNA modulators of trail-induced apoptosis pathway. *Cancer Res.* **2007**, *67*, 10782–10788.
76. Costinean, S.; Zanesi, N.; Pekarsky, Y.; Tili, E.; Volinia, S.; Heerema, N.; Croce, C.M. Pre-B cell proliferation and lymphoblastic leukemia/high-grade lymphoma in *e(mu)-miR155* transgenic mice. *Proc. Natl. Acad. Sci. USA* **2006**, *103*, 7024–7029.
77. Fornari, F.; Gramantieri, L.; Ferracin, M.; Veronese, A.; Sabbioni, S.; Calin, G.A.; Grazi, G.L.; Giovannini, C.; Croce, C.M.; Bolondi, L.; *et al. Mir-221* controls *cdkn1c/p57* and *cdkn1b/p27* expression in human hepatocellular carcinoma. *Oncogene* **2008**, *27*, 5651–5661.
78. Di Leva, G.; Gasparini, P.; Piovan, C.; Ngankeu, A.; Garofalo, M.; Taccioli, C.; Iorio, M.V.; Li, M.; Volinia, S.; Alder, H.; *et al.* MicroRNA cluster *221–222* and estrogen receptor alpha interactions in breast cancer. *J. Natl. Cancer Inst.* **2010**, *102*, 706–721.

79. Felicetti, F.; Errico, M.C.; Bottero, L.; Segnalini, P.; Stoppacciaro, A.; Biffoni, M.; Felli, N.; Mattia, G.; Petrini, M.; Colombo, M.P.; *et al.* The promyelocytic leukemia zinc finger-*microRNA-221/-222* pathway controls melanoma progression through multiple oncogenic mechanisms. *Cancer Res.* **2008**, *68*, 2745–2754.
80. Pallante, P.; Visone, R.; Ferracin, M.; Ferraro, A.; Berlingieri, M.T.; Troncone, G.; Chiappetta, G.; Liu, C.G.; Santoro, M.; Negrini, M.; *et al.* MicroRNA deregulation in human thyroid papillary carcinomas. *Endovr.-Relat. Cancer* **2006**, *13*, 497–508.
81. le Sage, C.; Nagel, R.; Egan, D.A.; Schrier, M.; Mesman, E.; Mangiola, A.; Anile, C.; Maira, G.; Mercatelli, N.; Ciafre, S.A.; *et al.* Regulation of the p27(kip1) tumor suppressor by *miR-221* and *miR-222* promotes cancer cell proliferation. *EMBO* **2007**, *26*, 3699–3708.
82. Garofalo, M.; Quintavalle, C.; Di Leva, G.; Zanca, C.; Romano, G.; Taccioli, C.; Liu, C.G.; Croce, C.M.; Condorelli, G. MicroRNA signatures of trail resistance in human non-small cell lung cancer. *Oncogene* **2008**, *27*, 3845–3855.
83. Quintavalle, C.; Garofalo, M.; Zanca, C.; Romano, G.; Iaboni, M.; del Basso De Caro, M.; Martinez-Montero, J.C.; Incoronato, M.; Nuovo, G.; Croce, C.M.; *et al.* *MiR-221/222* overexpession in human glioblastoma increases invasiveness by targeting the protein phosphate ptpmu. *Oncogene* **2012**, *31*, 858–868.
84. Garofalo, M.; Di Leva, G.; Romano, G.; Nuovo, G.; Suh, S.S.; Ngankeu, A.; Taccioli, C.; Pichiorri, F.; Alder, H.; Secchiero, P.; *et al.* *Mir-221 & 222* regulate trail resistance and enhance tumorigenicity through pten and timp3 downregulation. *Cancer Cell* **2009**, *16*, 498–509.
85. Manni, I.; Artuso, S.; Careccia, S.; Rizzo, M.G.; Baserga, R.; Piaggio, G.; Sacchi, A. The microRNA *miR-92* increases proliferation of myeloid cells and by targeting p63 modulates the abundance of its isoforms. *FASEB* **2009**, *23*, 3957–3966.
86. Voorhoeve, P.M.; le Sage, C.; Schrier, M.; Gillis, A.J.; Stoop, H.; Nagel, R.; Liu, Y.P.; van Duijse, J.; Drost, J.; Griekspoor, A.; *et al.* A genetic screen implicates *miRNA-372* and *miRNA-373* as oncogenes in testicular germ cell tumors. *Cell* **2006**, *124*, 1169–1181.
87. Ma, L.; Teruya-Feldstein, J.; Weinberg, R.A. Tumour invasion and metastasis initiated by *microRNA-10b* in breast cancer. *Nature* **2007**, *449*, 682–688.
88. Petrocca, F.; Visone, R.; Onelli, M.R.; Shah, M.H.; Nicoloso, M.S.; de Martino, I.; Iliopoulos, D.; Pilozzi, E.; Liu, C.G.; Negrini, M.; *et al.* E2f1-regulated microRNAs impair tgfbeta-dependent cell-cycle arrest and apoptosis in gastric cancer. *Cancer Cell* **2008**, *13*, 272–286.
89. Cimmino, A.; Calin, G.A.; Fabbri, M.; Iorio, M.V.; Ferracin, M.; Shimizu, M.; Wojcik, S.E.; Aqeilan, R.I.; Zupo, S.; Dono, M.; *et al.* *MiR-15* and *miR-16* induce apoptosis by targeting *bcl2*. *Proc. Natl. Acad. Sci. USA* **2005**, *102*, 13944–13949.
90. Bonci, D.; Coppola, V.; Musumeci, M.; Addario, A.; Giuffrida, R.; Memeo, L.; D'Urso, L.; Pagliuca, A.; Biffoni, M.; Labbaye, C.; *et al.* *The miR-15a-miR-16-1* cluster controls prostate cancer by targeting multiple oncogenic activities. *Nat. Med.* **2008**, *14*, 1271–1277.
91. Bhattacharya, R.; Nicoloso, M.; Arvizo, R.; Wang, E.; Cortez, A.; Rossi, S.; Calin, G.A.; Mukherjee, P. *MiR-15a* and *miR-16* control *bmi-1* expression in ovarian cancer. *Cancer Res.* **2009**, *69*, 9090–9095.

92. Sun, C.Y.; She, X.M.; Qin, Y.; Chu, Z.B.; Chen, L.; Ai, L.S.; Zhang, L.; Hu, Y. *MiR-15a* and *miR-16* affect the angiogenesis of multiple myeloma by targeting *vegf*. *Carcinogenesis* **2013**, *34*, 426–435.
93. Pouliot, L.M.; Chen, Y.C.; Bai, J.; Guha, R.; Martin, S.E.; Gottesman, M.M.; Hall, M.D. Cisplatin sensitivity mediated by wee1 and chk1 is mediated by *miR-155* and the *miR-15* family. *Cancer Res.* **2012**, *72*, 5945–5955.
94. Takamizawa, J.; Konishi, H.; Yanagisawa, K.; Tomida, S.; Osada, H.; Endoh, H.; Harano, T.; Yatabe, Y.; Nagino, M.; Nimura, Y.; *et al.* Reduced expression of the *let-7* microRNAs in human lung cancers in association with shortened postoperative survival. *Cancer Res.* **2004**, *64*, 3753–3756.
95. Zhang, H.H.; Wang, X.J.; Li, G.X.; Yang, E.; Yang, N.M. Detection of *let-7a* microRNA by real-time PCR in gastric carcinoma. *World J. Gastroenterol.* **2007**, *13*, 2883–2888.
96. Akao, Y.; Nakagawa, Y.; Naoe, T. *Let-7* microRNA functions as a potential growth suppressor in human colon cancer cells. *Biol. Pharm. Bull.* **2006**, *29*, 903–906.
97. Sampson, V.B.; Rong, N.H.; Han, J.; Yang, Q.; Aris, V.; Soteropoulos, P.; Petrelli, N.J.; Dunn, S.P.; Krueger, L.J. MicroRNA *let-7a* down-regulates *myc* and reverts *myc*-induced growth in burkitt lymphoma cells. *Cancer Res.* **2007**, *67*, 9762–9770.
98. Johnson, S.M.; Grosshans, H.; Shingara, J.; Byrom, M.; Jarvis, R.; Cheng, A.; Labourier, E.; Reinert, K.L.; Brown, D.; Slack, F.J. Ras is regulated by the *let-7* microRNA family. *Cell* **2005**, *120*, 635–647.
99. Mayr, C.; Hemann, M.T.; Bartel, D.P. Disrupting the pairing between *let-7* and *hmga2* enhances oncogenic transformation. *Science* **2007**, *315*, 1576–1579.
100. Wang, Y.; Lu, Y.; Toh, S.T.; Sung, W.K.; Tan, P.; Chow, P.; Chung, A.Y.; Jooi, L.L.; Lee, C.G. *Lethal-7* is down-regulated by the hepatitis B virus x protein and targets signal transducer and activator of transcription 3. *J. Hepatol.* **2010**, *53*, 57–66.
101. Hu, X.; Guo, J.; Zheng, L.; Li, C.; Zheng, T.M.; Tanyi, J.L.; Liang, S.; Benedetto, C.; Mitidieri, M.; Katsaros, D.; *et al.* The heterochronic microRNA *let-7* inhibits cell motility by regulating the genes in the actin cytoskeleton pathway in breast cancer. *Mol. Cancer Res.* **2013**, *11*, 240–250.
102. Wiggins, J.F.; Ruffino, L.; Kelnar, K.; Omotola, M.; Patrawala, L.; Brown, D.; Bader, A.G. Development of a lung cancer therapeutic based on the tumor suppressor *microRNA-34*. *Cancer Res.* **2010**, *70*, 5923–5930.
103. Li, N.; Fu, H.; Tie, Y.; Hu, Z.; Kong, W.; Wu, Y.; Zheng, X. *Mir-34a* inhibits migration and invasion by down-regulation of c-met expression in human hepatocellular carcinoma cells. *Cancer Lett.* **2009**, *275*, 44–53.
104. Lodygin, D.; Tarasov, V.; Epanchintsev, A.; Berking, C.; Knyazeva, T.; Korner, H.; Knyazev, P.; Diebold, J.; Hermeking, H. Inactivation of *miR-34a* by aberrant CpG methylation in multiple types of cancer. *Cell Cycle* **2008**, *7*, 2591–2600.
105. Tazawa, H.; Tsuchiya, N.; Izumiya, M.; Nakagama, H. Tumor-suppressive *miR-34a* induces senescence-like growth arrest through modulation of the e2f pathway in human colon cancer cells. *Proc. Natl. Acad. Sci. USA* **2007**, *104*, 15472–15477.

106. Welch, C.; Chen, Y.; Stallings, R.L. *MicroRNA-34a* functions as a potential tumor suppressor by inducing apoptosis in neuroblastoma cells. *Oncogene* **2007**, *26*, 5017–5022.
107. Corney, D.C.; Hwang, C.I.; Matoso, A.; Vogt, M.; Flesken-Nikitin, A.; Godwin, A.K.; Kamat, A.A.; Sood, A.K.; Ellenson, L.H.; Hermeking, H.; *et al.* Frequent downregulation of *miR-34* family in human ovarian cancers. *Clin. Cancer Res.* **2010**, *16*, 1119–1128.
108. Chen, X.; Hu, H.; Guan, X.; Xiong, G.; Wang, Y.; Wang, K.; Li, J.; Xu, X.; Yang, K.; Bai, Y. CpG island methylation status of miRNAs in esophageal squamous cell carcinoma. *Int. J. Cancer* **2012**, *130*, 1607–1613.
109. Mraz, M.; Malinova, K.; Kotaskova, J.; Pavlova, S.; Tichy, B.; Malcikova, J.; Stano Kozubik, K.; Smardova, J.; Brychtova, Y.; Doubek, M.; *et al. Mir-34a*, *miR-29c* and *miR-17–5p* are downregulated in cll patients with tp53 abnormalities. *Leukemia* **2009**, *23*, 1159–1163.
110. He, L.; He, X.; Lim, L.P.; de Stanchina, E.; Xuan, Z.; Liang, Y.; Xue, W.; Zender, L.; Magnus, J.; Ridzon, D.; *et al.* A microRNA component of the p53 tumour suppressor network. *Nature* **2007**, *447*, 1130–1134.
111. Garofalo, M.; Jeon, Y.J.; Nuovo, G.J.; Middleton, J.; Secchiero, P.; Joshi, P.; Alder, H.; Nazaryan, N.; Di Leva, G.; Romano, G.; *et al. Mir-34a/c*-dependent pdgfr-alpha/beta downregulation inhibits tumorigenesis and enhances trail-induced apoptosis in lung cancer. *PLoS One* **2013**, *8*, e67581.
112. Park, S.M.; Gaur, A.B.; Lengyel, E.; Peter, M.E. The *miR-200* family determines the epithelial phenotype of cancer cells by targeting the e-cadherin repressors zeb1 and zeb2. *Genes Dev.* **2008**, *22*, 894–907.
113. Liu, S.; Tetzlaff, M.T.; Cui, R.; Xu, X. *Mir-200c* inhibits melanoma progression and drug resistance through down-regulation of *bmi-1*. *Am. J. Pathol.* **2012**, *181*, 1823–1835.
114. Iorio, M.V.; Ferracin, M.; Liu, C.G.; Veronese, A.; Spizzo, R.; Sabbioni, S.; Magri, E.; Pedriali, M.; Fabbri, M.; Campiglio, M.; *et al.* MicroRNA gene expression deregulation in human breast cancer. *Cancer Res.* **2005**, *65*, 7065–7070.
115. Li, Y.; Wang, F.; Xu, J.; Ye, F.; Shen, Y.; Zhou, J.; Lu, W.; Wan, X.; Ma, D.; Xie, X. Progressive miRNA expression profiles in cervical carcinogenesis and identification of *hpv*-related target genes for *miR-29*. *J. Pathol.* **2011**, *224*, 484–495.
116. Xiong, Y.; Fang, J.H.; Yun, J.P.; Yang, J.; Zhang, Y.; Jia, W.H.; Zhuang, S.M. Effects of *microRNA-29* on apoptosis, tumorigenicity, and prognosis of hepatocellular carcinoma. *Hepatology* **2010**, *51*, 836–845.
117. Fabbri, M.; Garzon, R.; Cimmino, A.; Liu, Z.; Zanesi, N.; Callegari, E.; Liu, S.; Alder, H.; Costinean, S.; Fernandez-Cymering, C.; *et al. MicroRNA-29* family reverts aberrant methylation in lung cancer by targeting DNA methyltransferases 3a and 3b. *Proc. Natl. Acad. Sci. USA* **2007**, *104*, 15805–15810.
118. Pekarsky, Y.; Santanam, U.; Cimmino, A.; Palamarchuk, A.; Efanov, A.; Maximov, V.; Volinia, S.; Alder, H.; Liu, C.G.; Rassenti, L.; *et al. Tcl1* expression in chronic lymphocytic leukemia is regulated by *miR-29* and *miR-181*. *Cancer Res.* **2006**, *66*, 11590–11593.
119. Mott, J.L.; Kobayashi, S.; Bronk, S.F.; Gores, G.J. *Mir-29* regulates mcl-1 protein expression and apoptosis. *Oncogene* **2007**, *26*, 6133–6140.

120. Zhang, X.; Zhao, X.; Fiskus, W.; Lin, J.; Lwin, T.; Rao, R.; Zhang, Y.; Chan, J.C.; Fu, K.; Marquez, V.E.; *et al.* Coordinated silencing of *myc*-mediated *miR-29* by hdac3 and ezh2 as a therapeutic target of histone modification in aggressive B-cell lymphomas. *Cancer Cell* **2012**, *22*, 506–523.
121. Smith, C.M.; Steitz, J.A. Sno storm in the nucleolus: New roles for myriad small rnps. *Cell* **1997**, *89*, 669–672.
122. Bortolin, M.L.; Kiss, T. Human u19 intron-encoded snoRNA is processed from a long primary transcript that possesses little potential for protein coding. *RNA (New York, N.Y.)* **1998**, *4*, 445–454.
123. Kiss-Laszlo, Z.; Henry, Y.; Bachellerie, J.P.; Caizergues-Ferrer, M.; Kiss, T. Site-specific ribose methylation of preribosomal RNA: A novel function for small nucleolar RNAs. *Cell* **1996**, *85*, 1077–1088.
124. Lafontaine, D.L.; Tollervey, D. Birth of the snornps: The evolution of the modification-guide snoRNAs. *Trends Biochem. Sci.* **1998**, *23*, 383–388.
125. Terns, M.P.; Terns, R.M. Small nucleolar RNAs: Versatile trans-acting molecules of ancient evolutionary origin. *Gene Expr.* **2002**, *10*, 17–39.
126. Weinstein, L.B.; Steitz, J.A. Guided tours: From precursor snoRNA to functional snornp. *Curr. Opin. Cell Biol.* **1999**, *11*, 378–384.
127. Kiss, A.M.; Jady, B.E.; Bertrand, E.; Kiss, T. Human box h/aca pseudouridylation guide RNA machinery. *Mol. Cell. Biol.* **2004**, *24*, 5797–5807.
128. Bachellerie, J.P.; Cavaille, J.; Huttenhofer, A. The expanding snoRNA world. *Biochimie* **2002**, *84*, 775–790.
129. Coughlin, D.J.; Pleiss, J.A.; Walker, S.C.; Whitworth, G.B.; Engelke, D.R. Genome-wide search for yeast RNase p substrates reveals role in maturation of intron-encoded box c/d small nucleolar RNAs. *Proc. Natl. Acad. Sci. USA* **2008**, *105*, 12218–12223.
130. Brameier, M.; Herwig, A.; Reinhardt, R.; Walter, L.; Gruber, J. Human box c/d snoRNAs with miRNA like functions: Expanding the range of regulatory RNAs. *Nucleic Acids Res.* **2011**, *39*, 675–686.
131. Li, R.; Wang, H.; Bekele, B.N.; Yin, Z.; Caraway, N.P.; Katz, R.L.; Stass, S.A.; Jiang, F. Identification of putative oncogenes in lung adenocarcinoma by a comprehensive functional genomic approach. *Oncogene* **2006**, *25*, 2628–2635.
132. Schneider, C.; King, R.M.; Philipson, L. Genes specifically expressed at growth arrest of mammalian cells. *Cell* **1988**, *54*, 787–793.
133. Mourtada-Maarabouni, M.; Pickard, M.R.; Hedge, V.L.; Farzaneh, F.; Williams, G.T. Gas5, a non-protein-coding RNA, controls apoptosis and is downregulated in breast cancer. *Oncogene* **2009**, *28*, 195–208.
134. Nakamura, Y.; Takahashi, N.; Kakegawa, E.; Yoshida, K.; Ito, Y.; Kayano, H.; Niitsu, N.; Jinnai, I.; Bessho, M. The *gas5* (growth arrest-specific transcript 5) gene fuses to bcl6 as a result of t (1; 3) (q25; q27) in a patient with B-cell lymphoma. *Cancer Genet. Cytogenet.* **2008**, *182*, 144–149.

135. Gee, H.E.; Buffa, F.M.; Camps, C.; Ramachandran, A.; Leek, R.; Taylor, M.; Patil, M.; Sheldon, H.; Betts, G.; Homer, J.; *et al.* The small-nucleolar RNAs commonly used for microRNA normalisation correlate with tumour pathology and prognosis. *Br. J. Cancer* **2011**, *104*, 1168–1177.

136. Tanaka, R.; Satoh, H.; Moriyama, M.; Satoh, K.; Morishita, Y.; Yoshida, S.; Watanabe, T.; Nakamura, Y.; Mori, S. Intronic u50 small-nucleolar-RNA (snoRNA) host gene of no protein-coding potential is mapped at the chromosome breakpoint t (3; 6) (q27; q15) of human B-cell lymphoma. *Genes Cells* **2000**, *5*, 277–287.

137. Mei, Y.P.; Liao, J.P.; Shen, J.; Yu, L.; Liu, B.L.; Liu, L.; Li, R.Y.; Ji, L.; Dorsey, S.G.; Jiang, Z.R.; *et al. Small nucleolar RNA 42* acts as an oncogene in lung tumorigenesis. *Oncogene* **2012**, *31*, 2794–2804.

138. Liao, J.; Yu, L.; Mei, Y.; Guarnera, M.; Shen, J.; Li, R.; Liu, Z.; Jiang, F. Small nucleolar RNA signatures as biomarkers for non-small-cell lung cancer. *Mol. Cancer* **2010**, *9*, 198.

139. Ruggero, D.; Grisendi, S.; Piazza, F.; Rego, E.; Mari, F.; Rao, P.H.; Cordon-Cardo, C.; Pandolfi, P.P. Dyskeratosis congenita and cancer in mice deficient in ribosomal RNA modification. *Science* **2003**, *299*, 259–262.

140. Cabili, M.N.; Trapnell, C.; Goff, L.; Koziol, M.; Tazon-Vega, B.; Regev, A.; Rinn, J.L. Integrative annotation of human large intergenic noncoding RNAs reveals global properties and specific subclasses. *Genes Dev.* **2011**, *25*, 1915–1927.

141. Orom, U.A.; Derrien, T.; Beringer, M.; Gumireddy, K.; Gardini, A.; Bussotti, G.; Lai, F.; Zytnicki, M.; Notredame, C.; Huang, Q.; *et al.* Long noncoding RNAs with enhancer-like function in human cells. *Cell* **2010**, *143*, 46–58.

142. Ponting, C.P.; Oliver, P.L.; Reik, W. Evolution and functions of long noncoding RNAs. *Cell* **2009**, *136*, 629–641.

143. Khalil, A.M.; Guttman, M.; Huarte, M.; Garber, M.; Raj, A.; Rivea Morales, D.; Thomas, K.; Presser, A.; Bernstein, B.E.; van Oudenaarden, A.; *et al.* Many human large intergenic noncoding RNAs associate with chromatin-modifying complexes and affect gene expression. *Proc. Natl. Acad. Sci. USA* **2009**, *106*, 11667–11672.

144. Guttman, M.; Amit, I.; Garber, M.; French, C.; Lin, M.F.; Feldser, D.; Huarte, M.; Zuk, O.; Carey, B.W.; Cassady, J.P.; *et al.* Chromatin signature reveals over a thousand highly conserved large non-coding RNAs in mammals. *Nature* **2009**, *458*, 223–227.

145. Birney, E.; Stamatoyannopoulos, J.A.; Dutta, A.; Guigo, R.; Gingeras, T.R.; Margulies, E.H.; Weng, Z.; Snyder, M.; Dermitzakis, E.T.; Thurman, R.E.; *et al.* Identification and analysis of functional elements in 1% of the human genome by the encode pilot project. *Nature* **2007**, *447*, 799–816.

146. Brosnan, C.A.; Voinnet, O. The long and the short of noncoding RNAs. *Curr. Opin. Cell Biol.* **2009**, *21*, 416–425.

147. Amaral, P.P.; Mattick, J.S. Noncoding RNA in development. *Mamm. Genome* **2008**, *19*, 454–492.

148. Pang, K.C.; Frith, M.C.; Mattick, J.S. Rapid evolution of noncoding RNAs: Lack of conservation does not mean lack of function. *Trends Genet.* **2006**, *22*, 1–5.

149. Chen, L.L.; Carmichael, G.G. Decoding the function of nuclear long non-coding RNAs. *Curr. Opin. Cell Biol.* **2010**, *22*, 357–364.
150. Amaral, P.P.; Clark, M.B.; Gascoigne, D.K.; Dinger, M.E.; Mattick, J.S. Lncrnadb: A reference database for long noncoding RNAs. *Nucleic Acids Res.* **2011**, *39*, D146–D151.
151. Dinger, M.E.; Pang, K.C.; Mercer, T.R.; Crowe, M.L.; Grimmond, S.M.; Mattick, J.S. NRED: A database of long noncoding RNA expression. *Nucleic Acids Res.* **2009**, *37*, D122–D126.
152. Derrien, T.; Johnson, R.; Bussotti, G.; Tanzer, A.; Djebali, S.; Tilgner, H.; Guernec, G.; Martin, D.; Merkel, A.; Knowles, D.G.; *et al.* The gencode v7 catalog of human long noncoding RNAs: Analysis of their gene structure, evolution, and expression. *Genome Res.* **2012**, *22*, 1775–1789.
153. Wang, X.; Arai, S.; Song, X.; Reichart, D.; Du, K.; Pascual, G.; Tempst, P.; Rosenfeld, M.G.; Glass, C.K.; Kurokawa, R. Induced ncRNAs allosterically modify RNA-binding proteins in *cis* to inhibit transcription. *Nature* **2008**, *454*, 126–130.
154. Schwartz, J.C.; Younger, S.T.; Nguyen, N.B.; Hardy, D.B.; Monia, B.P.; Corey, D.R.; Janowski, B.A. Antisense transcripts are targets for activating small RNAs. *Nat. Struct. Mol. Biol.* **2008**, *15*, 842–848.
155. Tasheva, E.S.; Roufa, D.J. Regulation of human *rps14* transcription by intronic antisense RNAs and ribosomal protein s14. *Genes Dev.* **1995**, *9*, 304–316.
156. Kino, T.; Hurt, D.E.; Ichijo, T.; Nader, N.; Chrousos, G.P. Noncoding RNA *gas5* is a growth arrest- and starvation-associated repressor of the glucocorticoid receptor. *Science Signal.* **2010**, *3*, doi:10.1126/scisignal.2000568.
157. Lin, D.; Pestova, T.V.; Hellen, C.U.; Tiedge, H. Translational control by a small RNA: Dendritic bc1 RNA targets the eukaryotic initiation factor 4a helicase mechanism. *Mol. Cell. Biol.* **2008**, *28*, 3008–3019.
158. Young, T.L.; Matsuda, T.; Cepko, C.L. The noncoding RNA taurine upregulated gene 1 is required for differentiation of the murine retina. *Curr. Biol.* **2005**, *15*, 501–512.
159. Wang, K.C.; Chang, H.Y. Molecular mechanisms of long noncoding RNAs. *Mol. Cell* **2011**, *43*, 904–914.
160. Yochum, G.S.; Cleland, R.; McWeeney, S.; Goodman, R.H. An antisense transcript induced by wnt/beta-catenin signaling decreases e2f4. *J. Biol. Chem.* **2007**, *282*, 871–878.
161. Mourtada-Maarabouni, M.; Hedge, V.L.; Kirkham, L.; Farzaneh, F.; Williams, G.T. Growth arrest in human t-cells is controlled by the non-coding RNA growth-arrest-specific transcript 5 (*gas5*). *J. Cell Sci.* **2008**, *121*, 939–946.
162. Rinn, J.L.; Kertesz, M.; Wang, J.K.; Squazzo, S.L.; Xu, X.; Brugmann, S.A.; Goodnough, L.H.; Helms, J.A.; Farnham, P.J.; Segal, E.; *et al.* Functional demarcation of active and silent chromatin domains in human hox loci by noncoding RNAs. *Cell* **2007**, *129*, 1311–1323.
163. Yu, W.; Gius, D.; Onyango, P.; Muldoon-Jacobs, K.; Karp, J.; Feinberg, A.P.; Cui, H. Epigenetic silencing of tumour suppressor gene *p15* by its antisense RNA. *Nature* **2008**, *451*, 202–206.
164. Mercer, T.R.; Dinger, M.E.; Mattick, J.S. Long non-coding RNAs: Insights into functions. *Nat. Rev. Genet.* **2009**, *10*, 155–159.
165. Nagano, T.; Fraser, P. No-nonsense functions for long noncoding RNAs. *Cell* **2011**, *145*, 178–181.

166. Clark, M.B.; Mattick, J.S. Long noncoding RNAs in cell biology. *Semin. Cell Dev. Biol.* **2011**, *22*, 366–376.
167. Mattick, J.S.; Amaral, P.P.; Dinger, M.E.; Mercer, T.R.; Mehler, M.F. RNA regulation of epigenetic processes. *BioEssays* **2009**, *31*, 51–59.
168. Nagano, T.; Mitchell, J.A.; Sanz, L.A.; Pauler, F.M.; Ferguson-Smith, A.C.; Feil, R.; Fraser, P. The air noncoding RNA epigenetically silences transcription by targeting g9a to chromatin. *Science* **2008**, *322*, 1717–1720.
169. Pandey, R.R.; Mondal, T.; Mohammad, F.; Enroth, S.; Redrup, L.; Komorowski, J.; Nagano, T.; Mancini-Dinardo, D.; Kanduri, C. Kcnq1ot1 antisense noncoding RNA mediates lineage-specific transcriptional silencing through chromatin-level regulation. *Mol. Cell* **2008**, *32*, 232–246.
170. Prensner, J.R.; Iyer, M.K.; Balbin, O.A.; Dhanasekaran, S.M.; Cao, Q.; Brenner, J.C.; Laxman, B.; Asangani, I.A.; Grasso, C.S.; Kominsky, H.D.; *et al.* Transcriptome sequencing across a prostate cancer cohort identifies *pcat-1*, an unannotated lincRNA implicated in disease progression. *Nat. Biotechnol.* **2011**, *29*, 742–749.
171. Gupta, R.A.; Shah, N.; Wang, K.C.; Kim, J.; Horlings, H.M.; Wong, D.J.; Tsai, M.C.; Hung, T.; Argani, P.; Rinn, J.L.; *et al.* Long non-coding RNA hotair reprograms chromatin state to promote cancer metastasis. *Nature* **2010**, *464*, 1071–1076.
172. Lee, J.T. Gracefully ageing at 50, X-chromosome inactivation becomes a paradigm for RNA and chromatin control. *Nat. Rev. Mol. Cell Biol.* **2011**, *12*, 815–826.
173. Zhao, J.; Sun, B.K.; Erwin, J.A.; Song, J.J.; Lee, J.T. Polycomb proteins targeted by a short repeat RNA to the mouse x chromosome. *Science* **2008**, *322*, 750–756.
174. Morris, K.V.; Santoso, S.; Turner, A.M.; Pastori, C.; Hawkins, P.G. Bidirectional transcription directs both transcriptional gene activation and suppression in human cells. *PLoS Genet.* **2008**, *4*, e1000258.
175. Vanneste, R.; Smith, E.; Graham, G. Multiple neurofibromas as the presenting feature of familial atypical multiple malignant melanoma (fammm) syndrome. *Am. J. Med. Genet.* **2013**, *161*, 1425–1431.
176. Yap, K.L.; Li, S.; Munoz-Cabello, A.M.; Raguz, S.; Zeng, L.; Mujtaba, S.; Gil, J.; Walsh, M.J.; Zhou, M.M. Molecular interplay of the noncoding RNA anril and methylated histone h3 lysine 27 by polycomb cbx7 in transcriptional silencing of ink4a. *Mol. Cell* **2010**, *38*, 662–674.
177. Gutschner, T.; Hammerle, M.; Eissmann, M.; Hsu, J.; Kim, Y.; Hung, G.; Revenko, A.; Arun, G.; Stentrup, M.; Gross, M.; *et al.* The noncoding RNA *malat1* is a critical regulator of the metastasis phenotype of lung cancer cells. *Cancer Res.* **2013**, *73*, 1180–1189.
178. Pickard, M.R.; Mourtada-Maarabouni, M.; Williams, G.T. Long non-coding RNA *gas5* regulates apoptosis in prostate cancer cell lines. *Biochim. Biophys. Acta* **2013**, *1832*, 1613–1623.
179. Huarte, M.; Guttman, M.; Feldser, D.; Garber, M.; Koziol, M.J.; Kenzelmann-Broz, D.; Khalil, A.M.; Zuk, O.; Amit, I.; Rabani, M.; *et al.* A large intergenic noncoding RNA induced by *p53* mediates global gene repression in the *p53* response. *Cell* **2010**, *142*, 409–419.
180. Hung, T.; Wang, Y.; Lin, M.F.; Koegel, A.K.; Kotake, Y.; Grant, G.D.; Horlings, H.M.; Shah, N.; Umbricht, C.; Wang, P.; *et al.* Extensive and coordinated transcription of noncoding RNAs within cell-cycle promoters. *Nat. Genet.* **2011**, *43*, 621–629.

181. Poliseno, L.; Salmena, L.; Zhang, J.; Carver, B.; Haveman, W.J.; Pandolfi, P.P. A coding-independent function of gene and pseudogene mRNAs regulates tumour biology. *Nature* **2010**, *465*, 1033–1038.

182. Melo, C.A.; Drost, J.; Wijchers, P.J.; van de Werken, H.; de Wit, E.; Oude Vrielink, J.A.; Elkon, R.; Melo, S.A.; Leveille, N.; Kalluri, R.; *et al.* ERNAs are required for *p53*-dependent enhancer activity and gene transcription. *Mol. Cell* **2013**, *49*, 524–535.

183. Brantl, S. Antisense-RNA regulation and RNA interference. *Biochim. Biophys. Acta* **2002**, *1575*, 15–25.

184. Prescott, E.M.; Proudfoot, N.J. Transcriptional collision between convergent genes in budding yeast. *Proc. Natl. Acad. Sci. USA* **2002**, *99*, 8796–8801.

185. Thrash-Bingham, C.A.; Tartof, K.D. *Ahif*: A natural antisense transcript overexpressed in human renal cancer and during hypoxia. *J. Natl. Cancer Inst.* **1999**, *91*, 143–151.

186. Cayre, A.; Rossignol, F.; Clottes, E.; Penault-Llorca, F. *Ahif* but not *hif-1alpha* transcript is a poor prognostic marker in human breast cancer. *Breast Cancer Res.* **2003**, *5*, R223–230.

187. Zhang, H.; Gao, S.; De Geyter, C. A natural antisense transcript, bokas, regulates the pro-apoptotic activity of human bok. *Int. J. Oncol.* **2009**, *34*, 1135–1138.

188. Beltran, M.; Puig, I.; Pena, C.; Garcia, J.M.; Alvarez, A.B.; Pena, R.; Bonilla, F.; de Herreros, A.G. A natural antisense transcript regulates *zeb2/sip1* gene expression during snail1-induced epithelial-mesenchymal transition. *Genes Dev.* **2008**, *22*, 756–769.

189. Fabbri, M.; Paone, A.; Calore, F.; Galli, R.; Gaudio, E.; Santhanam, R.; Lovat, F.; Fadda, P.; Mao, C.; Nuovo, G.J.; *et al.* MicroRNAs bind to toll-like receptors to induce prometastatic inflammatory response. *Proc. Natl. Acad. Sci. USA* **2012**, *109*, E2110–2116.

190. Mitchell, P.S.; Parkin, R.K.; Kroh, E.M.; Fritz, B.R.; Wyman, S.K.; Pogosova-Agadjanyan, E.L.; Peterson, A.; Noteboom, J.; O'Briant, K.C.; Allen, A.; *et al.* Circulating microRNAs as stable blood-based markers for cancer detection. *Proc. Natl. Acad. Sci. USA* **2008**, *105*, 10513–10518.

191. Rai, K.; Takigawa, N.; Ito, S.; Kashihara, H.; Ichihara, E.; Yasuda, T.; Shimizu, K.; Tanimoto, M.; Kiura, K. Liposomal delivery of microRNA-7-expressing plasmid overcomes epidermal growth factor receptor tyrosine kinase inhibitor-resistance in lung cancer cells. *Mol. Cancer Ther.* **2011**, *10*, 1720–1727.

192. Kota, J.; Chivukula, R.R.; O'Donnell, K.A.; Wentzel, E.A.; Montgomery, C.L.; Hwang, H.W.; Chang, T.C.; Vivekanandan, P.; Torbenson, M.; Clark, K.R.; *et al.* Therapeutic microRNA delivery suppresses tumorigenesis in a murine liver cancer model. *Cell* **2009**, *137*, 1005–1017.

193. Su, J.; Baigude, H.; McCarroll, J.; Rana, T.M. Silencing microRNA by interfering nanoparticles in mice. *Nucleic Acids Res.* **2011**, *39*, e38.

7

Downregulation of miR-17~92 Expression Increase Paclitaxel Sensitivity in Human Ovarian Carcinoma SKOV3-TR30 Cells via BIM Instead of PTEN

Ting Shuang [1], Chunxue Shi [1,2,†], Shuang Chang [1,†], Min Wang [1,*] and Cui Hong Bai [1]

1 Department of Obstetrics and Gynecology, Shengjing Hospital of China Medical University, Shenyang 110004, China; E-Mails: shuangting87@163.com (T.S.); c.shuang-0519@163.com (S.C.); myearth126@126.com (C.H.B.)

2 Shenyang Maternity and infant Hospital, 90 Qingnian Road, Shenhe District, Shenyang 110014, China; E-Mail: shichunxue_502@163.com

† These authors contributed equally to this work.

* Author to whom correspondence should be addressed; E-Mail: wm21st@hotmail.com;

Abstract: To better understand the molecular mechanisms of paclitaxel resistance in ovarian carcinoma, we evaluated the expression of miRNAs using miRNA microarray between human ovarian carcinoma SKOV3 cells and paclitaxel resistant SKOV3-TR30 cells. Results showed that 69 miRNAs were upregulated while 102 miRNAs were downregulated in SKOV3-TR30 cells. Using real-time PCR, we further clarified that miR-17~92 was overexpressed in SKOV3-TR30 cells compared with that in SKOV3 cells. We then established stable virally transduced SKOV3-TR30-m-PTIP-Sponge all SKOV3-TR30 cells and its vector-only control SKOV3-TR30-m-PTIP-GFP cells. Real time-PCR revealed that SKOV3-TR30-m-PTIP-Sponge all cells expressed approximately 6.18-fold lower levels of miR-17~92 compared with the control group. Decreased expression of miR-17~92 resulted in cell cycle arrest in the G2/M phase and growth inhibition. After the transduction, the BIM protein level was increased in SKOV3-TR30 cells and luciferase reporter assays revealed that miR-17~92 binds directly to the 3'-UTR of BIM. Results of luciferase reporter assays accompanied with Western Blot showed that although miR-17~92 binds directly to the 3'-UTR of PTEN, the PTEN protein expression level was upregulated slightly while the result is of no statistical significance. Our results showed that miR-17~92 could be a causal factor of the downregulation of BIM in SKOV3-TR30 cells and thus induce the paclitaxel resistance in SKOV3-TR30 cells.

Keywords: miR-17~92 cluster; transduction; ovarian carcinoma; paclitaxel resistance; PTEN; BIM

---

1. Introduction

Ovarian carcinoma is one of the most common gynecological malignant tumors. Its incidence is ranked second in malignant tumors of the female reproductive system and is gradually increasing in recent years. The most recommended treatment worldwide is surgery, along with paclitaxel and platinum–based (first-line chemotherapy) adjuvant chemotherapy; however, its mortality rate is still about 70%, which is the highest in gynecological malignancies [1]. One of the most significant reasons for such high morality rate is that about 30%~40% of patients with ovarian carcinoma are resistant to chemotherapy and, moreover, 60% of first-line chemotherapy-sensitive patients are resistant to chemotherapy after six months. Therefore, clarification of the exact mechanism of resistance and resistance reversal of ovarian carcinoma has become extremely urgent and important research topic.

MicroRNAs (miRNAs) contain 18–24 nucleotides and are small, non-coding RNAs that post transcriptionally regulate gene expression through translational repression and mRNA degradation [2]. Recent research has shown the tumor suppressant and oncogenic potential of a number of miRNAs, underscoring their importance in human cancer therapeutic and diagnostic applications [3–7].

BIM is a member of the BH3-only family of pro-apoptotic proteins and is expressed in a wide variety of tissues. BIM can initiate apoptosis by directly activating Bax through interaction with the Bcl-2/Bax herterodimer complex, which can further induces mitochondrial cell death [8,9]. It plays a critical role in tumor cell biology, including the regulation of tumorigenesis through activities as a tumor suppressor, tumor metastasis, and tumor cell survival. Therefore, it has gradually become an interesting target for cancer chemotherapy. PTEN plays a well-established role in the negative regulation of the PI3K pathway, which is frequently activated in several cancer types, including ovarian cancer. PTEN loss of function occurs in a wide spectrum of human cancers through mutations, deletions and transcriptional silencing. The expression of BIM and PTEN is highly regulated by its transcriptional and post-translational levels.

Overexpression of miR-17~92 has been observed in lymphomas and solid tumors [10] and is related to cell proliferation. The gene cluster of miR-17~92 resides with intron 3 of cl3orf25 non-protein-coding gene at 13q31.3 gene [8,11]. Studies by Lewis, BP, and others [12] have shown that miR-17~92 may play a role in PTEN and BIM. In recent years, more and more studies have reported that abnormal expression of PTEN and BIM has participated in the formation of tumor drug-resistance [13–15].

Yet it is not clear if miR-17~92 gene cluster has an impact on the paclitaxel resistance of ovarian carcinoma through affecting the expression of the BIM or PTEN protein. Furthermore, the effect of both BIM and PTEN on paclitaxel resistant in ovarian cancer cells has not been thoroughly researched;

particularly, the mechanism involved in their expression regulation has barely been studied in ovarian carcinoma chemoresistance.

In this study, we investigated whether BIM or PTEN gene was post-transcriptionally regulated by miR-17~92 and the contribution of miR-17~92 to BIM or PTEN protein levels in SKOV3-TR30 cells. In addition, we also investigated the impact of miR-17~92 on SKOV3-TR30 cell proliferation and cell cycle. We aim to get a better understanding of the molecular mechanisms of paclitaxel resistance in SKOV3-TR30 cells, to provide a clue for further investigation of the function of miR-17~92 and its target genes, and their correlation in ovarian carcinoma paclitaxel resistance.

2. Results

2.1. MiRNA Expression Profiling

A microarray platform was used to analyze and compare the pattern of miRNA expression between the parental SKOV3 cell line and its counterpart paclitaxel resistant SKOV3-TR30 cell line. The expression profiles of 171 miRNAs changed significantly: 69 upregulated miRNAs (miR-17, miR-19b, miR-92-1) and 102 downregulated miRNAs (miR-134, miR-34, miR-196b) in SKOV3-TR30 cells as compared with SKOV3 cells (Figure 1).

Figure 1. Analysis of miRNAs Expression Profile of Sensitive and Drug-resistant Human Ovarian carcinoma Cells. Data were analyzed by the CLUSTER analysis software CLUSTER 3.0. Yellow means high expression, blue represents the low expression.

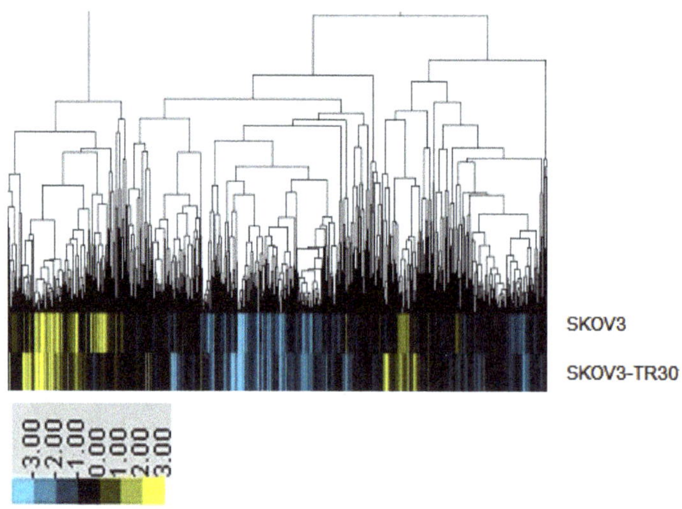

2.2. miR-17~92 Was Overexpressed in SKOV3-TR30 Cells Compared with SKOV3 Cells

We next studied the expression level of miR-17~92 in SKOV3 and SKOV3-TR30 cells using real-time PCR. Amplification Curve and Melting Curve of miR-17~92 and internal control β-actin see Figure 2. The expression level of miR-17~92 in the SKOV3-TR30 cells is indicated as "1" as control group. Compared to SKOV3-TR30 cells, the expression level of miRNA-17~92 in SKOV3 cells is 0.557, and the difference is of statistically significant ($t = 9.193$, $p < 0.05$) (Table 1).

The result revealed that the expression level of miR-17~92 was markedly increased in paclitaxel resistant SKOV3-TR30 cells compared with paclitaxel sensitive SKOV3 cells.

Figure 2. Amplification Curve and Melting Curve of miR-17~92 Expression in SKOV3, SKOV3-TR30, SKOV3-TR30-m-PTIP-Sponge all cells and the SKOV3-TR30-m-PTIP-GFP cells Detected by real-time PCR. (**a**) Amplification Curve of Expression of miR-17~92; (**b**) Melting Curve of Expression of miR-17~92.

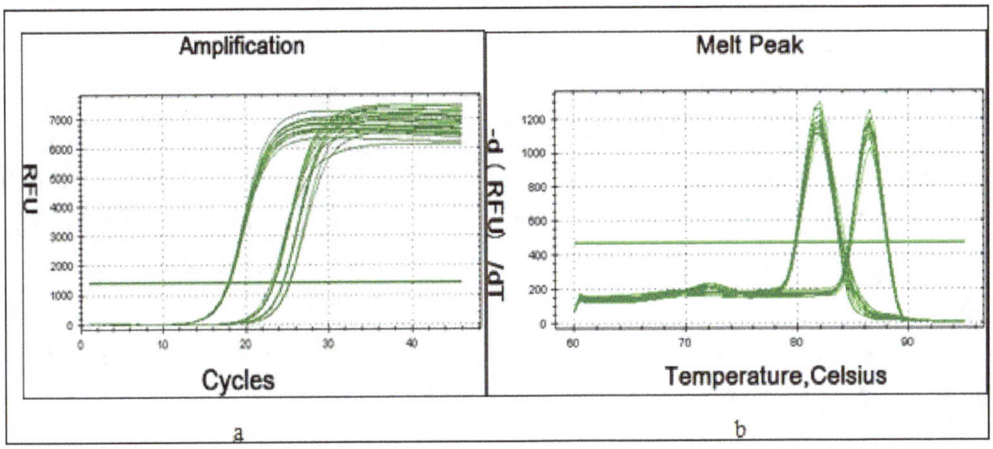

Table 1. Expression level of miR-17~92 in Different Groups.

Cells line	Δct	ΔΔct	2-ΔΔct
SKOV3	9.85 ± 0.198	0.845 ± 0.198	0.55 (0.483–0.642) *
SKOV3-TR30	10.695 ± 0.488	0.0 ± 0.488	1.0 (0.711–1.412)
SKOV3-TR30-m-PTIP-GFP	9.715 ± 0.007	0.98 ± 0.007	0.507 (0.642–0.801) *
SKOV3-TR30-m-PTIP-Sponge all	7.085 ± 0.304	3.61 ± 0.304	0.082 (0.075–0.09) *

* represents compared with the expression level of miRNA-17~92 in SKOV3, the different expression level of miRNA-17~92 in SKOV3-TR30, SKOV3-TR30-m-PTIP-GFP and SKOV3-TR30-m-PTIP-Sponge all is of great significance, $p < 0.05$.

2.3. Comparison of miR-17~92 Expression Levels in SKOV3-TR30-m-PTIP-Sponge all and the SKOV3-TR30-m-PTIP-GFP

In order to determine whether the resistance to paclitaxel could be caused by miR-17~92 in SKOV3-TR30, we established stable virally transduced SKOV3-TR30 cell line SKOV3-TR30-m-PTIP-Sponge all cells and its vector-only control SKOV3-TR30-m-PTIP-GFP cells. The m-PTIP-Sponge all inhibit plasmids were identified using restriction enzyme digestion shown in (Figure 3c). Using Flow Cytometry, the purity of stable transfectants identified by the expression of plasmid green fluorescent protein was greater than 95% (Figure 3a,b). We next studied the expression level of miR-17~92 in SKOV3-TR30 -m-PTIP-GFP cells and SKOV3-TR30-m-PTIP-Sponge all cells using real-time PCR. Real-time PCR results revealed that SKOV3-TR30-m-PTIP-Sponge all expressed approximately 6.18-fold lower levels of miR-17~92 (Table 1) compared with which in SKOV3-TR30-m-PTIP-GFP cells.

Amplification Curve and Melting Curve of miR-17~92 and internal control β-actin see Figure 2. The expression level of miR-17~92 in the SKOV3-TR30 cells is indicated as "1" as control. Compared with SKOV3-TR30 cells, the expression level of miRNA-17~92 in SKOV3-TR30-m-PTIP-GFP cells and SKOV3-TR30-m-PTIP-Sponge all cells is 0.507 ($t = 9.417$, $p < 0.05$) and 0.082 ($t = 23.659$, $p < 0.05$), respectively. Contrast the expression level of miR-17~92 in SKOV3-TR30-m-PTIP-GFP cells, which is 0.507 and in SKOV3-TR30-m-PTIP-Sponge all cells which 0.082, the difference is of statistical significance ($t = 10.726$, $p < 0.05$).

Figure 3. (**a**) Determination of stable virally transduced SKOV3-TR30-m-PTIP-Sponge all cell line and its vector-only control SKOV3-TR30-m-PTIP-GFP by Fluorescence Microscope. (100×) 1, 2, 3: transduced by m-PTIP-GFP empty plasmid 1, white field, 2, with fluorescence, 3, the integration of 1 and 2. 4,5,6: transduced by m-PTIP-Sponge all inhibit plasmid, 4, white field, 5, with fluorescence, 6, the integration of 4 and 5; (**b**) Determination of stable virally transduced SKOV3-TR30-m-PTIP-Sponge all cell line and its vector-only control SKOV3-TR30-m-PTIP-GFP using Flow Cytometry. Ratio of cells expression green fluorescent protein for SKOV3-TR30, SKOV3-TR30-m-PTIP-GFP and SKOV3-TR30-m-PTIP-Sponge all were 0.00, 97.48% and 99.79%; (**c**) Plasmid Restriction Enzyme Digestion Results 1: m-PTIP-GFP empty plasmid, 2: m-PTIP-Sponge all inhibit plasmid, M: Marker.

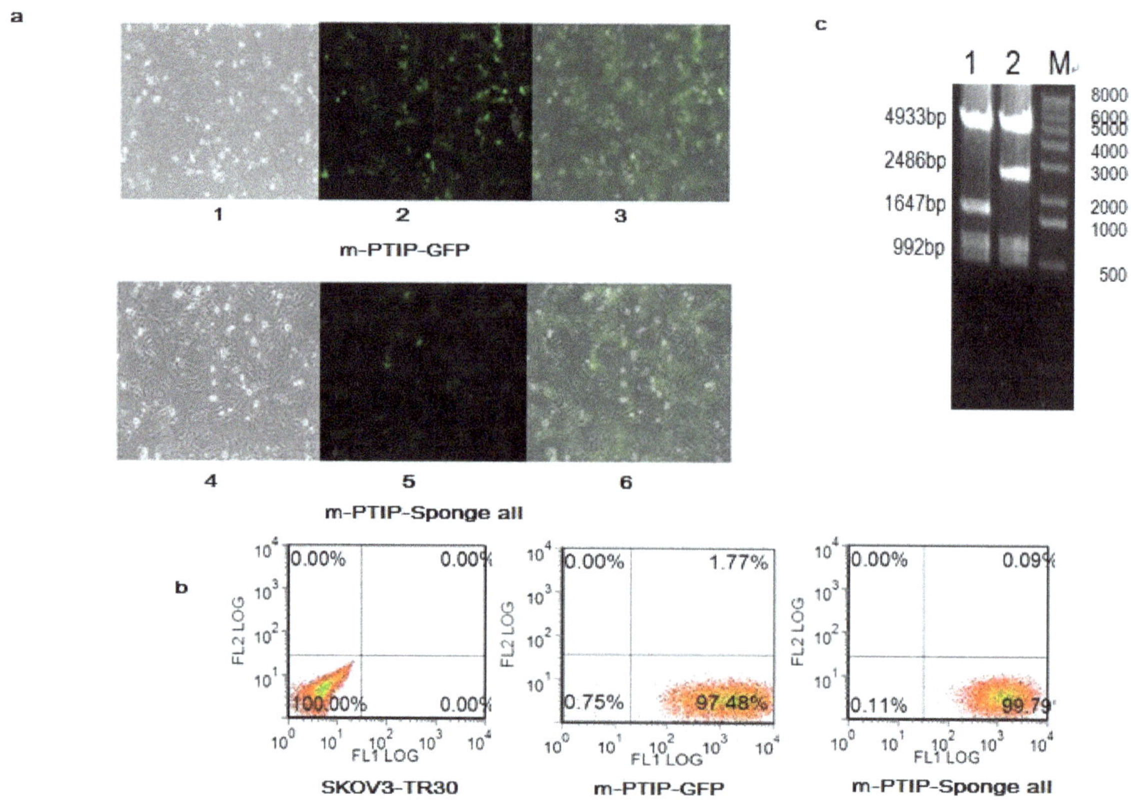

2.4. Inhibition of miR-17~92 could Inhibit the Proliferation of SKOV3-TR30 Cells

We further study the effects of miR-17~92 expression on the proliferation of SKOV3-TR30 using MTT assay. Paclitaxel inhibited cell proliferation in both SKOV3-TR30-m-PTIP-Sponge all and SKOV3-TR30-m-PTIP-GFP, and with increased concentration of paclitaxel the inhibitory effects was more obvious. Further more, SKOV3-TR30-m-PTIP-Sponge all cells had significantly reduced cell proliferation ($p < 0.05$) compared with SKOV3-TR30-m-PTIP-GFP (Figure 4) and when the concentration of paclitaxel was 100 nm, the inhibition rate difference was most obvious, which is 40.407% of SKOV3-TR30-m-PTIP-Sponge all cells and 20.934% of SKOV3-TR30-m-PTIP-GFP. IC50 of paclitaxel of SKOV3-TR30-m-PTIP-Sponge all cells and SKOV3-TR30-m-PTIP-GFP cells is 470.80 nM and 170.13 nM respectively.

Figure 4. MTT assay was used to detect the effects of miR-17~92 on cell growth in SKOV3-TR30 cells after transduced with either m-PTIP-Sponge all inhibit plasmid or m-PTIP-GFP empty plasmid. Inhibition rate of stable virally transduced SKOV3-TR30-m-PTIP-Sponge all cells and its vector-only control SKOV3-TR30-m-PTIP-GFP cells is of great difference, which has statistical significance ($p < 0.05$). The difference is most obvious when the concentration of paclitaxel is 100 nM (* represents that under the same concentration of paclitaxel, SKOV3-TR30 cells transduced by m-PTIP-Sponge all inhibit plasmid *vs.* SKOV3-TR30 cells transduced by m-PTIP-GFP empty plasmid, $p < 0.05$).

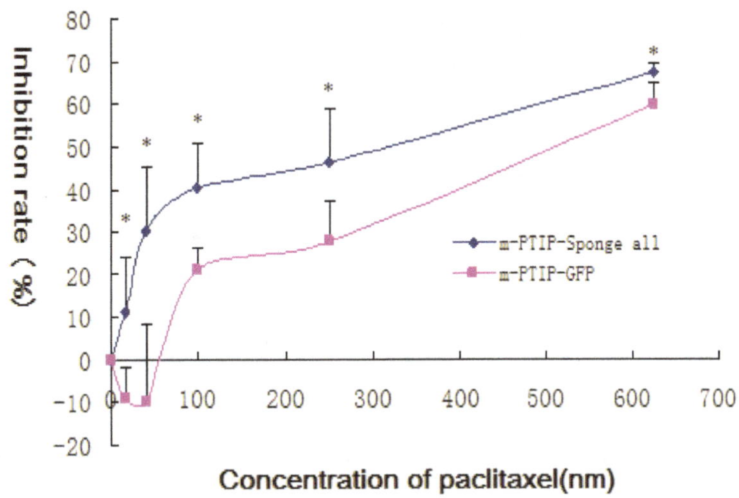

2.5. Decreased Expression of miR-17~92 Resulted in Cell Cycle Arrest in the G2/M Phase

We accessed cell cycle distribution profiles after the transduction of SKOV3-TR30 with either miR-17~92-PTIP-Sponge all plasmids or negative control miR-17~92-PTIP-GFP (m-PTIP-GFP) plasmids. Both SKOV3-TR30-m-PTIP-Sponge all and SKOV3-TR30-m-PTIP-GFP cells were treated with paclitaxel (the concentration was 0 nM, 20 nM, 100 nM, 200 nM). Decreased expression of miR-17~92 in SKOV3-TR30-m-PTIP-Sponge all cells resulted in a significant increase in the fraction of cells arrested at G2/M as well as a concomitant decrease in the fraction of cells arrested at S phase compared with SKOV3-TR30-m-PTIP-GFP. These effects were obvious when the concentration of paclitaxel was 100 nm (Figure 5).

Figure 5. Results of Cell Cycle of SKOV3-TR30 after transduction. (**a**) The Proportion of G0/G1 Phase; (**b**) The Proportion of S Phase; (**c**) The Proportion of G2/M Phase (* respects the group transduced by m-PTIP-Sponge all inhibit plasmid *vs.* the group transduced by m-PTIP-GFP empty plasmid in the same concentration of paclitaxel, $p < 0.05$; ** respects the group transduced by m-PTIP-Sponge all inhibit plasmid *vs.* the group transduced by m-PTIP-GFP empty plasmid in the same concentration of paclitaxel, $p < 0.01$).

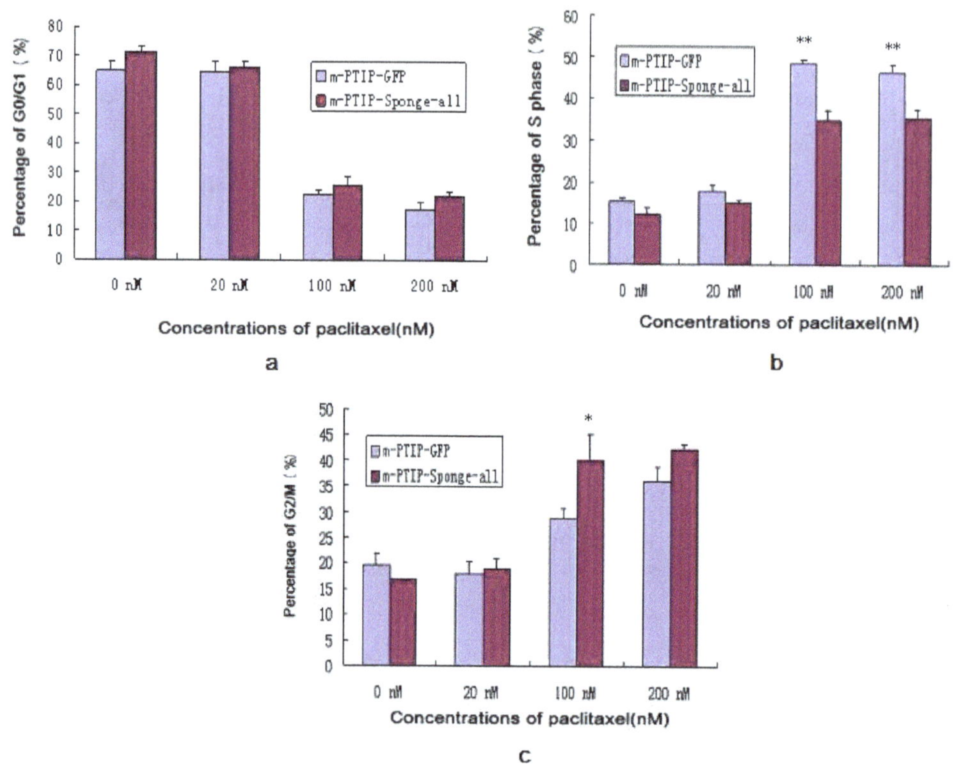

2.6. BIM, a BH3-only Propoapoptotic Protein, Is a Direct Target of the miR-17~92

As result shows miR-17~92 expression is significantly higher in SKOV3-TR30 than in SKOV3. To demonstrate the BIM protein level was directly mediated by miR-17~92 through biding to 3'-UTR of BIM, we co-transfected the BIM 3'-UTR (B1,B2,B3) along with TMP2-miR-17~92 into HEK293 cells and than performed luciferase reporter assays. The luciferase reporter plasmid contains a segment of the BIM 3'-UTR, among the three cloned fragments of the BIM 3'-UTR. B2 containing binding sites for miR-17-5p/-20a and miR-92, and B3 contains binding sites for miR-19 and miR-92 while there is not binding site of B1 with miR-17~92. The luciferase activity of the reporter plasmid containing the B1, B2 or B3 was decreased to 91.5%, 23.75% and 6.25% separately compared with the control construct (Figure 6b). We then used SKOV3 and SKOV3-TR30 cells to clarify if miR-17~92 has an effect on BIM protein expression which then trigger paclitaxel resistance in ovarian carcinoma cells. In contrast to SKOV3 cells, SKOV3-TR30 cells shows significantly decreased expression of BIM (Figure 7a, Table 2). We further applied SKOV3-TR30-m-PTIP-Sponge all cells and the control SKOV3-TR30-m-PTIP-GFP cells for a better understanding about the effect of miR-17~92 on BIM protein. Both of the cell lines were treated with paclitaxel (the concentration was 0 nM, 20 nM,

100 nM, 200 nM). In contrast to SKOV3-TR30-m-PTIP-GFP cells, SKOV3-TR30-m-PTIP-Sponge all cells with a significant lower expression of miR-17~92 showed a increase of the BIM protein level both at the steady-state condition without any paclitaxel (Figure 7b, Table 3) and after the treatment with paclitaxel in different concentrations (Figure 7c, Table 3). These findings suggest that miR-17~92 plays an important role in BIM protein expression decrease through the post-transcriptional regulation.

Figure 6. (**a**) Overexpression of miR-17~92 in HEK293 cells inhibited the luciferase activity of the PTEN 3'-UTR reporter plasmid. The 3'-UTR reporter plasmids PTEN or PTEN mut were co-transfected with TMP2-miR-17~92 into HEK293 cells. The activity of firefly luciferase was normalized to that of Renilla luciferase. (**b**) Overexpression of miR-17~92 in HEK293 cells inhibited the luciferase activity of the BIM 3'-UTR reporter plasmid. The 3'-UTR reporter plasmids B1, B2 orB3 were co-transfected with TMP2-miR-17~92 into the HEK293 cells. The activity of firefly luciferase was normalized to that of Renilla luciferase. * $p < 0.05$ compared with the control.

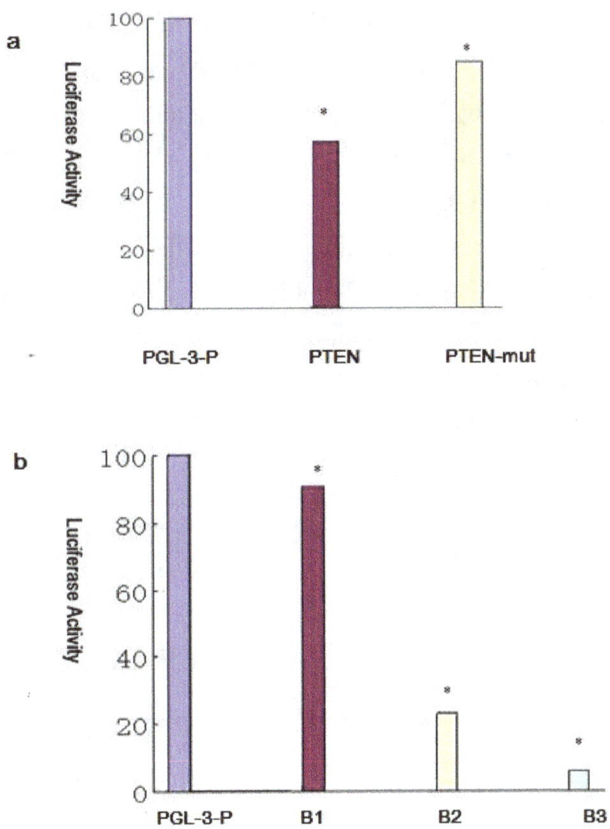

Table 2. Relative Content of BIM Protein in SKOV3-TR30 and SKOV3 cells ($\bar{x} \pm s$).

Cell group	BIM/actin
SKOV3-TR30	0.2118 ± 0.0923
SKOV3	0.2735 ± 0.1233

($t = 3.983, p = 0.028$).

Table 3. Relative Content of BIM and PTEN Protein in Each Cell Group (Relative Gray Rate, $\bar{x} \pm s$).

Concentrations of Paclitaxel (nM)	BIM/GAPDH		PTEN/GAPDH	
	m-PTIP-GFP	m-PTIP-Sponge all	m-PTIP-GFP	m-PTIP-Sponge all
0	70.59 ± 3.4215	107.70 ± 2.3265 *	122.77 ± 2.3265	137.67 ± 1.2452
20	94.13 ± 5.3265	154.08 ± 2.2641 *	98.15 ± 5.8736	121.96 ± 1.4737
100	123.72 ± 3.1749	136.83 ± 1.5821 *	122.59 ± 3.5521	130.93 ± 1.2481
200	109.54 ± 1.1399	164.46 ± 3.7326 *	116.54 ± 1.3357	125.28 ± 3.2158

* represents under the same concentration of paclitaxel, the group infected by m-PTIP-Sponge all inhibit plasmid *vs.* the group infected by m-PTIP-GFP empty plasmid, $p < 0.05$.

Figure 7. (**a**) The samples from SKOV3 cells and SKOV3-TR30 cells were subjected to Western Blot analysis with a BIM and PTEN antibody while β-actin levels were used as loading control; (**b**, **c**) The samples from SKOV3-TR30-m-PTIP-Sponge all cells and its vector-only control SKOV3-TR30-m-PTIP-GFP cells were subjected to Western Blot with BIM and PTEN antibody while GAPDH levels were used as loading control; (**b**) Decreased expression of miR-17~92 upregulates the expression of BIM instead of PTEN in SKOV3-TR30 cells without treatment of paclitaxel. 1,3 Protein samples from SKOV3-TR30 cells transduced by m-PTIP-GFP empty plasmid. 2,4 Protein samples from SKOV3-TR30 cells transduced by m-PTIP-Sponge all inhibit plasmid; (**c**) Decreased expression of miR-17~92 upregulates the expression of BIM instead of PTEN in SKOV3-TR30 cells after the treatment of paclitaxel with different concentration.1, 3, 5, Protein samples from SKOV3-TR30 cells transduced by m-PTIP-Sponge all inhibit plasmid 2,4,6: Protein samples from SKOV3-TR30 cells transduced by m-PTIP-GFP empty plasmid. The concentration of paclitaxel: 1,2: 20 nM, 3,4: 100 nM, 5,6: 200 nM.

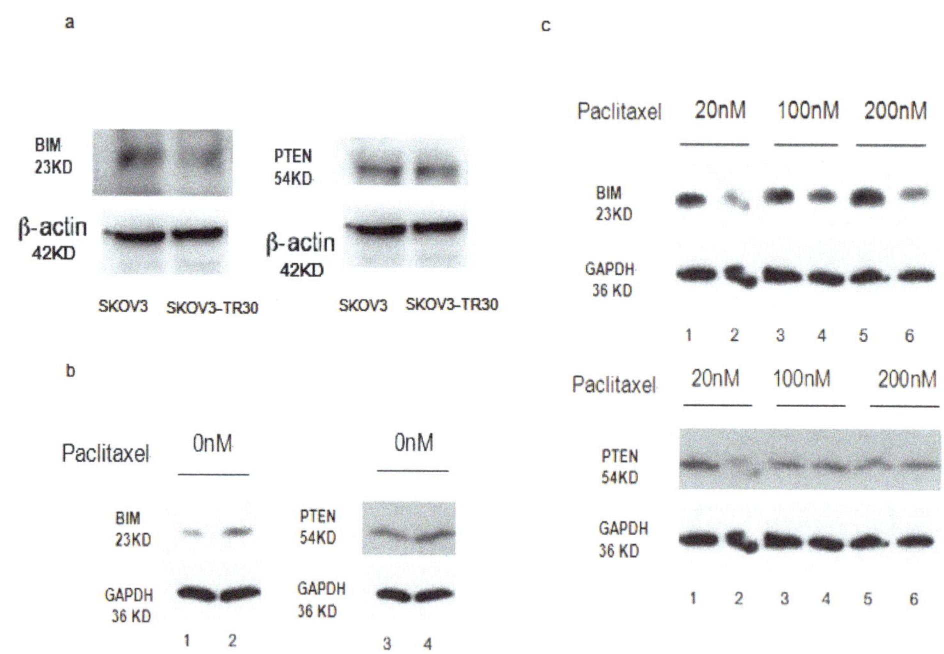

2.7. Effect of miR-17~92 on PTEN Expression

In order to clarify whether miR-17~92 has effect on PTEN expression, we constructed a luciferase reporter plasmid containing point mutations in the predicted miRNA binding sites within the PTEN 3'-UTR (PTEN mut). We then performed luciferase reporter assays in HEK293 cells. The luciferase activity of the luciferase reporter gene containing the wild-type PTEN 3'-UTR decreased 42.6% compared with the control construct pGL-3-P, but the decrease was not seen with the luciferase reporter gene PTEN mut (Figure 6a). That means miR-17~92 binds directly to 3'-UTR of PTEN. We then used SKOV3 and SKOV3-TR30 cells to see if there is a different expression of PTEN. In contrast to SKOV3 cells, expression of PTEN protein is decreased in SKOV3-TR30 cells but the result is of no statistical significance (Figure 7a, Table 4). We then use SKOV3-TR30 to make further clarifications. Results showed PTEN protein level is slightly upregulated in SKOV3-TR30-m-PTIP-Sponge all cells compared with SKOV3-TR30-m-PTIP-GFP cells while the difference is of no statistical significance.

To sum up, expression level of PTEN was not effected; even miR-17~92 binds directly to the 3'-UTR of PTEN. In addition, although miR-17~92 expression level was much lower in SKOV3-TR30-m-PTIP-Sponge all cells, there shows no significant increased expression of PTEN protein. These observations suggested that not only miR-17~92 but also other mechanisms may be responsible for the regulation of PTEN expression, including other miRNAs involved in the post-transcription of PTEN.

Table 4. Relative Content of PTEN Protein in SKOV3-TR30 and SKOV3 cells ($\bar{x}$ ± s).

Cell group	PTEN/actin
SKOV3-TR30	0.4043 ± 0.1266
SKOV3	0.4262 ± 0.1422

($t = 2.81, p = 0.067$).

3. Discussion

The current work shows that miR-17~92 cluster expression correlates with paclitaxel resistance of SKOV3-TR30 cells. It's also shown that BIM instead of PTEN is suppressed by miR-17~92 cluster via direct binding to the BIM 3'-UTR.

In this study, we established a stable virally transduced SKOV3-TR30 cell line SKOV3-TR30-m-PTIP-Sponge all, which expressed approximately 6.18-fold lower levels of miR-17~92 compared with its vector-only control SKOV3-TR30-m-PTIP-GFP, in order to seek an understanding of the molecular mechanisms of paclitaxel resistance with respect to miR-17~92. At first, we approached this issue by obtaining the miRNA differential expression profile between the parental SKOV3 cells and its paclitaxel resistance SKOV3-TR30 cells by using the array-based miRNA assay. Our results showed that 69 miRNAs are upregulated while 102 miRNAs were downregulated in SKOV3-TR30 cells. Among them miR-17~92 expression was significantly upregulated in paclitaxel resistance SKOV3-TR30 cells compared with that in the parental SKOV3 cells. Using real-time PCR we further clarified that miR-17~92 was overexpressed in SKOV3-TR30 cells compared with SKOV3 cells.

A miR-17~92 cluster comprising miR-17, miR-18a, miR-20a, miR-19a, miR-19b, and miR-92-1 is overexpressed in a large fraction of lymphomas [11]. Besides, miR-17~92 was detected as overexpressed also in multiple myeloma [16], breast cancer [17] and osteosarcoma [18]. Ohyashiki [19] states that miR-92a is not only used for the diagnosis of non-Hodgkin's lymphoma, but also used as an indicator for monitoring the recurrence of non-Hodgkin's lymphoma after chemotherapy. Other studies show that the tumorigenic effect of miR-17~92 may be due to synergism between its family members [20].

Significant decreased expression of miR-17~92 in SKOV3-TR30 cells by transduction with miR-17~92 inhibitory plasmids (miR-17-92-PTIP-Sponge all) markedly inhibited cell growth and the inhibition rate is most obvious (40.4%, $p < 0.05$) when the concentration of paclitaxel was 100 nm. Decreased expression of miR-17~92 also resulted in cell cycle arrest in the G2/M phase and is most obvious when the concentration of paclitaxel was 100 nm ($p < 0.05$). At that time, the expression of BIM protein was significantly increased. Thus, it is likely that the downregulation of miR-17~92 is closely associated with the sensitivity to paclitaxel through the upregulation BIM. To further clarify whether the BIM protein level was mediated through miR-17~92, we co-transfected the BIM 3'-UTR along with TMP2-miR-17~92 plasmid into HEK293 cells, than performed luciferase reporter assays. The reporter assays revealed that the firefly luciferase activity from HEK293 cells transfected with the reporter gene containing the wild-type B1, B2 or B3 was decreased to 91.5%, 23.75% and 6.25% separately compared with the control construct (Figure 6b).These findings suggest that BIM is likely to be direct target of miR-17~92 and that BIM protein was regulated at the post-transcriptional level in SKOV-TR30 cells.

So far, the research of the influence of miRNA-17~92 on ovarian carcinoma drug-resistance is rarely reported. Using bioinformatics analysis software, we predict that PTEN and BIM are most likely potential target genes of miR17~92. The newest transgenic animal experiments in 2008 shows that miR-17~92 gene clusters down regulate the expression of PTEN and BIM, which are tumor suppressor factors [21]. In recent years, more and more studies have shown that abnormal expression of PTEN and BIM is mostly related to the formation of tumor's drug resistance [13–15]. There are also studies show down regulation of PTEN and BIM in certain ovarian carcinoma cells with feature of chemoresistance [15,22], besides studies of Lewis and BP [10] show that miR-17~92 takes effect through PTEN and BIM. So in order to further prove whether miR-17~92 cause paclitaxel resistance through PTEN in SKOV3-TR30,we tested the expression of PTEN protein in paclitaxel resistant ovarian carcinoma SKOV3-TR30 cells after transduced with miR-17~92-PTIP-Sponge all. Different from the significant upregulation of BIM protein, PTEN was upregulated slightly while the result is of no statistical significance. The luciferase activity of the reporter gene containing the wild-type PTEN 3'-UTR was decreased 42.6% compared with the control construct pGL-3-P, but the decrease was not seen with the reporter gene PTEN mut (Figure 6a). These observations altogether suggested that PTEN expression is responsible for not only miR-17~92 but also other mechanisms regulating PTEN expression including other miRNAs or other mechanism involved in post-transcription.

4. Experimental Section

4.1. Cell Culture and Plasmids

The ovarian carcinoma cell line SKOV3 was provided by Tumor Cell Bank Research Institute of the Chinese Academy of Medical Sciences. The paclitaxel-resistant ovarian carcinoma cell line, SKOV3-TR30 with the resistant ability of 27-fold greater than its parental cell line, was derived from

SKOV3 cell line and provided by Zhejiang University affiliated Obstetrics and Gynecology Hospital. SKOV3 cells were maintained in McCoy's 5a medium containing fetal bovine serum with 10%, penicillin with 100 μg/mL and streptomycin with 100 μg/mL. SKOV3-TR30 cells were maintained in McCoy's 5a medium supplemented with 10% fetal bovine serum, penicillin with 100 μg/mL and streptomycin with 100 μg/mL and paclitaxel with 30 nmol/L, paclitaxel was withdrawn a week before the experiment. All cells were maintained in a humidified atmosphere containing 5% CO_2 at 37 °C. Cells in the logarithmic phase of growth were used for all studies described. HEK293 cells were maintained in a DMEM medium containing fetal bovine serum with 10% in a humidified atmosphere containing 5% CO_2 at 37 °C.

MiR-17~92 inhibitory plasmids miR-17~92-PTIP-Sponge all (m-PTIP-Sponge all), empty plasmids miR-17~92-PTIP-GFP (m-PTIP-GFP) and TMP2-miR-17~92 were provided by the Zoology Institute of Chinese Academy of Sciences.

Three plasmids named as B1, B2, B3 are cloned from the 3'-UTR of BIM. These vectors based on the pGL3-P vector were kind gifts from Professor Fu kai. B2 fragment contains miR-17-5p/-20a and miR-92 binding site and B3 fragment contains miR-19 and miR-92 binding site. There is not binding site of B1 with miR-17~92.

PTEN 3'-UTR and PTEN mut 3'-UTR were also gifts from Professor Fu kai. PTEN 3'-UTR contains putative binding sites for both miR20a/17-5p and miR-19.

4.2. MicroRNA Gene Chip

We analyze the expression profiles of miRNA on the sensitive and resistant ovarian carcinoma cell lines with using Affymetrix, the Gene Chip® analysis of miRNA chip.

4.3. Quantitative Real-Time PCR of miR-17~92 within SKOV3 and SKOV3-TR30 Cells

The expression level of mature miRNAs was determined using the TaqMan real-time quantitative PCR. Briefly, single-stranded cDNA was synthesized from 10ng of total RNA using the TaqMan MicroRNA Reverse Transcription Kit. Each cDNA generated was amplified by quantitative PCR using sequence-specific primers from the TaqMan MicroRNA Assays. PCR primers are as follows: (1) miR-17~92 gene: upstream: 5'-CAGTAAAGGTAAGGAGAGCTCAATCTG-3', downstream: 5'-CATACAACCACTAAGCTAAAGAATAATCTGA-3'; (2) internal control β-actin gene: upstream: 5'-GCAAAGACCTGTACGCCAACA-3', downstream: 5'-TGCATCCTGTCGGCAATG-3'. The relative quantity of the target miRNAs was estimated by the $2^{-\Delta\Delta CT}$ after normalizing to the expression level of β-actin, which was detected by a TaqMan gene expression Assay.

4.4. Establishment of Stable SKOV3-TR30 Cell Lines with Induced Expression of miR-17~92 Cluster

To establish miR-17~92-PTIP-Sponge all cell line, the HEK293T cell line was co-transfected with the miR-17~92-PTIP-Sponge all plasmids and pCL packaging plasmid or empty plasmids miR-17~92-PTIP-GFP (m-PTIP-GFP) and pCL packaging plasmid by the calcium phosphate method. The virus supernatant was collected and used to infect the SKOV3-TR30 cells. The stable cell lines which have decreased expression of miR-17~92 and cell lines transduced by empty plasmids miR-17~92-PTIP-GFP (m-PTIP-GFP) were maintained in the presence of 1 μg/mL doxycycline. Stable transfectants expressing green fluorescent protein were identified by Flow Cytometry and fluorescence microscopy.

4.5. Cell Proliferation Assays

Cells were plated at a density of 5×10^3 cells/well in a 96-well plate and incubated with indicated concentrations of paclitaxel (0 nM, 20 nM, 100 nM, 200 nM, quadrupled transfected wells were analyzed for each concentration) for 48 h. The cells were then incubated with MTT reagent and the absorbance at 490 nm was determined.

4.6. Cell Cycle Analysis

Cells in the logarithmic phase of growth were selected and paclitaxel were added to the cells with concentration of 0 nM, 20 nM, 100 nM, 200 nM separately. After 48 h, cells were collected and washed in cold phosphate-buffered saline (PBS). After centrifugation at 1000*g* for 5 min, cells were fixed in cold ethanol (70%) over night, washed with cold PBS and stained with 15 μL propidium iodide (50 μg/mL) in the presence of 15 μL of RNase A (10 mg/mL) .The cells were incubated for 30 min in the dark and analyzed using Flow Cytometery.

4.7. Western Bloting

Cells were collected and washed twice in cold phosphate-buffered saline (PBS). Then cells were solubilized with lysis buffer, on ice for 30 min, and then centrifuged at 12,000 rpm for 15 min at 4 °C. The supernatants were collected, and the protein concentrations were determined using a protein assay CBB kit with a BSA standard.

Samples were loaded for gel electrophoresis at 20 μg/sample, and after electrophoresis, they were blotted onto PVDF membranes. Membranes were blocked for 2 h with blocking buffer at room temperature and then were incubated overnight at 4 °C with rabbit primary antibodies to PTEN and BIM (both from Cell Signaling 1:1000), GAPDH (from Abcam) and then for 1 h at room temperature with HRP-conjugated anti-rabbit (1:2000) or antimouse IgG (1:2000) (both from Abcam). Membranes were then washed three times in TBS-Tween, and specific bands were visualized using the ECL system.

4.8. Luciferase Reporter Assays

HEK 293 cells were plated at 2×10^5 cells per well in a 24-well plate 24 h before transfection. The pGL-3-P promoter plasmids containing the wild type or mutated 3'-UTRs of PTEN and 3'-UTRBIM (B1,B2,B3)were co-transfected into HEK 293 cells with TMP-miR-17~92 according to the manufacturer's instructions using LIPOFECTAMINE 2000, pRL-SV40 (Promega, Madison, WI, USA) was also transfected as a normalization control. Luciferase Assays were performed 24 h after transfection using the Dual Luciferase Assay System. Firefly Luciferase activity was normalized to renilla luciferase activity for each reaction. Quintupled transfected wells were analyzed for each group.

4.9. Statistical Analysis

All values are in the expression of mean ± SD; Analysis was done using variance analysis, and comparisons among different groups were made using the Least Significant Difference (LSD) test. It is considered statistically significant when p value is less than 0.05.

5. Conclusions

In this study, we have demonstrated the significant contribution of miR-17~92/BIM to paclitaxel resistance in human ovarian carcinoma SKOV3-TR30 cells. However, we could not clearly clarify if there is a link from miR-17~92 and PTEN to paclitaxel resistance mechanisms. There are studies to show that decreased PTEN expression is partly responsible for ovarian carcinoma drug resistance [13]. In this present study, however, PTEN was not over-expressed in the resistant SKOV3-TR30 cells after transduced with miR-17~92-PTIP-Sponge all plasmid. Further studies to clarify the possible involvement of other miRNAs that might regulate the expression of PTEN in SKOV3-TR30 cells and the detailed molecular mechanism are needed in our future research.

Acknowledgments

This work is supported by the Natural Science Foundation of China (Grant Number: 30973189).

References

1. Jemal, A.; Siegel, R.; Xu, J.; Ward, E. Cancer statistics, 2010. *CA Cancer J. Clin.* **2010**, *60*, 277–300.
2. Filipowicz, W.; Bhattacharyya, S.N.; Sonenberg, N. Mechanisms of posttranscriptional regulation by microRNAs: Are the answers in sight? *Nat. Rev. Genet.* **2008**, *9*, 102–114.
3. Caldas, C.; Brenton, J.D. Sizing up miRNAs as cancer genes. *Nat. Med.* **2005**, *11*, 712–714.
4. Calin, G.A.; Croce, C.M. MicroRNA signatures in human cancers. *Nat. Rev. Cancer* **2006**, *6*, 857–866.
5. Esquela-Kerscher, A.; Slack, F.J. Oncomirs-microRNAs with a role in cancer. *Nat. Rev. Cancer* **2006**, *6*, 259–269.
6. Cho, W.C. MicroRNAs: Potential biomarkers for cancer diagnosis prognosis and targets for therapy. *Int. J. Biochem. Cell Biol.* **2010**, *42*, 1273–1281.
7. William, C.S.; Cho, W.C. MicroRNAs in cancer—from research to therapy. *Biochem. Biophys. Acta* **2010**, *1805*, 209–217.
8. Kuwana, T.; Bouchier-Hayes, L.; Chipuk, J.E.; Bonzon, C.; Sullivan, B.A. Green, D.R.; Newmeyer, D.D. BH3 domains of BH3-only proteins differentially regulate Bax-mediated mitochondrial membrane permeabilization both directly and indirectly. *Mol. Cell* **2005**, *17*, 525–535.
9. Letai, A.; Bassik, M.C.; Walensky, L.D.; Sorcinelli, M.D.; Weiler, S.; Korsmeyer, S.J. Distinct BH3 domains either sensitize or activate mitochondrial apoptosis, serving as prototype cancer therapeutics. *Cancer Cell* **2002**, *2*, 183–192.
10. Ota, A.; Tagawa, H.; Karnan, S.; Tsuzuki, S.; Karpas, A.; Kira, S.; Yoshida, Y.; Seto, M. Identification and characterization of a novel gene, C13orf25, as a target for 13q31-q32 amplification in malignant lymphoma. *Cancer Res.* **2004**, *64*, 3087–3095.

11. He, L.; Thomson, J.M.; Hemann, M.T.; Hernando-Monge, E.; Mu, D.; Goodson, S.; Powers, S.; Cordon-Cardo, C.; Lowe, S.W.; Hannon, G.J.; Hammond, S.M. A microRNA polycistron as a potential human oncogene. *Nature* **2005**, *435*, 828–833.
12. Lewis, B.P.; Shih, I.H.; Jones-Rhoades, M.W.; Bartel, D.P.; Burge, C.B. Prediction of mammalian microRNA targets. *Cell* **2003**, *115*, 787–798.
13. Willis, S.N.; Fletcher, J.I.; Kaufmann, T.; van Delft, M.F.; Chen, L.; Czabotar, P.E.; Ierino, H.; Lee, E.F.; Fairlie, W.D.; Bouillet, P.; *et al.* Apoptosis initiated when BH3 ligands engage multiple Bcl-2 homologs, not Bax or Bak. *Science* **2007**, *315*, 856–859.
14. Tanaka, M.; Grossman, H.B. *In vivo* gene therapy of human bladder cancer with PTEN suppresses tumor growth; down regulates phosphorylated Akt, and increases sensitivity to doxorubicin. *Gene Ther.* **2003**, *10*, 1636–1642.
15. Wu, H.; Cao, Y.; Weng, D.; Xing, H.; Song, X.; Zhou, J.; Xu, G.; Lu, Y.; Wang, S.; Ma, D. Effect of tumor suppressor gene PTEN on the resistance to cisplatin in human ovarian carcinoma cell lines and related mechanisms. *Cancer Lett.* **2008**, *271*, 260–271.
16. Gao, X.; Zhang, R.; Qu, X. MiR-15a, miR-16-1 and miR-17~92 cluster expression are linked to poor prognosis in multiple myeloma. *Leuk. Res.* **2012**, *36*, 1505–1509.
17. Ouchida, M.; Kanzaki, H. Novel direct targets of miR-19a identified in breast cancer cells by a quantitative proteomic approach. *PLoS One* **2012**, *7*, e44095.
18. Baumhoer, D.; Zillmer, S.; Unger, K.; Rosemann, M.; Atkinson, M.J.; Irmler, M.; Beckers, J.; Siggelkow, H.; von Luettichau, I.; Jundt, G.; *et al.* MicroRNA profiling with correlation to gene expression revealed the oncogenic miR-17~92 cluster to be up-regulated in osteosarcoma. *Cancer Genet.* **2012**, *205*, 212–219.
19. Ohyashiki, K.; Umezu, T.; Yoshizawa, S.; Ito, Y.; Ohyashiki, M.; Kawashima, H.; Tanaka, M.; Kuroda, M.; Ohyashiki, J.H. Clinical impact of down-regulated plasma miR-92a levels in non-hodgkin's lymphoma. *PLoS One* **2011**, *6*, e16408.
20. Grimson, A.; Farh, K.K.; Johnston, W.K.; Garrett-Engele, P.; Lim, L.P.; Bartel, D.P. MicroRNA targeting specificity in mammals: Determinants beyond seed pairing. *Mol. Cell.* **2007**, *27*, 91–105.
21. Xiao, C.; Srinivasan, L.; Calado, D.P; Patterson, H.C.; Zhang, B.; Wang, J.; Henderson, J.M.; Kutok, J.L.; Rajewsky, K. Lymphoproliferative disease and autoimmunity in mice with increased miR-17~92 expression in lymphocytes. *Nat. Immunol.* **2008**, *9*, 405–414.
22. Wang, J.; Zhou, J.Y.; Wu, G.S. Bim protein degradation contributes to cisplatin resistance. *J. Biol. Chem.* **2011**, *286*, 22384–22392.

8

Mechanisms of Lin28-Mediated miRNA and mRNA Regulation—A Structural and Functional Perspective

Florian Mayr [1,2] **and Udo Heinemann** [1,2,*]

[1] Crystallography, Max-Delbrück Center for Molecular Medicine, Robert-Rössle Straße 10, Berlin 13125, Germany; E-Mail: florian.mayr@mdc-berlin.de

[2] Institute for Chemistry and Biochemistry, Freie Universität Berlin, Takustraße 6, Berlin 14195, Germany

* Author to whom correspondence should be addressed; E-Mail: heinemann@mdc-berlin.de;

Abstract: Lin28 is an essential RNA-binding protein that is ubiquitously expressed in embryonic stem cells. Its physiological function has been linked to the regulation of differentiation, development, and oncogenesis as well as glucose metabolism. Lin28 mediates these pleiotropic functions by inhibiting *let-7* miRNA biogenesis and by modulating the translation of target mRNAs. Both activities strongly depend on Lin28's RNA-binding domains (RBDs), an N-terminal cold-shock domain (CSD) and a C-terminal Zn-knuckle domain (ZKD). Recent biochemical and structural studies revealed the mechanisms of how Lin28 controls *let-7* biogenesis. Lin28 binds to the terminal loop of pri- and pre-*let-7* miRNA and represses their processing by Drosha and Dicer. Several biochemical and structural studies showed that the specificity of this interaction is mainly mediated by the ZKD with a conserved GGAGA or GGAGA-like motif. Further RNA crosslinking and immunoprecipitation coupled to high-throughput sequencing (CLIP-seq) studies confirmed this binding motif and uncovered a large number of new mRNA binding sites. Here we review exciting recent progress in our understanding of how Lin28 binds structurally diverse RNAs and fulfills its pleiotropic functions.

Keywords: Lin28; *let-7* miRNA; miRNA processing; RNA-binding protein; cold-shock domain; zinc-knuckle domain; TUTase; oncogene; stem cell

---

1. Introduction

Lin28 (cell lineage abnormal 28) is a conserved RNA-binding protein in higher eukaryotes that regulates several important cellular functions associated with development, glucose metabolism, differentiation and pluripotency. It was first described as a heterochronic gene in *Caenorhabditis elegans (C. elegans)*, since mutations within *lin-28* disturbed the developmental timing of the worm and accelerated differentiation of hypodermal seam cells and vulva stem cells [1,2]. Subsequent experiments revealed that Lin28 is expressed early in nematode embryonic and larval development, but its expression is down-regulated by *lin-4* and *let-7* miRNA as differentiation proceeds [2,3].

A similar expression pattern and physiological function was also shown for *Drosophila*, *Xenopus* and mammalian Lin28 [4]. The human paralogs Lin28a (routinely termed simply Lin28) and Lin28b encode for basic 23- or 28-kDa proteins that are highly expressed in embryonic stem cells (ESC) but are down-regulated upon differentiation of ESCs into embryoid bodies [5]. Reciprocally, Yu and colleagues used Lin28a, in conjunction with Oct4, Sox2 and Nanog, to reprogram adult human fibroblasts to induced pluripotent stem cells (iPSCs) [6]. A knockdown of Lin28a expression in mouse ESCs led to loss of Oct4 and Nanog expression, indicating an impaired self-renewal potential [7]. Increased Lin28a/Lin28b expression was reported in various hepatocellular and other carcinomas and was associated with poor patient prognosis [8–13]. Recently, Lin28a was linked to the regulation of developmental and metabolic processes. After ectopic overexpression of Lin28a mice developed a bigger size and delayed sexual maturation, whereas Lin28 knockout mice were smaller and died shortly after birth [14]. In addition, Lin28a overexpression was associated with increased insulin sensitivity and glucose metabolism, while a depletion of Lin28a resulted in insulin resistance and glucose intolerance [15].

On the molecular level, Lin28a and Lin28b act as both negative regulator of *let-7* miRNA biogenesis and post-transcriptional regulator of mRNA translation. Both activities strongly depend on Lin28's two RNA-binding domains (RBDs): an N-terminal cold-shock domain (CSD) and a C-terminal Zn-knuckle domain (ZKD) composed of two tandemly arranged retroviral-type CCHC Zn knuckles. The individual domain combination of both RBDs is unique in animals with the RBDs being highly conserved. The human Lin28 paralogs share an overall sequence identity of 65% (FFAS, [16]) and contain low-complexity regions at the N-terminus, a putative bipartite nucleolar localization sequence (NoLS) as well as a C-terminal nuclear localization signal (NLS) in the case of Lin28b [17] (Figure 1). Lin28a and Lin28b can localize to both cytosol and nucleus [2,4,17–21] and interact with primary (pri-) or precursor (pre-) *let-7* miRNAs thereby preventing their maturation [20,22,23]. In addition, binding of Lin28a to messenger ribonucleoprotein complexes containing translation initiation (eIF3B, eI4E) and elongation factors (EF1α, EF1α2), poly(A) binding proteins, Igf2bps and RNA helicase A was reported in various studies [18,20,24,25]. Under stress conditions, Lin28a was shown to localize to cytoplasmic stress granules and P-bodies where mRNA translation is temporally stalled [18]. Since a mutation in Lin28a's ZKD caused Lin28a to accumulate in the nucleus, it was suggested that Lin28a exits the nucleus in a complex with bound RNA and thus regulates the post-transcriptional processing of its target RNAs [18].

Figure 1. Domain organization of Lin28. (**A**) *Caenorhabditis elegans* (*Cel*) and human (h) Lin28a/Lin28b contain two RNA-binding domains (RBDs): an N-terminal cold-shock domain (CSD) and a C-terminal Zn-knuckle domain (ZKD) comprised of two retroviral type CCHC Zn knuckles (ZnK). Additionally, Lin28 harbors low-complexity sequences, Lys/Arg (K/R)-rich stretches, bipartite nuclear localization signals (NLS) or putative nucleolar localization sequences (NoLS); (**B**) Sequence alignment of hLin28a and hLin28B. Amino acids belonging to CSD or ZKD are shaded in blue or green, respectively.

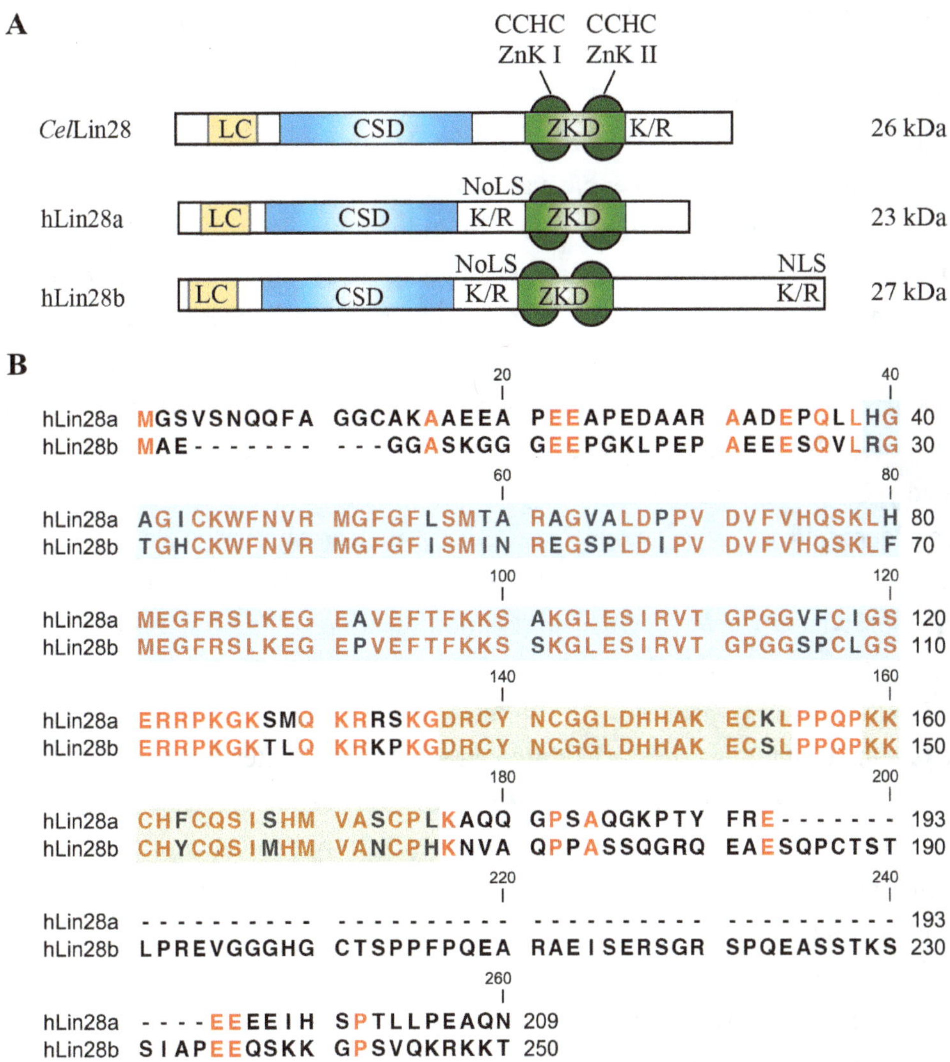

2. Lin28 Blocks *let-7* Processing

The opposing expression pattern of Lin28 and *let-7* miRNA became initially apparent when studying *C. elegans* larval development [2–4,26,27]. At an early stage in larval development both Lin28 and pri-*let-7* are present, however, no levels of either pre-*let-7* or mature *let-7* can be detected, indicating a regulation at a post-transcriptional level [28]. As larval development proceeds, a heterogenic cascade involving *lin-4* miRNA and the *let-7* sisters *mir-48/84/241* lead to a relief of pri-*let-7* processing inhibition and to a subsequent down-regulation of Lin28 expression (reviewed in [29]). This inverse relationship between Lin28 and *let-7* miRNA is also present in mammalian cells, where Lin28a/b are mainly expressed in undifferentiated cells, and mature *let-7* is only detectable upon differentiation or tissue development [5,20,23]. Furthermore, levels of pri-*let-7* remain constant throughout the entire differentiation or development process suggesting a negative regulation of *let-7* biogenesis by Lin28a/b in stem or progenitor cells [30–32]. Purification of pre-*let-7* bound complexes and subsequent analysis via mass spectrometry revealed that both human Lin28 paralogs specifically associate with pri- or pre-*let-7 in vivo* [20,22,23]. Moreover, *in vitro* purified Lin28a could inhibit pri- and pre-*let-7* processing by Drosha and Dicer by binding to the double-stranded stem close to the Dicer cleavage site and the pre-element (preE, terminal loop or hairpin) [33]. Mutations within Lin28's CSD and ZKD impaired pre-*let-7* binding and inhibition of Dicer processing, suggesting a competitive relationship between Lin28 and Dicer [33–35]. Moreover, recent studies provided evidence that Lin28a/b can induce a structural change within pre-*let-7*'s preE, thereby leading to an opening of the double-stranded stem including the Dicer cleavage site [34–36].

Heo and colleagues revealed an additional inhibition mode of *let-7* miRNA processing, which irreversibly targets pre-*let-7* to a decay pathway [7,37]. They demonstrated that Lin28a/b induce oligo-uridylation of pre-*let-7*'s 3' overhang. Oligo-uridylated pre-*let-7* is resistant to Dicer cleavage given that Dicer normally recognizes a 2-nt 3' overhang in miRNAs via its PAZ domain. Thus, Dicer is unable to recognize the elongated 3' overhang and to process pre-*let-7*. Furthermore, it was reported that oligo-uridylated RNAs recruit 3'–5' exonucleases and are targeted for decay [38,39]. Indeed, oligo-uridylated pre-*let-7* was more rapidly degraded than unmodified pre-*let-7* [37]. Recently, Chang and colleagues identified the 3'–5' exonuclease Dis3l2 that catalyzes the decay of oligo-uridylated pre-*let-7* in mouse ESCs [40]. Consistent with this, a knockdown of Dis3l2 in mouse ESCs caused an accumulation of uridylated pre-*let-7*. Oligo-uridylation of pre-*let-7* is catalyzed by the non-canonical poly(A) polymerase TUT4 (terminal uridyl transferase 4/Zcchc11) and to a minor extent by TUT7 (Zcchc6) in a Lin28-dependent manner [7,41,42]. Interestingly, these enzymes catalyze mono-uridylation of pre-miRNAs with a 1-nt 3' overhang (like most pre-*let-7* family members) in the absence of Lin28, thereby enhancing Dicer-mediated processing [43]. However, in the presence of Lin28, pre-*let-7* and other miRNAs containing a GGAG motif within their preE were subjected to oligo-uridylation. Upon mutation of this motif, both Lin28 binding and oligo-uridylation were impaired, indicating that the GGAG motif is essential for these processes [7].

In *C. elegans* a similar mechanism for inhibiting pre-*let-7* processing has been reported [44]. The poly(U) polymerase PUP-2 was shown to oligo-uridylate pre-*let-7* in a Lin28-dependent fashion, thereby suppressing premature expression of mature *let-7* during larval development. In addition, subsequent RNA and chromatin immunoprecipitation assays revealed a specific interaction between Lin28 and pri-*let-7* that co-transcriptionally inhibits pri-*let-7* processing by Drosha [28]. An interaction between Lin28 and endogenous pri-*let-7* was also described for human ESCs and neuronal stem/progenitor cells [28]. Here, a highly expressed RNA-binding protein called Musashi1 (Msi1) selectively recruits Lin28a to the nucleus and synergistically blocks the cropping step of pri-*let-7* [45]. Moreover, it was suggested that Lin28b predominantly localizes to the nucleolus where it sequesters pri-*let-7*, thereby preventing Drosha processing in the nucleus [17]. Thus, Lin28a/b seem to obviate precocious expression of mature *let-7* during early development and differentiation by interfering with both the Drosha and Dicer complexes and by targeting pre-*let-7* towards degradation. Conversely, upon differentiation of stem or progenitor cells, *let-7* ensures constant down-regulation of Lin28 by binding to the 3' UTR of Lin28 and its promoting transcription factor c-Myc [20] (Figure 2).

Figure 2. Lin28/*let-7* regulatory axis. In undifferentiated cells, Lin28 is highly expressed and blocks the biogenesis of *let-7* miRNA. By binding to the pre-element of pri- or pre-*let-7*, neither Drosha nor Dicer can process the corresponding *let-7* precursor. In addition, Lin28 recruits TUT4/TUT7 to pre-*let-7* and promotes its 3'-end oligo-uridylation. Oligo-uridylated pre-*let-7* cannot be cleaved by Dicer and thus serves as a signal for the cellular 3'–5' exoribonuclease Dis3l2. Upon differentiation, Lin28 expression is reduced, which leads to increased levels of mature *let-7*. The latter silences gene expression of proto-oncogenes (Ras, c-Myc, Hmga2), cell cycle progression factors (Cyclin D1 and D3, Cdk4), components of the insulin-PI3K-mTOR pathway and Lin28 itself, thereby establishing a positive feedback loop. Besides its role in differentiation, a Lin28/*let-7* regulatory network is apparently involved in several cellular processes such as proliferation, oncogenesis, development and physiology, as well as metabolism (recently reviewed in [42]).

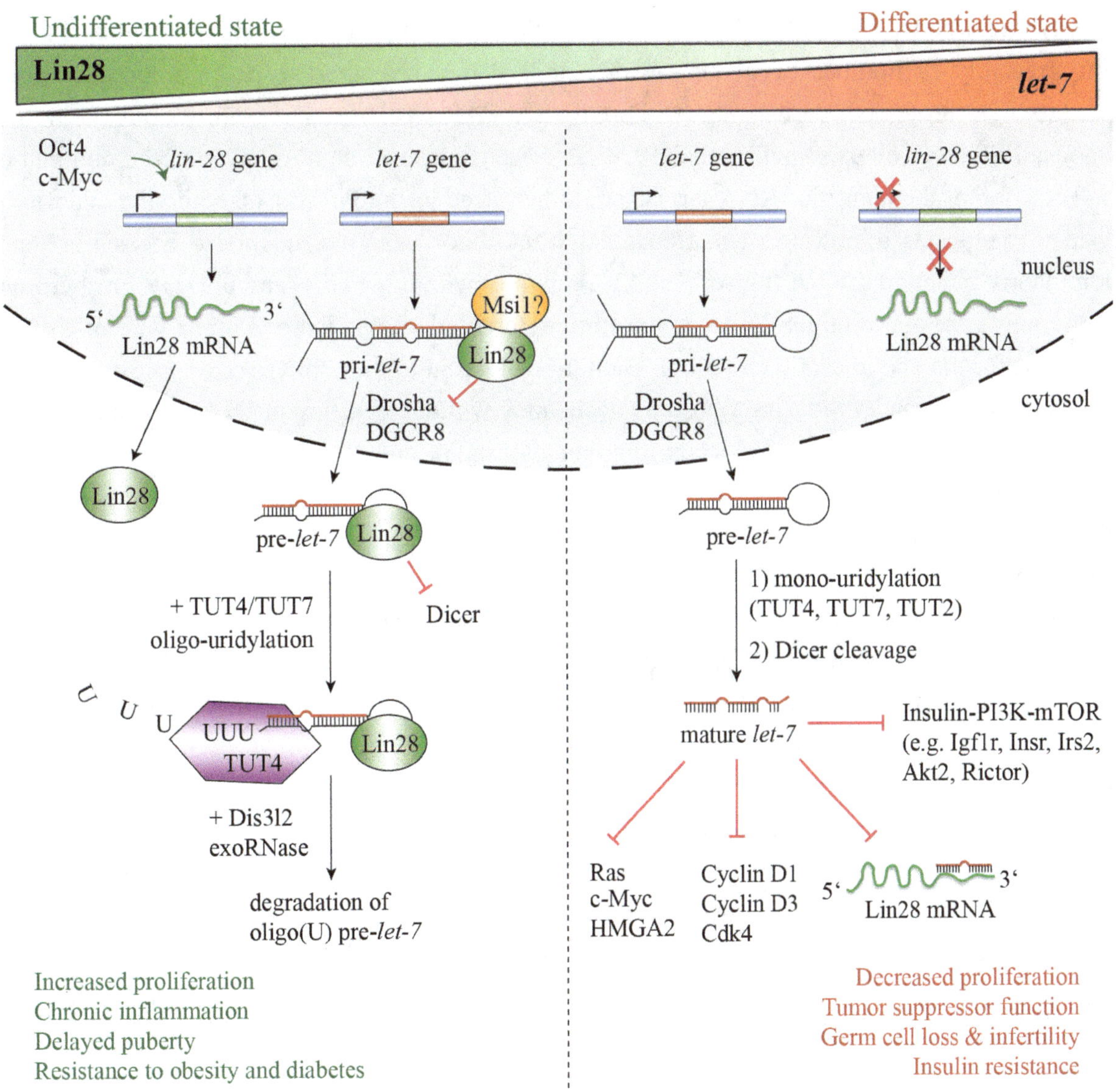

3. Lin28 Influences mRNA Translation

Besides regulation of *let-7* biogenesis, Lin28a/b can interact with various mRNAs and modulate their translation. Polesskaya and colleagues revealed that Lin28a can associate with polysomes and enhance translation of a number of mRNAs in differentiating myoblasts [25]. Among the first identified mRNA target was Igf2 (insulin-like growth factor 2), a major growth and differentiation factor in muscle tissue. Further evidence was provided that Lin28 recruits Igf2 mRNA to polysomes and enhances its translation via interactions with components of the translation initiation machinery. Subsequent studies revealed a number of additional mRNA targets of Lin28a in mouse ESCs such as H2a (histone 2a), Hmga1, Cyclin A, Cyclin B, Cdk4 and Oct4 [11,46–50]. An association of Lin28a with most of these mRNAs correlated with enhanced translation, suggesting that Lin28a maintains pluripotency by stimulating the translation of corresponding cell-cycle effectors. Further genome-wide studies revealed that Lin28a facilitates translation of genes important for growth and survival in human ESCs by recruiting RNA helicase A (RHA) to polysomes [24,51,52]. Additional mutagenesis studies revealed that the C-terminal part of Lin28a is required for RHA interactions, while mutations in the ZKD only impaired the stimulatory impact on translation, but not protein-protein interactions [46].

Very recently, a number of genome-wide Lin28 RNA crosslinking and immunoprecipitation coupled to high-throughput sequencing (HITS-Clip and PAR-CLIP) studies were conducted in human and mouse ESCs as well as somatic cells [19,49,53,54]. All of these studies have in common that only a small fraction of the identified RNA targets could be traced back to miRNAs, while the majority was mapped to thousands of mRNAs and ribosomal RNAs. For example, in mouse ESCs Lin28a was predominantly bound to mRNA transcripts (42%), mainly within the CDS and 3' UTR. Furthermore, a gene ontology analysis of target RNAs, revealed a preferential interaction of Lin28a with mRNAs that are destined for the endoplasmic reticulum. Binding of Lin28a to these mRNAs was associated with a translational repression by reducing ribosome occupancy without affecting mRNA abundance [49].

On contrary, in human HEK293 cells, binding of Lin28a and Lin28b to its mRNA targets was linked to a globally enhanced protein synthesis [19,53]. As before in mESCs, both human Lin28 paralogs predominantly bound within exonic regions of mRNAs, thereby mirroring the predominant cytosolic localization of Lin28a/b in HEK293 cells. Among the top RNA targets were mRNAs encoding for splicing factors and RNA-binding proteins, cell-cycle regulators as well as Lin28 itself. Binding of Lin28b to its own mRNA, indeed, correlated with increased levels of Lin28b protein, thereby suggesting a *let-7* independent feed-forward mechanism to maintain high levels of Lin28b in proliferative cell types [19,53,54]. Apart from their own expression Lin28a/b also seem to drive expression of important cell-cycle regulators of the ERK signaling cascade, such as Cdk1, N-Ras, Ran and ERK. This would explain the strong proliferative defects observed upon Lin28b knockdown [53]. Wilbert and colleagues further detected widespread changes in protein levels of splicing factors upon down-regulation of Lin28a and Lin28b in human ESCs. Whereas Lin28a binding to hnRNP F mRNA repressed translation, binding to TDP-43 and FUS/TLS mRNA was associated with an enhanced protein synthesis of the corresponding transcript. Consistent with Lin28's impact on alternative splicing factors, up-regulation of Lin28a in somatic HEK293 cells caused dramatic changes in alternative splicing patterns [54] (Figure 3).

Figure 3. Lin28 binds various mRNAs and modulates their translation. Both Lin28 paralogs were shown to influence mRNA processing on several levels. In the nucleus, Lin28 could regulate splicing of bound pre-mRNAs in concert with heterogeneous nuclear ribonucleoproteins (hnRNPs). In the cytosol, Lin28 was shown to interact with an RNA helicase A (RHA) thereby modulating the translation of target mRNAs via interactions with eukaryotic translation initiation factors (eIFS), elongation factors (eEFS) and poly(A)-binding proteins (PABP). Furthermore, Lin28 was found to shuttle mRNAs to poly-ribosomes and, under stress condition, to P-bodies and stress granules, thereby providing a direct link to the miRNA decay machinery. Lin28 binding to mRNAs was typically associated with a globally enhanced protein synthesis. However, in hESCs Lin28 binding repressed translation of bound mRNAs that were destined for the ER.

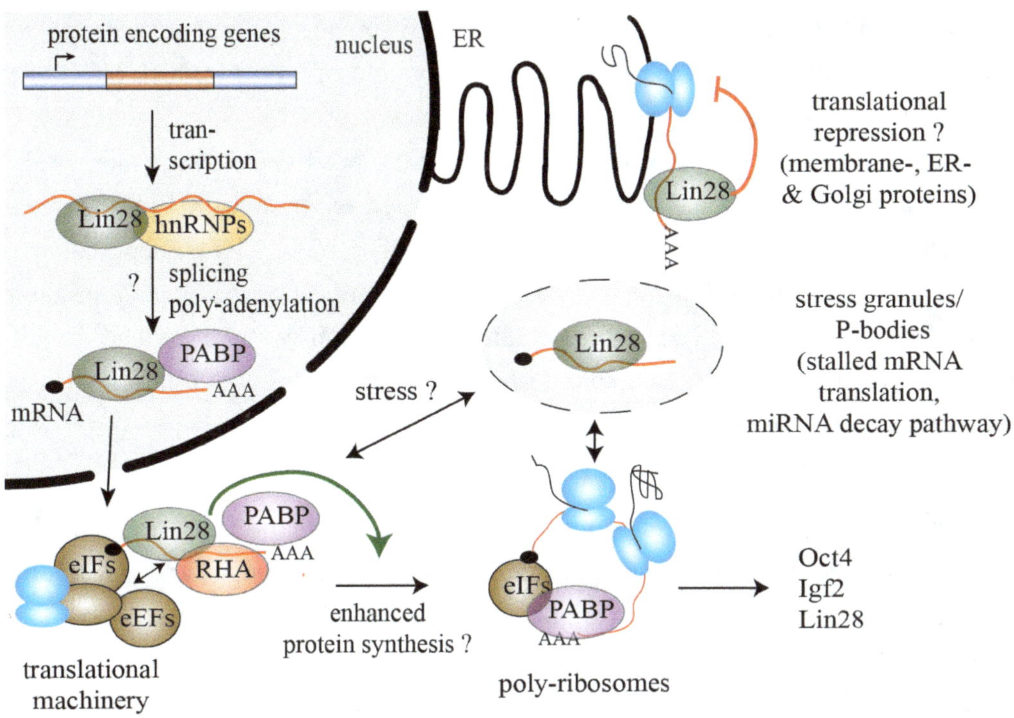

4. Functional Importance of Lin28-Mediated mRNA and miRNA Regulation for Stem Cell Maintenance, Cancer and Development

The functional importance of Lin28 in stem cell maintenance and reconstituting pluripotency becomes apparent when looking at the signaling pathways in which Lin28a/b are involved. Both paralogs are highly expressed in mammalian ESCs and are a central part of a conserved pluripotency network. For example, the expression of Lin28a is driven by the proto-oncogenic transcription factors Oct4, Sox2 and Nanog with Sox2 being most critical for an efficient Lin28a expression [55,56]. Once Lin28a is expressed, it antagonizes *let-7* and hence de-represses *let-7* targets such as c-Myc, Sal4, Igf2bps, Hmga2, various cyclins as well as Lin28 itself, thereby ensuring a constant expression of stemness factors and cell cycle regulator [57]. In addition, Lin28a directly or indirectly stimulates the translation of mRNAs encoding for cell-cycle regulators or growth-promoting factors such as Cyclin A/B, Oct4 and Igf2 [18,25,48,50]. Consequently, Lin28a/b up-regulate the expression of cell-cycle regulators and growth-promoting factors via *let-7* dependent and *let-7* independent mechanisms, thereby activating and maintaining signaling pathways that are important for self-renewal and proliferation. In agreement with this, Lin28a overexpression is not essential for reprogramming human fibroblast to iPSCs but strongly accelerates reprogramming by stimulating cell proliferation [58].

The strong effect of Lin28a/b on cell progression and proliferation [59] and the frequent re-activation of Lin28a/b in multiple cancers [12] supported the role of Lin28 as a potential oncogene. Indeed, Lin28a/b overexpressing in NIH/3T3 cells led to tumor formation in nude mice and was linked to depletion of mature *let-7*. As a consequence, oncogenic *let-7* targets such as c-Myc and N-Ras were de-repressed, and, since c-Myc itself transcriptionally activates various oncogenic miRNAs as well as Lin28b, a positive feed-forward loop is established [12,60]. Iliopoulus and colleagues revealed another positive feedback loop between NF-κB, Lin28b, *let-7* and IL-6 (Interleukin 6). Transient activation of Src tyrosine kinase in immortalized breast cells led to activation of NF-κB, which binds to the Lin28b promoter and induces its expression. As a result, Lin28b represses *let-7* processing, the *let-7* target IL-6 can be produced and activate NF-κB, thereby closing the positive feedback loop [61]. Similar to its role in reprogramming somatic cells to iPSCs, an elevated expression of Lin28a/b might also be important in the formation of cancer stem cells (CSCs) [62]. This subpopulation of tumor cells is thought to be essential for the propagation of some cancer cells and might arise in a reprogramming-like mechanism [63]. Hence, Lin28a/b reactivation would contribute to the formation of metastasis thereby explaining why Lin28a/b up-regulation correlates with tumor aggressiveness and an advanced tumor stage [12,62].

Given that *let-7* family members target numerous metabolic genes, it is not surprising that Lin28a/b overexpression also has an impact on growth, developmental timing and metabolism. Using genome-wide association studies, genetic variations within the LIN28B loci was linked to changes of human height, timing of puberty and the age of menopause [64–66]. Consistent with these studies, Lin28a overexpression in transgenic mice led to similar phenotypes and was associated with increased insulin sensitivity and increased glucose uptake [14]. On the molecular level, Lin28a/b act on multiple components of the insulin-P13K-mTOR pathway, thereby explaining why administration of the mTOR inhibitor rapamycin could rescue the Lin28a-mediated metabolic phenotype [15]. Further *in vitro* studies showed that Lin28a de-represses *let-7* targets of the insulin-P13K-mTOR pathway such as Igf1r, Insr, Irs2, Akt2, Tsc1 and Rictor [15,67]. The authors could not rule out that Lin28a/b associate with these mRNAs itself and enhance their translation. Recent genome-wide Clip-seq studies indeed suggested that Lin28a/b binds to mRNAs of insulin and Igf receptors, glycolytic and mitochondrial enzymes thereby modulating their translation directly [19,52,53]. Hence, Lin28a/b seem to regulate both mRNA translation and *let-7* maturation to coordinate proliferative signaling pathways and cellular metabolism in order to maintain the self-renewal potential of stem or progenitor cells. However, given the wealth of recently identified mRNA targets of Lin28a/b, their overlap with known *let-7* targets and the interwoven signaling pathways, it remains to be determined which of the identified targets indeed contribute to the observed physiological functions.

5. Structural Basis for the RNA-Binding Specificity of Lin28

5.1. The Lin28 Zinc-Knuckle Domain Specifically Recognizes GGAG or GGAG-Like Motifs

After identifying *let-7* precursors as major targets of Lin28a and Lin28b, several groups aimed to identify the specificity of this interaction. Using electrophoretic mobility shift assays with different pre-*let-7* sequences, it became initially apparent that the terminal loop of pre-*let-7* (also called pre-element or preE) is sufficient for Lin28a binding [33]. An alignment of stem-loop precursors of *let-7* revealed a highly conserved GGAG motif within vertebrates that is critical for Lin28 binding. Mutations within this motif (GGAG→AAAG and GGAG→GUAU) released the Lin28a-mediated block of pri- or pre-*let-7* processing and impaired TUT4-mediated oligo-uridylation of pre-*let-7* [7,22]. On contrary, introduction of the GGAG motif into preE of an unrelated miRNA (pre-*miR-16-1*) allowed Lin28a binding and TUT4-mediated uridylation of this chimeric pre-miRNA [7].

Due to the close homology between Lin28's ZKD and the ZKD of HIV-1 nucleocapsid protein (HIV NC), which was known to bind GGAG- or GGUG-containing loops within the HIV Ψ-RNA recognition element [68–70], it was suggested that Lin28's ZKD mediates a specific interaction with the conserved GGAG motif (see Figure 5A). Indeed, mutations with Lin28's ZKD specifically impaired pre-*let-7* binding as well as binding of the isolated Lin28 ZKD to GGAG-containing RNAs [7,34,35,71,72]. Co-crystal structures of a minimal mouse Lin28a construct with GGAG-containing oligonucleotides derived from preE-*let-7* (Figures 4 and 5) and a NMR solution structure of human Lin28a's ZKD bound to AGGAGAU provided the final proof for the supposed interaction (Figure 5B,C) [35,72].

For Lin28:mRNA binding, so far no structural data has been obtained. However, despite their discrepancies in individual mRNA targets, most of the above mentioned genome-wide HITS-CLIP and PAR-CLIP studies identified GGAG or GGAG-like consensus motifs within Lin28a/b binding sites. For example, Wilbert *et al.* found a highly enriched GGAGA(U) consensus sequence that was enriched within loop structures [54]. Cho *et al.* detected AAGNNG, AAGNG and UGUG motifs that are often located in terminal loops of small RNA hairpins [49]. Finally, Graf *et al.* detected GGSWG (S = G or C, W = A or T) or AAGRWG (R = A or G) motifs in Lin28b binding sites. Using individual domain PAR-CLIP (iDo-PAR-CLIP) Graf and colleagues further confirmed the GGGAG sequence as the top motif within Lin28 ZKD binding sites, whereas Lin28 CSD binding sites were rather U-rich [53]. These data indicate that the GGAG motif is indeed the major determinant of Lin28 RNA binding and is recognized by the ZKD. Even a mutation of the first or second guanosine only moderately impairs the interaction, thereby mirroring the overall flexibility of both ZKD and RNA. A recent study revealed that CCHC Zn knuckles can be used to design single-stranded nucleic-acid binding proteins that specifically recognize a number of guanosines [73]. This study further demonstrated that the length of the inter-knuckle linker affects spacing between specifically bound guanosines. Hence, Lin28 ZKD probably prefers GNNG motifs over NGNG motifs as seen for HIV-1 NC ZKD (see Figure 5C). Interestingly, TUT4 and TUT7 also contain CCHC Zn knuckles that are critical for pre-*let-7* oligo-uridylation [42,74]. Compared to Lin28, the distance between these knuckles is larger (37 aa), indicating that they act independently from each other.

Figure 4. Co-crystal structure of a minimal mouse Lin28a construct with preE-*let*-7d derived RNA (PDB ID 3TRZ). The ZKD specifically binds to the conserved GGAG motif, whereas Lin28 CSD establishes extensive interactions with the less conserved terminal hairpin loop.

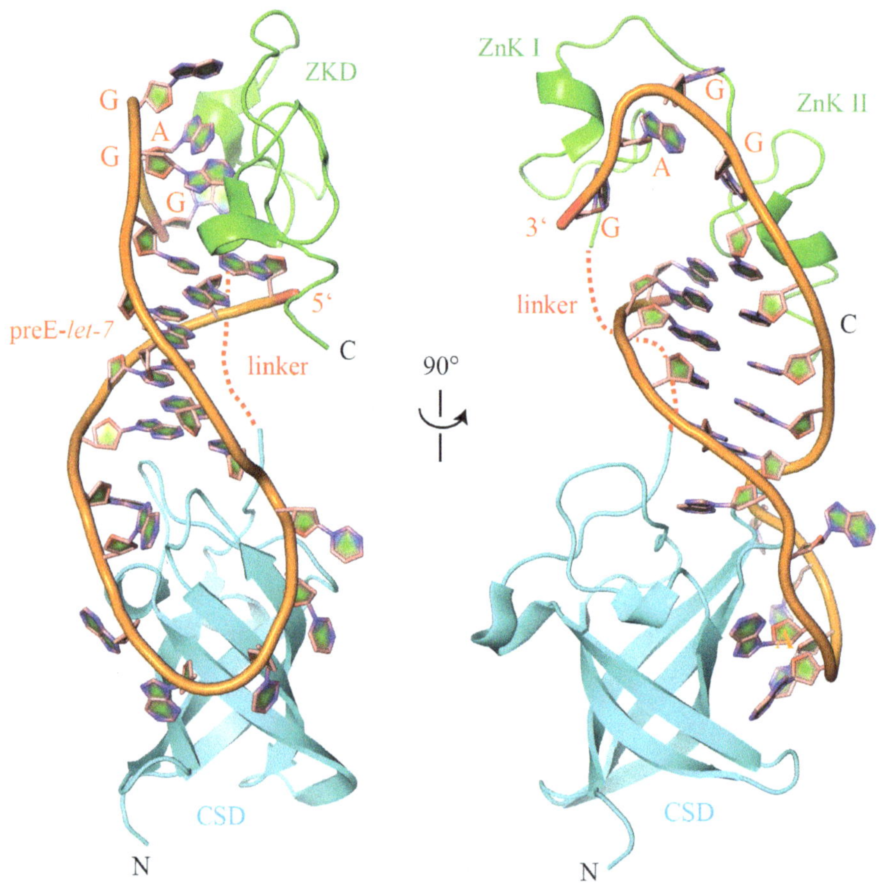

The second CCHC Zn knuckle undergoes a larger structural change upon RNA binding, whereby the central Zn^{2+} ion moves about 25 Å. Responsible for this large conformational shift is Pro158 within the Pro-rich linker region, since its ψ torsion angle performs a 130° rotation (Figure 5B) [72]. While HIV-1 NC ZKD specifically binds G-2 and G-4 of a GGAG tetraloop in a sequence-specific manner, each CCHC Zn knuckle of Lin28 specifically recognizes the first and fourth guanosine of the GGAG motif by sequence-specific hydrogen bonds to the bases. Hydrogen bonding is mediated by backbone carbonyl and amide groups of residues that are located within the rigid parts of the CCHC Zn knuckle. In addition to this sequence-specific interaction, both G-1 and G-4 are sandwiched in a hydrophobic pocket by one conserved Tyr and His in the first Zn knuckle and another conserved His and Met in the second Zn knuckle (Figure 5C). In the case of mLin28a:GGAG structures, G-2 is also bound in a sequence-specific manner via hydrogen bonds from backbone carbonyl groups and the N1 amino group of A-3. Even more, A-3 contributes to the formation of a strong kink within the RNA backbone, since it also contacts G-1 [35] (Figure 5D). Although such a strong bending of the RNA backbone was not observed in the hLin28a:AGGAGAU structure, the imposed structural changes within RNA and protein likely lead to a constant opening of neighboring double-stranded pre-*let*-7 stem thereby masking the Dicer cleavage site [34,35].

Figure 5. Lin28 ZKD specifically recognizes single-stranded GGAG or GGAG-like sequences. (**A**) Sequence alignment of HIV-1 NC, HIV-2 NC, hLin28a and hLin28b ZKDs. The chelating Cys and His residues of the CCHC Zn knuckles (ZnK) are shaded in red. Conserved residues are labeled from light red (100% type-conserved) to dark red (70% type-conserved); (**B**) Comparison between unbound hLin28a ZKD (green, PDB-ID 2CQF) and AGGAGAU-bound hLin28a ZKD (purple, PDB-ID 2LI8). Upon RNA binding, hLin28a ZKD undergoes a dramatic conformational shift mainly caused by a rotation of the Pro158 ψ angle; (**C**) In comparison to HIV-1 NC, the inter-knuckle linker of hLin28a ZKD harbors an additional Pro. As a consequence, the knuckles are further apart, thereby explaining why HIV-1 NC ZKD specifically binds G-2 and G-4 while hLin28a ZKD binds G-1 and G-4 of the GGAG motif in a hydrophobic pocket; (**D**) Structure of mLin28a ZKD bound to GGAG (derived from PDB-ID 3TSO). mLin28a is represented in green cartoon and the bound GGAG motif in purple (G) and pink (A). Tyr140 of the first and His162 of the second ZnK are key residues for the interaction, since they contact each other and stack with the bases, thereby establishing a kinked conformation in the RNA. All three guanosines are specifically recognized via various hydrogen bonds with backbone amide and carbonyl groups. In addition, G-1 and G-4 are bound in a hydrophobic pocket formed by His140, His162, Tyr140 and Met170.

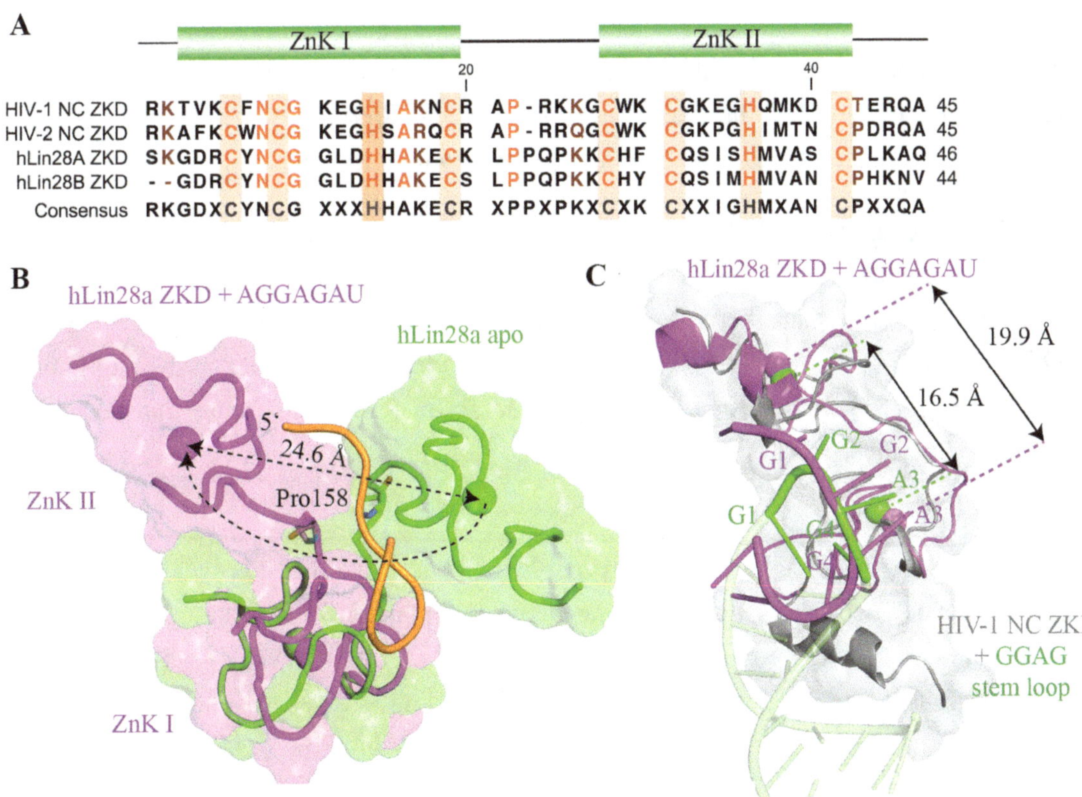

Figure 5. *Cont.*

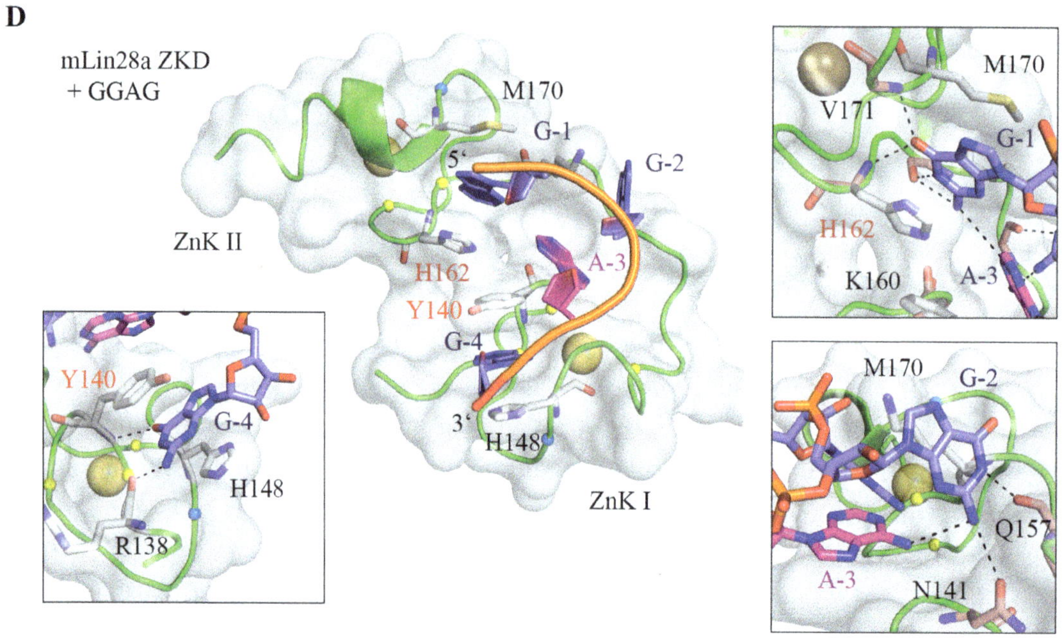

5.2. The Lin28 CSD Has Broad Sequence Specificity and Can Induce Local Structural Changes within RNAs

Despite Lin28's specificity for GGAG-containing RNAs, the isolated Lin28 ZKD is not sufficient for binding *let-7* precursors and blocking their processing [34,35]. So what is the contribution of Lin28's CSD with respect to sequence specificity, binding affinity and inhibition of pre-*let-7* processing?

CSDs are highly conserved RBDs that are widely distributed in bacteria, animals and plants and fulfill pleiotropic functions mainly related to RNA metabolism (reviewed in [75]). Bacterial major cold-shock proteins (Csps) share between 30% and 45% sequence identity to Lin28 CSDs and are known to bind pyrimidine-rich ssDNA/ssRNA oligonucleotides with affinities in the sub-nanomolar to micromolar range [76–81]. In addition to this, they can act as RNA chaperones that destabilize local RNA secondary structures [82–84]. Crystal and NMR structures of Csps have been known since the 1990s [85–89].

A systematic binding analysis with *Xenopus tropicalis* (*Xtr*) Lin28b CSD revealed that this domain has a broad sequence specificity and shows the highest binding affinities for pyrimidine-rich RNA octamers that contain at least one guanosine at the 5' end [34]. The observation was further confirmed by genome-wide PAR-CLIP studies, in which Lin28a/b binding sites were generally uridine-rich and flanked by one or more guanosines [19,53]. Moreover, these binding sites were typically located upstream of the corresponding ZKD binding sites, indicating a defined domain orientation of Lin28s' RBDs on RNA targets [53].

Co-crystal structures of Lin28 CSDs in complex with ssDNA and preE-*let-7* derived RNA stem loops provided valuable information about Lin28's specificity and function in pre-*let-7* and mRNA binding [34,35]. Lin28 CSDs bind to single-stranded nucleic acids via a conserved nucleic acid-binding platform mainly formed of exposed aromatic residues. Unlike for Lin28 ZKD, this binding platform is already pre-formed in the apo protein and, consequently, only subtle changes are observed upon nucleic-acid binding (Figure 6A). Binding of ssDNA and ssRNA are remarkably similar and dominated by π-stacking interactions with exposed aromatic residues (Figure 6B). Consistent with solution binding experiments and bacterial Csp:ssDNA/RNA structures [79,90], Lin28 CSD binds up to 8 nucleotides arranged in a curved single strand with defined orientation. In the case of the mLin28a:preE-*let-7* structures, an additional ninth nucleotide is visible that establishes hydrogen bonds with the first base, thereby closing the preE stem loop. Sequence-specific binding is mainly mediated at position 6, since the presence of a conserved Lys-Asp salt bridge limits the flexibility and, consequently, the size of the binding pocket and contributes to specific hydrogen bonds with the U/T base. In addition to this, at binding subsite 2 either a T, U or G is specifically recognized within a hydrophobic pocket. Despite the difference in size, the corresponding bases are recognized by similar hydrogen bonds. The lack of contacts with the CSD allows the DNA/RNA backbone to adopt slightly different conformations without disturbing hydrogen bonding.

Besides its contribution to binding affinity and specificity, Lin28 CSD can affect and reorganize secondary structures within RNA targets. The first hint in this direction came from a study that examined the effect of Lin28a binding on pre-*let-7g* secondary structure using enzymatic foot-printing [36]. Upon Lin28a binding, some regions of preE as well as a part of the double-stranded stem of pre-*let-7* became more susceptible to cleavage by single-strand-specific ribonucleases. Hence, the authors concluded that Lin28a is able to unwind the double-stranded stem of pre-*let-7*, thereby blocking the Dicer cleavage site. Second, Nam *et al.* provided evidence that Lin28a's CSD can partially melt double-stranded stem loops to generate an optimal binding interface [35]. Third, using site-directed mutagenesis in combination with a kinetic analysis of *Xtr*Lin28b-mediated remodeling of pre-*let-7g*, it was shown that Lin28's CSD first binds to pre-*let-7* and induces a structural change [34]. Consistent with earlier studies on bacterial Csps [83,91], highly conserved His (His68 in *Xtr*Lin28b-binding subsite 4) and Phe residues (Phe77-binding subsite 1,2) were crucial for the remodeling reaction. The CSD-induced remodeling might be important for proper recognition of the GGAG motif by Lin28's ZKD, since in most pre-*let-7* structures the conserved GGAG motif is involved in secondary structures and therefore not accessible for binding (Figure 7). Genome-wide PAR-CLIP and HITS-CLIP studies further supported this hypothesis, since Lin28a/b could recognize RNA binding sites that are predicted to be involved in stable secondary structures [19,49]. Such a chaperone-like function of Lin28 might be an important regulatory mechanism that allows downstream RBPs either to dissociate from or associate with RNPs and influence their processing. Most notably, a recent study provided evidence that Dis3l2 exoribonuclease degrades oligo-uridylated pre-*let-7*. This enzyme is composed of one ribonuclease II domain, two CSDs and one CSD-like S1 domain. Interestingly, both the CSDs and the S1 domain were essential for Dis3l2 binding and degradation of oligo-uridylated pre-*let-7*. Given the preference of Lin28 CSD's for U-rich binding sites, this suggests that the CSD might recognize the oligo(U) tail. In addition, it may assist in the exoribonucleolytic degradation of oligo(U)-pre-*let-7* by partially unwinding the double-stranded miRNA stem.

Figure 6. The Lin28 CSD can bind to a wide range of different RNA sequences. (**A**) Superimposition of unbound (skin color, PDB-ID 3ULJ) and heptathymidine-bound *Xtr*Lin28b CSD (green, PDB-ID 4A76). Both structures are highly conserved and reveal a pre-formed nucleic-acid binding platform with exposed aromatic residues; (**B**) Superimposition of *Xtr*Lin28b:dT_7 (green) and mLin28s:preE-*let-7f* (RNA: blue, protein: gray, PDB-ID 3TS0). Both Lin28 CSDs bind single-stranded nucleic acids predominantly via base stacking interactions in a defined orientation. The protein nucleic-acid interaction surface is similar for binding subsites 1 to 7. Binding of an additional eighth (U-8) and ninth (U-9) base in mLin28:preE-*let-7f* is triggered by the formation of a closed RNA loop; (**C**) Superimposition of bound nucleotides at binding subsite 6 derived from various bacterial and Lin28 CSDs in complex with ssDNA/ssRNA (PDB-IDs 4A76, 4A75, 3TS0, 3TS2, 3PF4, 2HAX). All structures contained T or U nucleotides at this binding pocket. A highly conserved Lys-Asp salt bridge limits the size of the pocket and establishes specific hydrogen bonds with the T/U base; (**D**) Since few interactions are formed with the sugar-phosphate backbone, the bound oligonucleotides can adopt different backbone conformations to optimize binding with Lin28 CSD. For example, at binding subsite 2, the sugar-phosphate backbone of mLin28a:preE-*let-7f* is farther displaced from the protein, thereby enabling binding of G (G-2) instead of T (T-2) without disrupting hydrogen bonds.

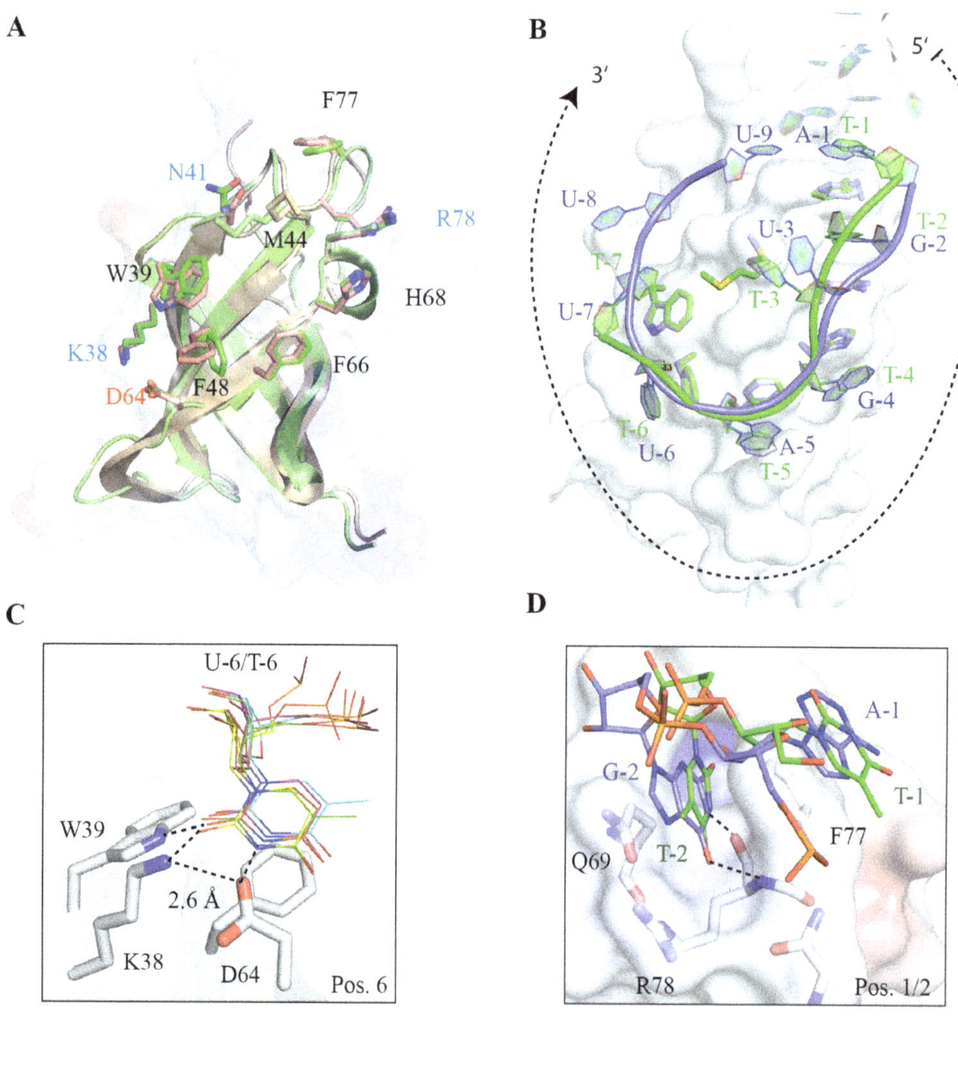

Figure 7. The pre-elements of *let-7* family members are structurally diverse. In six out of eleven human *let-7* family members, the conserved GGAG motif (blue) is inaccessible for ZKD binding in the lowest-energy folding state. Secondary structure predictions of human *let-7* family members (except miR-98 and miR-202) were calculated and visualized by CLC genomics workbench 3.65. All lowest-energy structures within a ΔΔG range of 1.5 kcal/mol are depicted. For simplicity, only 5 bp of the miRNA stem are shown (labeled in red).

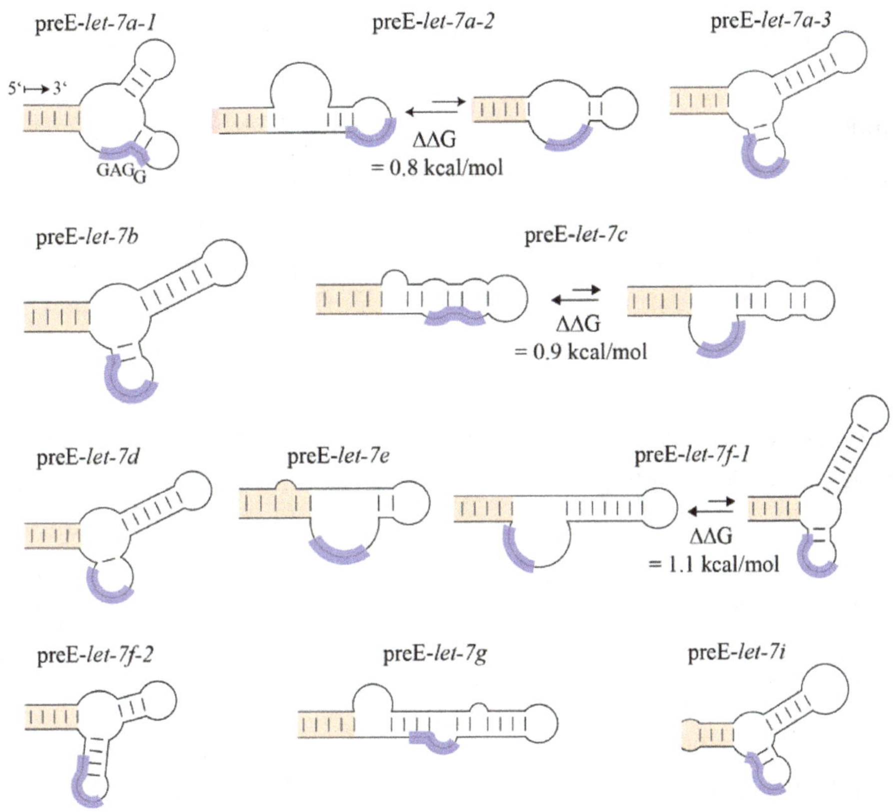

6. Summary and Conclusions

Recent structural and biochemical studies along with genome-wide CLIP-seq studies have greatly improved our understanding of how Lin28 recognizes target RNAs and fulfills its pleiotropic functions related to regulation of miRNA and mRNA processing as well as mRNA translation. In the case of *let-7* biogenesis, Lin28 ZKD specifically recognizes a conserved GGAG motif within preE-*let-7* and induces a strong bending of the bound RNA backbone. Thus, the adjacent Dicer cleavage site remains constantly unwound and pre-*let-7* cannot be processed anymore. Apart from a minor preference for pyrimidine-rich sequences with one flanking guanosine, the CSD did not reveal any clear sequence specificity, but was able to remodel local RNA secondary structures. This might be important in three ways. First, initial binding of the CSD to single-stranded RNA sequences can induce a conformation in which the GGAG motif is accessible for subsequent ZKD binding. Second, the CSD might trigger structural changes within target RNAs thereby stimulating downstream processes such as pre-*let-7* oligo-uridylation. Third, the wide RNA-binding specificity of the CSD enables Lin28 to recognize all *let-7* family precursors in a defined 5'–3' orientation despite the low sequence conservation within *let-7* preEs. Consequently, Lin28 impairs *let-7* biogenesis and irreversibly targets pre-*let-7* to degradation.

On the mRNA level the combination of both domains enables Lin28a/b to bind thousands of mRNAs. Although there is still no consensus how Lin28 mRNA binding influences mRNA processing and regulates translation, the observed principles with respect to sequence specificity and RNA remodeling are also valid here. Lin28a/b recognize GGAG or GGAG-like motifs and can access these motifs even if they are embedded in predicted secondary structures. The CSD typically binds to uridine-rich regions upstream of the ZKD binding site and is responsible for Lin28's RNA chaperone-like function. The defined orientation of Lin28 on bound RNA might allow a specific recruitment of downstream factors, such as RNA helicase A. However, downstream effects of Lin28:mRNA binding were quite distinct in recent genome-wide CLIP-seq studies and comprised translational stimulation of growth-promoting and alternative splicing factors, as wells as translational repression of ER-destined mRNAs [19,49,52–54]. In addition to the localization of binding sites (coding sequence, 3'UTR), Lin28-induced structural changes within mRNAs and/or direct protein-protein interactions might affect the translation efficiency of target mRNAs.

Therefore, it will be essential to understand how Lin28a/b regulate mRNA processing and translation in more detail. Ongoing studies will need to determine which cellular factors are involved in these processes and which factors regulate the activity of Lin28a/b within different cellular compartments and cells/tissues. Furthermore, it remains to be verified which of the huge number of novel mRNA targets are indeed directly regulated by Lin28a/b and what are their impact on stem cell maintenance/differentiation, development, metabolism and cancer. Last but not least, additional structural and functional data of how Lin28 binds mRNA targets and interacts with downstream components such as RNA helicase A may help to elucidate the mechanisms behind Lin28-mediated translational enhancement. Although much effort has been undertaken to unravel the molecular mechanisms that control the Lin28/*let-7* regulatory axis, there are still a number of issues that remain to be solved. Which precise mechanisms does Lin28 use to inhibit pri-*let-7* processing by the Microprocessor in the nucleus? How does Lin28 stimulate TUT4/TUT7-mediated oligo-uridylation of pre-*let-7*? And is the observed RNA-chaperone-like function of Lin28 mandatory for fulfilling its tasks? Understanding these issues might help us to exploit Lin28's function and manipulate the involved pathways for improved tissue re-engineering and novel treatments of cancer or metabolic diseases.

Acknowledgments

The authors express their gratitude to Yvette Roske and Anja Schütz (Max-Delbrück-Center, Berlin) for their help with some of the structural studies covered in this review and helpful discussions.

References

1. Ambros, V.; Horvitz, H. Heterochronic mutants of the nematode *Caenorhabditis elegans*. *Science* **1984**, *226*, 409–416.
2. Moss, E.G.; Lee, R.C.; Ambros, V. The cold shock domain protein LIN-28 controls developmental timing in *C. elegans* and is regulated by the lin-4 RNA. *Cell* **1997**, *88*, 637–646.
3. Seggerson, K.; Tang, L.; Moss, E.G. Two genetic circuits repress the *Caenorhabditis elegans* heterochronic gene lin-28 after translation initiation. *Dev. Biol.* **2002**, *243*, 215–225.
4. Moss, E.G.; Tang, L. Conservation of the heterochronic regulator Lin-28, its developmental expression and microRNA complementary sites. *Dev. Biol.* **2003**, *258*, 432–442.

5. Darr, H.; Benvenisty, N. Genetic analysis of the role of the reprogramming gene LIN-28 in human embryonic stem cells. *Stem Cells* **2009**, *27*, 352–362.
6. Yu, J.; Vodyanik, M.; Smuga-Otto, K.; Antosiewicz-Bourget, J.; Frane, J.; Tian, S.; Nie, J.; Jonsdottir, G.; Ruotti, V.; Stewart, R.; *et al.* Induced pluripotent stem cell lines derived from human somatic cells. *Science* **2007**, *318*, 1917–1920.
7. Heo, I.; Joo, C.; Kim, Y.-K.; Ha, M.; Yoon, M.-J.; Cho, J.; Yeom, K.-H.; Han, J.; Kim, V.N. TUT4 in concert with Lin28 suppresses microRNA biogenesis through pre-microRNA uridylation. *Cell* **2009**, *138*, 696–708.
8. Guo, Y.; Chen, Y.; Ito, H.; Watanabe, A.; Ge, X.; Kodama, T.; Aburatani, H. Identification and characterization of lin-28 homolog B (LIN28B) in human hepatocellular carcinoma. *Gene* **2006**, *384*, 51–61.
9. King, C.E.; Cuatrecasas, M.; Castells, A.; Sepulveda, A.R.; Lee, J.-S.; Rustgi, A.K. LIN28B promotes colon cancer progression and metastasis. *Cancer Res.* **2011**, *71*, 4260–4268.
10. King, C.E.; Wang, L.; Winograd, R.; Madison, B.B.; Mongroo, P.S.; Johnstone, C.N.; Rustgi, A.K. LIN28B fosters colon cancer migration, invasion and transformation through let-7-dependent and -independent mechanisms. *Oncogene* **2011**, *30*, 4185–4193.
11. Peng, S.; Maihle, N.J.; Huang, Y. Pluripotency factors Lin28 and Oct4 identify a sub-population of stem cell-like cells in ovarian cancer. *Oncogene* **2010**, *29*, 2153–2159.
12. Viswanathan, S.; Powers, J.; Einhorn, W.; Hoshida, Y.; Ng, T.; Toffanin, S.; O'Sullivan, M.; Lu, J.; Phillips, L.; Lockhart, V.; *et al.* Lin28 promotes transformation and is associated with advanced human malignancies. *Nat. Genet* **2009**, *41*, 843–848.
13. Wang, Y.-C.; Chen, Y.-L.; Yuan, R.-H.; Pan, H.-W.; Yang, W.-C.; Hsu, H.-C.; Jeng, Y.-M. Lin-28B expression promotes transformation and invasion in human hepatocellular carcinoma. *Carcinogenesis* **2010**, *31*, 1516–1522.
14. Zhu, H.; Shah, S.; Shyh-Chang, N.; Shinoda, G.; Einhorn, W.; Viswanathan, S.; Takeuchi, A.; Grasemann, C.; Rinn, J.; Lopez, M.; *et al.* Lin28a transgenic mice manifest size and puberty phenotypes identified in human genetic association studies. *Nat. Genet* **2010**, *42*, 626–630.
15. Zhu, H.; Shyh-Chang, N.; Segre, A.V.; Shinoda, G.; Shah, S.P.; Einhorn, W.S.; Takeuchi, A.; Engreitz, J.M.; Hagan, J.P.; Kharas, M.G.; *et al.* The Lin28/let-7 axis regulates glucose metabolism. *Cell* **2011**, *147*, 81–94.
16. Jaroszewski, L.; Li, Z.; Cai, X.H.; Weber, C.; Godzik, A. FFAS server: Novel features and applications. *Nucleic Acids Res.* **2011**, *39*, W38–W44.
17. Piskounova, E.; Polytarchou, C.; Thornton, J.E.; LaPierre, R.J.; Pothoulakis, C.; Hagan, J.P.; Iliopoulos, D.; Gregory, R.I. Lin28A and Lin28B inhibit let-7 microRNA biogenesis by distinct mechanisms. *Cell* **2011**, *147*, 1066–1079.
18. Balzer, E.; Moss, E.G. Localization of the developmental timing regulator Lin28 to mRNP complexes, P-bodies and stress granules. *RNA Biol.* **2007**, *4*, 16–25.
19. Hafner, M.; Max, K.E.; Bandaru, P.; Morozov, P.; Gerstberger, S.; Brown, M.; Molina, H.; Tuschl, T. Identification of mRNAs bound and regulated by human LIN28 proteins and molecular requirements for RNA recognition. *RNA* **2013**, *19*, 613–626.

20. Rybak, A.; Fuchs, H.; Smirnova, L.; Brandt, C.; Pohl, E.; Nitsch, R.; Wulczyn, F. A feedback loop comprising lin-28 and let-7 controls pre-let-7 maturation during neural stem-cell commitment. *Nat. Cell Biol.* **2008**, *10*, 987–993.
21. Gaytan, F.; Sangiao-Alvarellos, S.; Manfredi-Lozano, M.; Garcia-Galiano, D.; Ruiz-Pino, F.; Romero-Ruiz, A.; Leon, S.; Morales, C.; Cordido, F.; Pinilla, L.; Tena-Sempere, M. Distinct expression patterns predict differential roles of the miRNA-binding proteins, Lin28 and Lin28b, in the mouse testis: Studies during postnatal development and in a model of hypogonadotropic hypogonadism. *Endocrinology* **2013**, *154*, 1321–1336.
22. Newman, M.A.; Thomson, J.M.; Hammond, S.M. Lin-28 interaction with the Let-7 precursor loop mediates regulated microRNA processing. *RNA* **2008**, *14*, 1539–1549.
23. Viswanathan, S.R.; Daley, G.Q.; Gregory, R.I. Selective blockade of microRNA processing by Lin28. *Science* **2008**, *320*, 97–100.
24. Jin, J.; Jing, W.; Lei, X.-X.; Feng, C.; Peng, S.; Boris-Lawrie, K.; Huang, Y. Evidence that Lin28 stimulates translation by recruiting RNA helicase A to polysomes. *Nucleic Acids Res.* **2011**, *39*, 3724–3734.
25. Polesskaya, A.; Cuvellier, S.; Naguibneva, I.; Duquet, A.; Moss, E.; Harel-Bellan, A. Lin-28 binds IGF-2 mRNA and participates in skeletal myogenesis by increasing translation efficiency. *Genes Dev.* **2007**, *21*, 1125–1138.
26. Pasquinelli, A.E.; Reinhart, B.J.; Slack, F.; Martindale, M.Q.; Kuroda, M.I.; Maller, B.; Hayward, D.C.; Ball, E.E.; Degnan, B.; Müller, P.; *et al.* Conservation of the sequence and temporal expression of let-7 heterochronic regulatory RNA. *Nature* **2000**, *408*, 86–89.
27. Reinhart, B.; Slack, F.; Basson, M.; Pasquinelli, A.; Bettinger, J.; Rougvie, A.; Horvitz, H.; Ruvkun, G. The 21-nucleotide let-7 RNA regulates developmental timing in *Caenorhabditis elegans. Nature* **2000**, *403*, 901–906.
28. Van Wynsberghe, P.; Kai, Z.; Massirer, K.; Burton, V.; Yeo, G.; Pasquinelli, A. LIN-28 co-transcriptionally binds primary let-7 to regulate miRNA maturation in *Caenorhabditis elegans. Nat. Struct. Mol. Biol.* **2011**, *18*, 302–308.
29. Mondol, V.; Pasquinelli, A.E. Let's make it happen: The role of let-7 microRNA in development. *Curr. Top. Dev. Biol.* **2012**, *99*, 1–30.
30. Thomson, J.M.; Newman, M.; Parker, J.S.; Morin-Kensicki, E.M.; Wright, T.; Hammond, S.M. Extensive post-transcriptional regulation of microRNAs and its implications for cancer. *Genes Dev.* **2006**, *20*, 2202–2207.
31. Wulczyn, F.; Smirnova, L.; Rybak, A.; Brandt, C.; Kwidzinski, E.; Ninnemann, O.; Strehle, M.; Seiler, A.; Schumacher, S.; Nitsch, R. Post-transcriptional regulation of the let-7 microRNA during neural cell specification. *FASEB J.* **2007**, *21*, 415–426.
32. Yang, D.H.; Moss, E.G. Temporally regulated expression of Lin-28 in diverse tissues of the developing mouse. *Gene Expr. Patterns* **2003**, *3*, 719–726.
33. Piskounova, E.; Viswanathan, S.R.; Janas, M.; LaPierre, R.J.; Daley, G.Q.; Sliz, P.; Gregory, R.I. Determinants of microRNA processing inhibition by the developmentally regulated RNA-binding protein Lin28. *J. Biol. Chem.* **2008**, *283*, 21310–21314.
34. Mayr, F.; Schutz, A.; Doge, N.; Heinemann, U. The Lin28 cold-shock domain remodels pre-let-7 microRNA. *Nucleic Acids Res.* **2012**, *40*, 7492–7506.

35. Nam, Y.; Chen, C.; Gregory, R.I.; Chou, J.J.; Sliz, P. Molecular basis for interaction of let-7 microRNAs with Lin28. *Cell* **2011**, *147*, 1080–1091.
36. Lightfoot, H.L.; Bugaut, A.; Armisen, J.; Lehrbach, N.J.; Miska, E.A.; Balasubramanian, S. A LIN28-dependent structural change in pre-let-7g directly inhibits dicer processing. *Biochemistry* **2011**, *50*, 7514–7521.
37. Heo, I.; Joo, C.; Cho, J.; Ha, M.; Han, J.; Kim, V.N. Lin28 mediates the terminal uridylation of let-7 precursor MicroRNA. *Mol. Cell* **2008**, *32*, 276–284.
38. Mullen, T.E.; Marzluff, W.F. Degradation of histone mRNA requires oligouridylation followed by decapping and simultaneous degradation of the mRNA both 5' to 3' and 3' to 5'. *Genes Dev.* **2008**, *22*, 50–65.
39. Shen, B.; Goodman, H.M. Uridine addition after microRNA-directed cleavage. *Science* **2004**, *306*, 997.
40. Chang, H.M.; Triboulet, R.; Thornton, J.E.; Gregory, R.I. A role for the Perlman syndrome exonuclease Dis3l2 in the Lin28-let-7 pathway. *Nature* **2013**, *497*, 244–248.
41. Hagan, J.P.; Piskounova, E.; Gregory, R.I. Lin28 recruits the TUTase Zcchc11 to inhibit let-7 maturation in mouse embryonic stem cells. *Nat. Struct. Mol. Biol.* **2009**, *16*, 1021–1025.
42. Thornton, J.E.; Chang, H.M.; Piskounova, E.; Gregory, R.I. Lin28-mediated control of let-7 microRNA expression by alternative TUTases Zcchc11 (TUT4) and Zcchc6 (TUT7). *RNA* **2012**, *18*, 1875–1885.
43. Heo, I.; Ha, M.; Lim, J.; Yoon, M.J.; Park, J.E.; Kwon, S.C.; Chang, H.; Kim, V.N. Mono-uridylation of pre-microRNA as a key step in the biogenesis of group II let-7 microRNAs. *Cell* **2012**, *151*, 521–532.
44. Lehrbach, N.J.; Armisen, J.; Lightfoot, H.L.; Murfitt, K.J.; Bugaut, A.; Balasubramanian, S.; Miska, E.A. LIN-28 and the poly(U) polymerase PUP-2 regulate let-7 microRNA processing in Caenorhabditis elegans. *Nat. Struct. Mol. Biol.* **2009**, *16*, 1016–1020.
45. Kawahara, H.; Okada, Y.; Imai, T.; Iwanami, A.; Mischel, P.S.; Okano, H. Musashi1 cooperates in abnormal cell lineage protein 28 (Lin28)-mediated let-7 family microRNA biogenesis in early neural differentiation. *J. Biol. Chem.* **2011**, *286*, 16121–16130.
46. Lei, X.X.; Xu, J.; Ma, W.; Qiao, C.; Newman, M.A.; Hammond, S.M.; Huang, Y. Determinants of mRNA recognition and translation regulation by Lin28. *Nucleic Acids Res.* **2012**, *40*, 3574–3584.
47. Xu, B.; Huang, Y. Histone H2a mRNA interacts with Lin28 and contains a Lin28-dependent posttranscriptional regulatory element. *Nucleic Acids Res.* **2009**, *37*, 4256–4263.
48. Xu, B.; Zhang, K.; Huang, Y. Lin28 modulates cell growth and associates with a subset of cell cycle regulator mRNAs in mouse embryonic stem cells. *RNA* **2009**, *15*, 357–361.
49. Cho, J.; Chang, H.; Kwon, S.C.; Kim, B.; Kim, Y.; Choe, J.; Ha, M.; Kim, Y.K.; Kim, V.N. LIN28A is a suppressor of ER-associated translation in embryonic stem cells. *Cell* **2012**, *151*, 765–777.
50. Qiu, C.; Ma, Y.; Wang, J.; Peng, S.; Huang, Y. Lin28-mediated post-transcriptional regulation of Oct4 expression in human embryonic stem cells. *Nucleic Acids Res.* **2010**, *38*, 1240–1248.
51. Winter, J.; Jung, S.; Keller, S.; Gregory, R.I.; Diederichs, S. Many roads to maturity: microRNA biogenesis pathways and their regulation. *Nat. Cell Biol.* **2009**, *11*, 228–234.

52. Peng, S.; Chen, L.-L.; Lei, X.-X.; Yang, L.; Lin, H.; Carmichael, G.G.; Huang, Y. Genome-wide studies reveal that Lin28 enhances the translation of genes important for growth and survival of human embryonic stem cells. *Stem Cells* **2011**, *29*, 496–504.
53. Graf, R.; Munschauer, M.; Mastrobuoni, G.; Mayr, F.; Heinemann, U.; Kempa, F.; Rajewski, N.; Landthaler, M. Identification of LIN28-bound mRNAs reveals features of target recognition and regulation. *RNA Biol.* **2013**, *10*, 1146–1159.
54. Wilbert, M.L.; Huelga, S.C.; Kapeli, K.; Stark, T.J.; Liang, T.Y.; Chen, S.X.; Yan, B.Y.; Nathanson, J.L.; Hutt, K.R.; Lovci, M.T.; *et al.* LIN28 binds messenger RNAs at GGAGA motifs and regulates splicing factor abundance. *Mol. Cell* **2012**, *48*, 195–206.
55. Cox, J.; Mallanna, S.; Luo, X.; Rizzino, A. Sox2 uses multiple domains to associate with proteins present in Sox2-protein complexes. *PLoS One* **2010**, *5*, e15486.
56. Marson, A.; Levine, S.S.; Cole, M.F.; Frampton, G.M.; Brambrink, T.; Johnstone, S.; Guenther, M.G.; Johnston, W.K.; Wernig, M.; Newman, J.; *et al.* Connecting microRNA genes to the core transcriptional regulatory circuitry of embryonic stem cells. *Cell* **2008**, *134*, 521–533.
57. Melton, C.; Judson, R.L.; Blelloch, R. Opposing microRNA families regulate self-renewal in mouse embryonic stem cells. *Nature* **2010**, *463*, 621–626.
58. Hanna, J.; Saha, K.; Pando, B.; van Zon, J.; Lengner, C.J.; Creyghton, M.P.; van Oudenaarden, A.; Jaenisch, R. Direct cell reprogramming is a stochastic process amenable to acceleration. *Nature* **2009**, *462*, 595–601.
59. Li, N.; Zhong, X.; Lin, X.; Guo, J.; Zou, L.; Tanyi, J.L.; Shao, Z.; Liang, S.; Wang, L.P.; Hwang, W.T.; *et al.* Lin-28 homologue A (LIN28A) promotes cell cycle progression via regulation of cyclin-dependent kinase 2 (CDK2), cyclin D1 (CCND1), and cell division cycle 25 homolog A (CDC25A) expression in cancer. *J. Biol. Chem.* **2012**, *287*, 17386–17397.
60. Chang, T.C.; Zeitels, L.R.; Hwang, H.W.; Chivukula, R.R.; Wentzel, E.A.; Dews, M.; Jung, J.; Gao, P.; Dang, C.V.; Beer, M.A.; *et al.* Lin-28B transactivation is necessary for Myc-mediated let-7 repression and proliferation. *Proc. Natl. Acad. Sci. USA* **2009**, *106*, 3384–3389.
61. Iliopoulos, D.; Hirsch, H.A.; Struhl, K. An epigenetic switch involving NF-kappaB, Lin28, Let-7 MicroRNA, and IL6 links inflammation to cell transformation. *Cell* **2009**, *139*, 693–706.
62. Yang, X.; Lin, X.; Zhong, X.; Kaur, S.; Li, N.; Liang, S.; Lassus, H.; Wang, L.; Katsaros, D.; Montone, K.; *et al.* Double-negative feedback loop between reprogramming factor LIN28 and microRNA let-7 regulates aldehyde dehydrogenase 1-positive cancer stem cells. *Cancer Res.* **2010**, *70*, 9463–9472.
63. Krizhanovsky, V.; Lowe, S.W. Stem cells: The promises and perils of p53. *Nature* **2009**, *460*, 1085–1086.
64. Lettre, G.; Jackson, A.U.; Gieger, C.; Schumacher, F.R.; Berndt, S.I.; Sanna, S.; Eyheramendy, S.; Voight, B.F.; Butler, J.L.; *et al.* Identification of ten loci associated with height highlights new biological pathways in human growth. *Nat. Genet.* **2008**, *40*, 584–591.
65. Ong, K.K.; Elks, C.E.; Li, S.; Zhao, J.H.; Luan, J.; Andersen, L.B.; Bingham, S.A.; Brage, S.; Smith, G.D.; Ekelund, U.; *et al.* Genetic variation in LIN28B is associated with the timing of puberty. *Nat. Genet.* **2009**, *41*, 729–733.

66. Perry, J.R.; Stolk, L.; Franceschini, N.; Lunetta, K.L.; Zhai, G.; McArdle, P.F.; Smith, A.V.; Aspelund, T.; Bandinelli, S.; Boerwinkle, E.; *et al.* Meta-analysis of genome-wide association data identifies two loci influencing age at menarche. *Nat. Genet.* **2009**, *41*, 648–650.
67. Frost, R.J.; Olson, E.N. Control of glucose homeostasis and insulin sensitivity by the Let-7 family of microRNAs. *Proc. Natl. Acad. Sci. USA* **2011**, *108*, 21075–21080.
68. Amarasinghe, G.K.; de Guzman, R.N.; Turner, R.B.; Chancellor, K.J.; Wu, Z.R.; Summers, M.F. NMR structure of the HIV-1 nucleocapsid protein bound to stem-loop SL2 of the psi-RNA packaging signal. Implications for genome recognition. *J. Mol. Biol.* **2000**, *301*, 491–511.
69. De Guzman, R.N.; Wu, Z.R.; Stalling, C.C.; Pappalardo, L.; Borer, P.N.; Summers, M.F. Structure of the HIV-1 nucleocapsid protein bound to the SL3 psi-RNA recognition element. *Science* **1998**, *279*, 384–388.
70. Pappalardo, L.; Kerwood, D.J.; Pelczer, I.; Borer, P.N. Three-dimensional folding of an RNA hairpin required for packaging HIV-1. *J. Mol. Biol.* **1998**, *282*, 801–818.
71. Desjardins, A.; Yang, A.; Bouvette, J.; Omichinski, J.G.; Legault, P. Importance of the NCp7-like domain in the recognition of pre-let-7g by the pluripotency factor Lin28. *Nucleic Acids Res.* **2012**, *40*, 1767–1777.
72. Loughlin, F.E.; Gebert, L.F.; Towbin, H.; Brunschweiger, A.; Hall, J.; Allain, F.H. Structural basis of pre-let-7 miRNA recognition by the zinc knuckles of pluripotency factor Lin28. *Nat. Struct. Mol. Biol.* **2012**, *19*, 84–89.
73. Guerrerio, A.L.; Berg, J.M. Design of single-stranded nucleic acid binding peptides based on nucleocapsid CCHC-box zinc-binding domains. *J. Am. Chem. Soc.* **2010**, *132*, 9638–9643.
74. Mayr, F. *Structural and Functional Analysis of Lin28-mediated Inhibition of let-7 miRNA Biogenesis*; Freie Universität Berlin: Berlin, Germany, 2013.
75. Mihailovich, M.; Militti, C.; Gabaldón, T.; Gebauer, F. Eukaryotic cold shock domain proteins: Highly versatile regulators of gene expression. *Bioessays* **2010**, *32*, 109–118.
76. Lopez, M.M.; Makhatadze, G.I. Major cold shock proteins, CspA from *Escherichia coli* and CspB from *Bacillus subtilis*, interact differently with single-stranded DNA templates. *Biochim. Biophys. Acta* **2000**, *1479*, 196–202.
77. Lopez, M.M.; Yutani, K.; Makhatadze, G.I. Interactions of the cold shock protein CspB from *Bacillus subtilis* with single-stranded DNA. Importance of the T base content and position within the template. *J. Biol. Chem.* **2001**, *276*, 15511–15518.
78. Max, K.E.A.; Zeeb, M.; Bienert, R.; Balbach, J.; Heinemann, U. T-rich DNA single strands bind to a preformed site on the bacterial cold shock protein Bs-CspB. *J. Mol. Biol.* **2006**, *360*, 702–714.
79. Max, K.E.A.; Zeeb, M.; Bienert, R.; Balbach, J.; Heinemann, U. Common mode of DNA binding to cold shock domains. Crystal structure of hexathymidine bound to the domain-swapped form of a major cold shock protein from *Bacillus caldolyticus*. *FEBS J.* **2007**, *274*, 1265–1279.
80. Morgan, H.P.; Estibeiro, P.; Wear, M.A.; Max, K.E.A.; Heinemann, U.; Cubeddu, L.; Gallagher, M.P.; Sadler, P.J.; Walkinshaw, M.D. Sequence specificity of single-stranded DNA-binding proteins: A novel DNA microarray approach. *Nucleic Acids Res.* **2007**, *35*, e75.
81. Zeeb, M.; Max, K.E.A.; Weininger, U.; Löw, C.; Sticht, H.; Balbach, J. Recognition of T-rich single-stranded DNA by the cold shock protein Bs-CspB in solution. *Nucleic Acids Res.* **2006**, *34*, 4561–4571.

82. Phadtare, S. Unwinding activity of cold shock proteins and RNA metabolism. *RNA Biol.* **2011**, *8*, 394–397.
83. Phadtare, S.; Severinov, K. Nucleic acid melting by *Escherichia coli* CspE. *Nucleic Acids Res.* **2005**, *33*, 5583–5590.
84. Phadtare, S.; Tyagi, S.; Inouye, M.; Severinov, K. Three amino acids in *Escherichia coli* CspE surface-exposed aromatic patch are critical for nucleic acid melting activity leading to transcription antitermination and cold acclimation of cells. *J. Biol. Chem.* **2002**, *277*, 46706–46711.
85. Newkirk, K.; Feng, W.; Jiang, W.; Tejero, R.; Emerson, S.D.; Inouye, M.; Montelione, G.T. Solution NMR structure of the major cold shock protein (CspA) from *Escherichia coli*: Identification of a binding epitope for DNA. *Proc. Natl. Acad. Sci. USA* **1994**, *91*, 5114–5118.
86. Schindelin, H.; Herrler, M.; Willimsky, G.; Marahiel, M.A.; Heinemann, U. Overproduction, crystallization, and preliminary X-ray diffraction studies of the major cold shock protein from *Bacillus subtilis*, CspB. *Proteins* **1992**, *14*, 120–124.
87. Schindelin, H.; Jiang, W.; Inouye, M.; Heinemann, U. Crystal structure of CspA, the major cold shock protein of *Escherichia coli*. *Proc. Natl. Acad. Sci. USA* **1994**, *91*, 5119–5123.
88. Schindelin, H.; Marahiel, M.A.; Heinemann, U. Universal nucleic acid-binding domain revealed by crystal structure of the *B. subtilis* major cold-shock protein. *Nature* **1993**, *364*, 164–168.
89. Schnuchel, A.; Wiltscheck, R.; Czisch, M.; Herrler, M.; Willimsky, G.; Graumann, P.; Marahiel, M.A.; Holak, T.A. Structure in solution of the major cold-shock protein from Bacillus subtilis. *Nature* **1993**, *364*, 169–171.
90. Sachs, R.; Max, K.E.; Heinemann, U.; Balbach, J. RNA single strands bind to a conserved surface of the major cold shock protein in crystals and solution. *RNA* **2012**, *18*, 65–76.
91. Phadtare, S.; Inouye, M.; Severinov, K. The mechanism of nucleic acid melting by a CspA family protein. *J. Mol. Biol.* **2004**, *337*, 147–155.

9

Long Non-Coding RNAs in Haematological Malignancies

Andoni Garitano-Trojaola [1], Xabier Agirre [1], Felipe Prósper [1,2] and Puri Fortes [3,*]

1 Laboratory of Myeloproliferative Syndromes, Oncology Area, Foundation for Applied Medical Research, University of Navarra, Pamplona 31008, Spain; E-Mails: agaritano@alumni.unav.es (A.G.-T.); xaguirre@unav.es (X.A.); fprosper@unav.es (F.P.)

2 Hematology Service and Area of Cell Therapy, University of Navarra Clinic, University of Navarra, Pamplona 31008, Spain

3 Department of Hepatology and Gene Therapy, Foundation for Applied Medical Research, University of Navarra, Pamplona 31008, Spain

* Author to whom correspondence should be addressed; E-Mail: pfortes@unav.es;

Abstract: Long non-coding RNAs (lncRNAs) are functional RNAs longer than 200 nucleotides in length. LncRNAs are as diverse as mRNAs and they normally share the same biosynthetic machinery based on RNA polymerase II, splicing and polyadenylation. However, lncRNAs have low coding potential. Compared to mRNAs, lncRNAs are preferentially nuclear, more tissue specific and expressed at lower levels. Most of the lncRNAs described to date modulate the expression of specific genes by guiding chromatin remodelling factors; inducing chromosomal loopings; affecting transcription, splicing, translation or mRNA stability; or serving as scaffolds for the organization of cellular structures. They can function in *cis*, cotranscriptionally, or in *trans*, acting as decoys, scaffolds or guides. These functions seem essential to allow cell differentiation and growth. In fact, many lncRNAs have been shown to exert oncogenic or tumor suppressor properties in several cancers including haematological malignancies. In this review, we summarize what is known about lncRNAs, the mechanisms for their regulation in cancer and their role in leukemogenesis, lymphomagenesis and hematopoiesis. Furthermore, we discuss the potential of lncRNAs in diagnosis, prognosis and therapy in cancer, with special attention to haematological malignancies.

Keywords: lncRNAs; leukemia; hematologic malignancies

1. Introduction

Transcriptome analysis by tiling arrays and RNA sequencing has led to the amazing conclusion that while 70%–90% of the genome is transcribed, only 2% is dedicated to the transcription of protein coding sequences [1]. This result has caused a great impression in a scientific community that is deeply proteocentric, *i.e.*, is dedicated to the study of proteins and generally does not pay much attention to other molecules such as lipids or RNAs.

Most cellular RNA is composed of highly expressed non-coding RNAs whose relevance in cell functionality has been well-known for years. However, their transcription requires a relatively small proportion of the genome. These housekeeping non-coding RNAs include transfer RNAs (tRNAs) and ribosomal RNAs (rRNAs), required for mRNA translation; small nuclear RNAs (snRNAs), essential for splicing; and small nucleolar RNAs (snoRNAs), involved in RNA modification. More recently, several small RNAs have been described as playing essential roles in gene expression and transposon silencing. These include microRNAs (miRNAs), small interfering RNAs (siRNAs) and piwi interacting RNAs (piRNAs). Less clear is the role and the molecular mechanisms involved in the function of other small RNAs derived from retrotransposons or 3' untranslated regions or associated with transcription start sites, promoters, termini or repeats. All these non-coding RNAs, with the exception of some of the housekeeping RNAs (some rRNAs and a few snRNAs and snoRNAs), share the common characteristic of being smaller than 200 nts. Therefore the remaining non-coding RNAs, longer than 200 nts, have been grouped under the name of long non-coding RNAs (lncRNAs).

LncRNAs have a terrible name. They are not really long, just longer than the limit of 200 nts imposed by small RNAs. In fact, the average size of coding mRNAs is near 2500 nts while the average length of all the lncRNAs recently described by the Encode project is less than 600 nts [2]. Thus, most of the long non-coding RNAs are shorter than the coding mRNAs, even if some of the lncRNAs may be longer than 100 kbs. Apart from not being really long, it is difficult to determine whether lncRNAs are indeed non-coding. Traditionally, lncRNAs have been characterized by what they do not have: they lack open reading frames (ORFs) longer than 100 amino acids, conserved codons and homology to protein databases [3,4]. Therefore, they have poor coding potential, although they could still code for small open reading frames or non-conserved peptides. Some authors have also analyzed coding capacities of specific lncRNAs by matching their sequences with ribosome footprints or peptide fragments from mass spectrometry analysis. Hits would indicate translation [5–8]. In spite of these efforts, it should be borne in mind that what makes lncRNAs interesting for most scientists is not whether they can encode for proteins or not but the fact that they are functional as RNA molecules. The demonstration of function as an RNA should be required for annotation as an lncRNA, as a functional long RNA is the best definition for lncRNAs. To complicate things further, there are several cases of coding mRNAs that contain regulatory RNA elements and act as bifunctional RNAs; on one hand they code for a protein (p53, for instance) and on the other hand they have a function as RNAs [9–14]. Furthermore, several coding genes are transcribed to non-coding alternative splicing variants.

Functional or lncRNA genes are very similar to coding genes at the DNA and chromatin level as they share the same epigenetic marks. Similar to mRNAs, most lncRNAs are transcribed from RNA polymerase II, are capped at the 5' end, contain introns and approximately 40% are polyadenylated at the 3' end [15]. The lncRNAs recently described by Encode show a bias for having just one intron and a trend for less-efficient cotranscriptional splicing [8,16]. It has been estimated that there could be as many lncRNA genes as coding genes, but the number of lncRNAs is still growing and some authors consider that it could increase from ~20,000 to ~200,000 [17,18]. Compared to mRNAs, most lncRNAs localize preferentially to the nucleus, are more cell type specific and are expressed at lower levels [19]. In fact, there is less than one copy per cell of many lncRNAs. The low expression levels and the fact that the sequence of lncRNAs is poorly conserved have convinced many scientists that they are not relevant for cell functionality. However, although lncRNAs are under lower selective pressure than protein-coding genes, sequence analysis shows that lncRNAs are under higher selective pressure than ancestral repeat sequences with neutral selection. Moreover, promoters of lncRNAs have similar selection levels than promoters of protein coding genes [8]. Even in the absence of strong sequence conservation, the genomic location and structure of many lncRNAs is conserved together with short stretches of sequences, suggesting that lncRNAs could be under selective pressure to maintain a functional RNA structure rather than a linear sequence [8].

Recent publications in the field have led to the hypothesis that many lncRNAs may be key regulators of development and may play relevant roles in cell homeostasis and proliferation. In fact, several lncRNAs have been described that function as oncogenes or tumor suppressors [20]. It is expected that for cell biology the role of lncRNAs could be as revolutionary as the role of small non-coding RNAs such as miRNAs. miRNA studies have highlighted the relevance of gene regulation in cell homeostasis, differentiation and proliferation and may impact the clinic with new therapies and new diagnostic and prognostic tools for many diseases. The relevance of miRNAs has been clearly established for haematological malignancies [21,22]. In this review we will summarize what is known about lncRNAs in normal haematopoiesis and in haematological tumors. Even though many more studies need to be done, the results obtained thus far suggest that several lncRNAs may be key molecules in haematopoiesis and in the pathogenesis of haematological malignancies.

2. Classes of lncRNAs and lncRNA Functionality

2.1. Classification by Genomic Location

Under the name of lncRNAs there are RNAs with many different characteristics, which complicates classification. Therefore a well accepted method is based on genomic location rather than on functionality, conservation or origin. From a genetic point of view lncRNAs can be classified into one or more of the following categories: (a) sense, when overlapping with one or more exons of another transcript in the same strand; (b) antisense, when overlapping with one or more exons of another transcript in the opposite strand; (c) intronic, when derived from an intron of another transcript; (d) divergent or bidirectional, when they share a promoter with another transcript in the opposite strand and therefore are coregulated; (e) intergenic, when they are independent, located in between two other genes. Long intergenic non coding RNAs (lincRNAs) are a special class of intergenic lncRNAs whose genes have histone mark signatures of active transcription (trimethylation in lysine 4 and lysine 36 of histone 3: H3K4m3, H3K36m3) [23].

In the case of antisense transcripts, classification based on genomic location helps to predict functionality. 50%–70% of sense transcripts have natural antisense partners (NATs) [24–26]. NATs are generally involved in the regulation in *cis* of the corresponding sense RNA by mechanisms that act at the transcriptional and posttranscriptional level. NATs can induce transcriptional interference or recruit chromatin modifiers and remodelers to establish a local transcriptionally active or inactive chromatin conformation [27]. Posttranscriptionally, examples of NATs exist that regulate imprinting, RNA editing, splicing, by blocking binding of the spliceosome to the 5' splice site of an intron leading to intron retention [28–32] or translation and stability by forming a duplex with the sense RNA that masks the binding site for miRNAs [33]. Thus, NATs can modify processing and induce or reduce the expression or the translation of their sense counterpart. Some intronic lncRNAs also regulate the expression of their genomic partners. Intronic lncRNAs may be generated by stabilization of the intron after splicing of the host gene but, more commonly, they are produced from independent transcription. Some intronic non-coding RNAs are associated with polycomb-related repressive histone marks along the promoter region and gene body of their host gene, which results in local transcriptional silencing [34].

2.2. Classification by Specific Characteristics

Most lncRNAs with special characteristics cannot be easily classified into a single group according to genomic location. These include enhancer RNAs (eRNAs), lncRNA-activating (lncRNA-a) genes, transcribed ultraconserved regions (T-UCRs), pseudogenes, telomere-associated ncRNAs (TERRAs), circular RNAs, *etc.*

eRNAs are transcribed by RNA polymerase II at active enhancer regions, characterized by H3 Lys4 monomethylation or Lys27 acetylation and binding of the regulatory protein p300 [35–39]. eRNAs are not polyadenylated. Many are bidirectional and poorly expressed [38,40], but expression of several eRNAs seems to be tightly regulated [38,39]. Although many eRNAs were thought to be by-products of the presence of RNA pol II in enhancers, recent evidence suggests that some may function to control the expression of neighbouring genes [41]. eRNAs are also important in the formation of the chromosomal loopings that bring enhancers closer to promoters [39], and in the induction of, for example, p53-dependent enhancer activity and transcription [42].

lncRNA-a genes generally transcribe intergenic RNAs which are involved in the expression of neighbouring genes [41]. Thus, downregulation of the lncRNA-a results in downregulation of the neighbour gene. This effect requires expression of the Mediator complex and it has been shown that interaction of the lncRNA-a with Mediator is required for the upregulation of nearby genes [43].

T-UCRs and pseudogenes are lncRNAs that share sequence similarity to other mammalian genomes or other regions of the same genome, respectively. There are 481 UCRs longer than 200 bp that are absolutely conserved between human, rat, and mouse genomes [44]. Most are transcribed or T-UCRs in normal human tissues, both ubiquitously and tissue specifically. The high degree of conservation across species implies that T-UCRs may be essential, but deletion of some of these regions in knockout mice has not been associated with a detectable phenotype [45]. One possible function of some T-UCRs is miRNA control, as many T-UCRs have significant antisense complementarity with particular miRNAs and there is a negative correlation between expression of specific T-UCRs and predicted antisense miRNAs targets [46,47]. In fact, some T-URCs have been shown to be targeted by miRNAs.

Pseudogenes originated from duplication of ancestor or parental coding genes (duplicated pseudogenes) or through retrotransposition of processed RNAs transcribed from ancestor genes (processed pseudogenes). Subsequently, they have lost their coding capacity as a result of the accumulation of mutations. When pseudogenes are expressed, they may regulate the expression and function of their parental gene by several mechanisms [48,49]. For instance, pseudogenes may act as miRNA decoys that lead to increased stability and translation of their parental gene [50–53].

Circular RNAs, newcomers to the RNA list, can also function as RNA decoys [54–56]. It is generally accepted that circular RNAs originate from reverse splicing, where the acceptor splice site located downstream binds to an upstream donor splice site. This causes the circularization of the RNA and a tremendous increase in RNA stability, as circular RNAs lack 5' or 3' ends and therefore, are resistant to exonucleases. The increased stability of circular RNAs may lead to long-term functionality by miRNA sequestration [57].

2.3. Classification as cis *or* trans-*Acting Molecules*

LncRNAs can also be classified according to their functionality as *cis* and/or *trans* acting molecules (Figure 1). *Trans*-acting lncRNAs function away from the site of synthesis while *cis*-acting lncRNAs function at the site of transcription to affect the expression of neighbouring genes. Several *cis*-acting lncRNAs guide epigenetic regulators to their site of transcription while they are being transcribed. Thus, lncRNA transcription is critical and rapidly creates an anchor to recruit proteins involved in chromatin remodelling [58–61]. This molecular mechanism has tremendous advantages: (i) it responds very fast, as it only requires transcription of an RNA and a proper accumulation of nuclear chromatin remodelers; (ii) it is very specific, as the targeting does not involve RNA-DNA interactions other than those required for lncRNA transcription and (iii) it may function with just a single molecule of lncRNA per locus. This may explain the low abundance of *cis*-acting lncRNAs and the relatively high concentration of lncRNAs close to developmental genes whose expression is strictly controlled [62]. Thus, *cis*-acting lncRNAs control the epigenetic regulation of some imprinted genes. Imprinting depends on the parental origin of the imprinted genes, which play critical roles in mammalian development and therefore, their expression must be tightly regulated [63]. Many imprinted gene loci express lncRNAs that appear to regulate the expression of neighbouring imprinted protein-coding genes in *cis*, allele specifically [64]. The lncRNA *AIR*, for example, silences the neighbouring imprinted genes *SLC22A3*, *SLC22A2* and *IGF2R* [65].

The clear division between *cis* and *trans* acting lncRNAs has been blurred by recent experiments, where exogenously expressed lncRNAs that normally work in *cis*, are able to find their target sites. Thus, even *cis*-acting lncRNAs may have the capacity to act in *trans* [65]. Furthermore, when considering *cis*-acting lncRNAs, the 3D organization of the genome should be taken into consideration. A *cis*-acting lncRNA may control the expression of neighbour genes brought into proximity by chromosome looping.

Figure 1. Schematic representation of *cis* and *trans*-acting lncRNAs. *cis*-acting lncRNAs function at the site of transcription and affect the expression of neighbouring genes. *Trans*-acting lncRNAs function away from the site of synthesis.

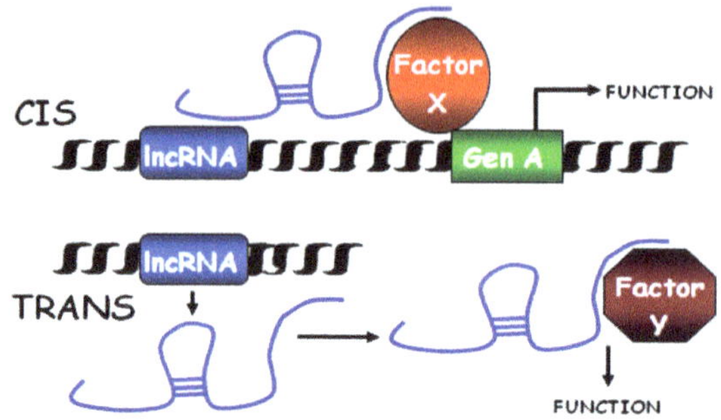

Trans-acting lncRNAs regulate gene expression on a genome-wide scale. A good example is *HOTAIR*, which binds the chromatin-modifying complexes PRC2, LSD1 and CoREST/REST [66–69]. Guiding chromatin remodelers to specific sites is easier to conceive for *cis*-acting lncRNAs. Targeting mediated by *trans*-acting lncRNAs would probably require RNA:DNA:DNA triplex formation via Hoogsteen base-pairing, as has been shown *in vitro* for a promoter-associated lncRNA [70]. However, such interactions may expose the genome to deamination and damage [71,72]. Furthermore, lncRNAs could form secondary and tertiary structures that behave similary to DNA-binding domains from proteins or that bind proteins that mediate DNA binding. This is what has been described for the *XIST* lncRNA, which binds YY1 transcription factor to reach specific sites in the X chromosome [73]. Theoretically, lncRNAs could also form an RNA:DNA hybrid that displaces a single strand of DNA (the so-called R-loop) or an RNA:RNA hybrid of lncRNA with a nascent transcript [74–76].

2.4. lncRNA Functionality

Guiding chromatin remodelling factors seems to be the predominant function exerted by lncRNAs. In fact, it has been estimated that 20% of all lncRNAs may bind PRC2 [66]. Several lncRNAs have also been shown to bind to PRC1, the CoREST/REST repressor complex [66], the histone methyltransferase associated with the activating trithorax complex, MLL1 [77,78], and H3-K9 methyltransferase, G9a [65,79]. However, lncRNAs have also been shown to exert several other functions in the cell nucleus and cytoplasm, including regulation of DNA bending and insulation, RNA transcription, splicing, translation and stability, organization of subnuclear structures and protein localization, among others.

DNA looping. CTCF can induce chromosomal bending and protect specific genes from the effects of distal enhancers and regulatory elements. The lncRNA *SRA* can interact with and enhance the function of CTCF [80]. Also, endogenous but not exogenous nascent *HOTTIP* lncRNA, binds target genes via chromosomal looping [81].

Transcription. LncRNAs may activate or inhibit transcription of specific targets. Some lncRNAs act as coactivators that bind transcription factors and enhance their transcriptional activity [82–84]. This is the function of *SRA* lncRNA in the progestin steroid hormone receptor [85,86]. However, some lncRNAs act as decoys of transcription factors [87] and may move them to the cytoplasm to keep them away from their nuclear targets [88]. Thus, p53-induced lncRNA *PANDA* binds transcription factor

NF-YA and prevents NF-YA activation of cell death genes [89]. *DHFR* lncRNA forms a triplex structure which sequesters the general transcription factor IIB and prevents transcription of the *DHFR* coding gene [90]. Finally, the act of lncRNA transcription may interfere with transcription initiation, elongation or termination of another sense or antisense gene [91]. Transcriptional interference can also lead to activation of gene expression by inhibiting the action of repressor elements.

Organization of subnuclear structures. LncRNAs can recruit protein factors to nuclear structures. This is the case of lncRNA *MALAT1* and *NEAT-1*. *MALAT1* recruits serine/arginine–rich splicing factors to nuclear speckles [92]. More importantly, *NEAT-1* is an essential structural component of paraspeckles, subnuclear structure implicated in RNA splicing and editing [93,94]. Depletion of *NEAT-1* leads to loss of paraspeckles while overexpression of *NEAT-1* causes an increase in the number of paraspeckles [95–97]. *MALAT1* and *NEAT-1* are genomic neighbours overexpressed in several tumors compared to healthy tissues. Surprisingly the mouse knockouts of either *NEAT-1* or *MALAT1* had no detectable phenotype, suggesting that there could be redundant or compensatory molecules [98–101].

Splicing. Splicing can be inhibited by lncRNAs antisense to intron sequences that impede spliceosome binding causing intron retention [28–32]. Furthermore, alternative splicing can be altered by lncRNA-mediated sequestration or modification of splicing factors. Thus, *MALAT1* binds splicing factors present in nuclear speckles and modulates the activity of SR proteins, involved in the selection of splice sites, and therefore regulates the splicing of many pre-mRNAs [92]. Some snoRNA-containing lncRNAs (sno-lncRNAs) are retained close to their sites of transcription where the splicing factor Fox2 is enriched. Changes in the level of the sno-lncRNA lead to a nuclear redistribution of Fox2 and to changes in alternative splicing. Thus, the sno-lncRNAs could function as a regulator of splicing in specific subnuclear domains [102].

Translation. LncRNAs have been described that increase or inhibit translation of specific targets [103,104]. Expression of antisense *UCHL1* lncRNA leads to an increase in Uchl1 protein level without any change at the Uchl1 mRNA level. A repetitive SINEB2 sequence is required for this function. Under cap dependent translation inhibition due to stress, *UCHL1* lncRNA moves from the nucleus to the cytoplasm, binds to Uchl1 mRNA and allows its cap-independent translation. Thus, *UCHL1* lncRNA could behave as a mobile internal ribosomal entry sequence.

Stability. LncRNAs have been described that increase or decrease stability of specific targets [105,106]. Binding of lncRNAs containing ancestral Alu repeats to complementary Alu sequences in the 3′UTR of coding mRNAs forms a dsRNA recognized by the dsRNA binding protein Stau1, which induces Stau-mediated RNA decay [106]. Instead, lncRNA *TINCR* localizes to the cytoplasm, where it interacts with Stau1 and promotes the stability of mRNAs containing the TINCR box motif [105].

miRNA binding. LncRNAs can regulate mRNA stability and translation by binding to miRNAs and preventing their action. Besides the already described role of some pseudogenes and circular lncRNAs in miRNA sequestration, other lncRNAs such as *linc-MD1*, have been shown to serve as "sponge" for miRNAs. *Linc-MD*1 binds two miRNAs, which downregulate transcription factors involved in muscle differentiation and therefore muscle differentiation is induced upon *Linc-MD*1 expression [107].

LncRNAs have been implicated in many other different functions. LncRNA *NRON* is a repressor of NFAT by binding β-importins and regulating the nuclear trafficking of NFAT [88]. *TERC* is a well-known telomerase-associated lncRNA that serves as a template for the synthesis of chromosome ends. The dsRNA-protein kinase PKR may be activated by binding to a lncRNA [108]. It is expected that in the near future novel and unexpected mechanisms of lncRNA functionality will be discovered. For instance, to date few lncRNAs have been described to have catalytic properties.

The high number of lncRNAs and their heterogeneity helps them to exert such a myriad of functions. In fact, all lncRNA functions respond to just three different mechanisms: decoys, scaffolds and guides [109]. Decoy-acting lncRNAs impede the access of proteins such as transcription factors and RNAs such as miRNAs to their targets. LncRNAs *MD-1* and *PANDA* act as decoys for miRNAs and transcription factors, respectively [89,107]. Scaffold-acting lncRNAs serve as adaptors to bring two or more factors into discrete ribonucleoproteins (RNPs) [110]. LncRNA *TERC*, *HOTAIR* or *NEAT-1* act as scaffolds to form the telomerase complex [111], a silencing complex [69] or the paraspeckle, respectively [93,94]. Guide-acting lncRNAs are required to localize protein complexes at specific positions. *XIST* or *AIR* lncRNAs act as guides to target gene silencing activity in an allele-specific manner. Guide lncRNAs such as *HOTAIR*, can also behave as scaffolds.

It is conceivable that lncRNAs may function through linear or structured domains. Linear domains may bind proteins but also RNA or, possibly, DNA sequences by perfect (e.g., antisense lncRNAs with their sense counterpart) or imperfect complementarity. Novel linear domains able to bind and regulate mRNAs, miRNAs or other lncRNAs could be very easily created evolutionarily. In many cases though, the secondary and tertiary structure of lncRNAs dictates their function. Thus, lncRNAs generally have complex structures with higher folding energies than those observed in mRNAs [112]. Proteins are expected to be the major partners of lncRNAs to form functional RNP particles. RNA binding proteins represent more than 15% of the total amount of proteins [113]. In several cases studied to date, interaction between proteins and RNAs results in conformational changes to the protein, the RNA or both, which could endow the complex with a novel ability.

LncRNA function impacts cell behaviour. LncRNAs have specially emerged as regulators of development. Some transcription factors involved in pluripotency bind promoter regions of more than 100 mouse lncRNAs [15]. 26 lincRNAs have already been described as being required for the maintenance of pluripotency in mouse [114]. Two lncRNAs regulated by pluripotency transcription factors such as Oct4 and Nanog are essential for pluripotency maintenance, as they, in turn, control the expression of Oct4 and Nanog [115]. Therefore, these lncRNAs participate in positive regulatory loops. Similarly, several lncRNAs have been implicated in human disease, including several cancers [116]. Dysregulated lncRNAs have been described in heart disease, Alzheimer disease, psoriasis, spinocerebellar ataxia and fragile X syndrome [33,117–121] and in several tumours including breast, brain, lung, colorectal, prostate and liver cancers, melanoma, leukaemia and others [46,68,116,122–128]. LncRNAs have been described that function as oncogenes [129], tumour suppressors [23,130] or drivers of metastatic transformation, such as *HOTAIR* in breast cancer [68]. In this review we will concentrate on those lncRNAs whose expression is altered in haematological malignancies.

3. LncRNAs Deregulated in Haematological Malignancies

The impact of non-coding RNAs on haematological malignancies has been well described for microRNAs [131,132]. The list of lncRNAs involved in the initiation and progression of blood tumors is still very short and expected to grow exponentially in the near future. Some of the lncRNAs that play a role in haematological malignancies (Table 1) are in fact host genes of miRNAs with oncogenic or tumour suppressor properties. Others endow oncogenic or tumour suppressor properties in the long non-coding RNA molecule. The mechanism of action of few of them has been studied in some detail.

3.1. Host Genes of Small RNAs

3.1.1. BIC and C13ORF25

Some lncRNAs were described to have oncogenic properties in blood cells before the discovery of miRNAs. This is the case of the B cell Integration cluster (*BIC*) or host gene *mir-155* (*MIR155HG*) (Figure 2A). *BIC* and *miR-155* expression is increased in Hodgkin lymphoma, Acute Myeloid Leukemia (AML) and Chronic Lymphocytic Leukemia (CLL) but it is not detected in healthy samples [133]. Increased expression of BIC and *miR-155* results from transcription activation by the MYB transcription factor [134] and leads to *miR-155*-mediated downregulation of several tumor suppressor genes [135]. In this case, the lncRNA *BIC* plays an important role in the regulation of *miR-155* which is directly involved in the lymphomagenesis or leukemogenesis. Similarly, *C13ORF25* or host gene *mir-17* (*MIR17HG*) encodes the *miR-17-92* cluster and its expression is increased in B-cell lymphoma [136], Mantle Cell Lymphoma (MCL) [137] and other tumors [138,139].

3.1.2. nc886 or *vtRNA2-1*

vtRNA2-1, previously known as *pre-miR-886*, is a short ncRNA suppressed in a wide range of cancer cells that inhibits activation of protein kinase R (PKR) [140]. Even if nc886 is shorter than 200 nts and therefore is not a lncRNA, its relevance in AML merits a short description. *vtRNA2-1* is transcribed from the long arm of chromosome 5 region whose deletion is associated with poor outcome in AML. Furthermore, decreased expression by monoallelic or biallelic DNA methylation correlates with a worse outcome in AML patients [141]. Thus, *vtRNA2-1* could be a tumour suppressor for AML and its role could be mediated by PKR.

Table 1. lncRNAs in hematopoiesis and hematological malignancies.

LncRNAs	LOCATION	HEMATOLOGIC DISEASE/SYSTEM	FUNCTION	MOLECULAR MECHANISM	MECHANISMS INVOLVED IN DYSREGULATION	CITATIONS
MIR155HG BIC	21q21.3	Burkitt, Hodgkin lymphoma, AML, CLL	Host of miRNAs	miR-155	Target MYB and NFKB	[134]
MIR17HG	13q31.3	B-cell lymphoma, MCL	Host of miRNAs	miR-17-92	Target MYC	[136,137,142]
vtRNA2-1	5q31.1	AML (poor prognosis)		PKR inhibition	DNA methylation Deletion 5q	[140,141]
PVT1	8q24.21	MM, Burkitt Lymphoma, T-cell Leukemia, CLL	Oncogene and host of miRNAs	miR-1204 MYC activation	Translocation t(8;14)(q24;q11) t(2;8)(p11;q24) t(8;22)(q24;q11)	[143–148]
CDKN2B-AS1/ANRIL	9p21.3	AML, ALL	Oncogene	PRC1 and PRC2 targeting	rs3731217-G SNP Deletion, hypermethylation	[26,128,149–151]
MEG3	14q32.2	AML,MM	Tumor suppressor	PRC2 binding to control DLK1 imprinting. p53 activation.	DNA methylation	[149,150,152–155]
DLEU1/DLEU2	13q14.2	CLL, MM, Lymphoma	Tumor suppressor	*hsa-miR-16-1* and *15a* BCL2 targeting. NFKB activation	Histone modification, DNA methylation, deletion	[156]
GAS5	1q25.1	B-cell Lymphoma, Leukemia	Tumor suppressor	Glucorticoid receptor repression. Regulated by mTOR pathway.	Translocation (1;3)(q25;q27)	[87]
H19	11p15.5	AML, CML, MPN, T-cell Leukemia, Lymphoma	Oncogene/tumor suppressor	Activated by Myc and down-regulated by p53. miR-675 targeting Rb	Loss of imprinting	[157]
T-UCRs		CLL (prognosis marker)	Oncogene/tumor suppressor	miR control		[46]
lincRNA-p21	Not annotated in human	ALL, CML	Tumor suppressor	Activated by p53 binds hnRNP K to induce apoptosis	Not known	[158]
TCL-6	14q32.13	T cell leukemia	Poorly characterized	Not described	Translocation and inversions with TCR	[151]
WT1-AS	11p13	AML, ALL	Poorly characterized	WT-1 control	Not known	[159]
CRNDE	16q12.2	AML, MM, T-cell leukemia	Oncogene	PRC2 and COREST binding	Not known	[160]
RMRP	9p13.3	Non-Hodgkin lymphoma	Poorly characterized	Not described	Mutation	[161]
SNHG5	6q14.3	B-cell Lymphoma	Poorly characterized	snoRNA host	Translocation (1;3)(q25;q27)	[162]
HOXA-AS2	7p15.2	APL	Poorly characterized		Not known	[163]
HOTAIRM1	7p15.2	Hematopoietic regulator	Regulator of myelopoiesis	HOX A genes.		[164]
EGOT	3p26.1	Hematopoietic regulator	Regulator of eosinophil development			[165]
PU.1-AS	11p11.2 Non annotated	Hematopoietic regulator	*PU.1-AS* regulate the hematopoiesis regulator PU.1	PU.1 control		[166]
EPS	Mouse 4qC7	Hematopoietic regulator	Regulator of erytropoyesis	Pycard repression		[167]
ThyncR1	1q23.1	Hematopoietic regulator	Regulator of T cell selection and maduration.	Riboregulator		[168]

Figure 2. Schematic representation of the function of lncRNAs deregulated in haematological malignancies. (**A**) *BIC*. Myb transcription factor increases the expression of *BIC* in several leukemias and lymphomas. This results in increased levels of *miR-155* and *miR-155*-mediated downregulation of several tumor suppressor genes; (**B**) *ANRIL*. The INK4 *p15INK4b-p14ARF-p16INK4a* cluster transcribes for an antisense transcript named *ANRIL*; PcG complex (PRC2) is targeted to the INK4 locus by *ANRIL*, and locus expression is inhibited; (**C**) *MEG3*. *MEG3*, among other functions, stimulates p53-dependent tumor suppressor pathways by several mechanisms. *MEG3* down-regulates MDM2 expression, therefore decreasing p53 MDM2-mediated degradation. *MEG3* increases p53 protein levels and stimulates p53-dependent transcription. *MEG3* enhances p53 binding to some target promoters such as *GDF15*; (**D**) *GAS5*. *GAS5* binds the DNA binding domain of glucocorticoid receptors (GR) and impedes GR binding to DNA and induction of GR-dependent genes such as *cIAP2*.

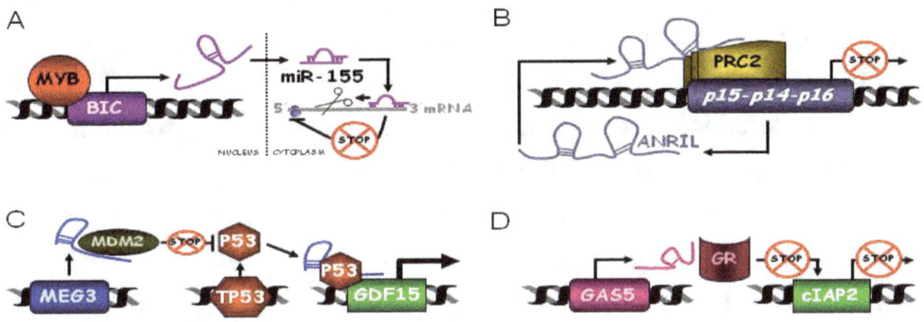

3.1.3. *PVT1*

It is not clear whether the role of Plasmacytoma variant translocation 1 (*PVT1*) lncRNA in haematological malignancies depends exclusively on being a miRNA host gene. The *PVT1* gene is transcribed to several mature RNAs by alternative splicing, including a cluster of seven miRNAs, six of them annotated in the miRBase as *miR-1204*, *miR-1205*, *miR-1206*, *miR-1207-5p*, *miR-1207-3p*, and *miR-1208*. The function of these miRNAs is unknown with the exception of *miR-1204*. *miR-1204* has been involved in different roles related to development, differentiation and senescence [146,169]. On one hand *miR-1204* has been described as increasing p53 levels and causing cell death [148]. In fact *PVT1* expression is induced in response to p53 [148]. On the other hand, *miR-1204* has been shown to activate Myc and cell proliferation in mouse pre- B cell lines [146,147].

PVT1 is located in chromosome region 8q24.21, relatively close to the transcription factor c-Myc. Translocations within c-Myc or *PVT1*, which cause the overexpression of these two oncogenes compared to healthy cells, are characteristics associated with B cell malignancies including Burkitt Lymphoma (BL), AIDs, Non-Hodgkin lymphoma, mouse plasmacytoma (Pct) and multiple myeloma (MM) [147]. Furthermore, *PVT1* is in a susceptibility locus for classical Hodgkin's lymphoma [145] and a SNP that causes increased *PVT1* expression is associated with prostate cancer risk [170]. *PVT1* is overexpressed, compared to healthy tissues, in breast and ovarian cancer, pediatric malignant astrocytomas, AML and Hodgkin lymphoma [171], suggesting that *PVT1* could be an oncogene. In fact, upregulation of *PVT1* contributes to tumor survival and chemoresistance [171–174] while its downregulation inhibits cell proliferation and induces a strong apoptotic response [171]. It has been proposed that *PVT1* regulates c-Myc expression but also that *PVT1* is regulated by c-Myc [175]. However, some authors suggest that Myc and *PVT1* contribute to cancer by different mechanisms [147,171]. Further studies are required to understand the role of *PVT1* in tumorigenesis and to determine whether the miRNAs encoded by *PVT1* mediate its functionality.

3.2. LncRNAs with Oncogenic Properties

ANRIL or *CDKN2B-AS1*

Antisense Non-coding RNA in the INK4 Locus (*ANRIL*) or *CDKN2B-AS1* is transcribed antisense to the *p15INK4b-p14ARF-p16INK4a* cluster, whose members are key effectors of oncogene-induced senescence (Figure 2B). The INK4 proteins are induced during aging and in premalignant lesions, limiting tumor progression. Therefore, expression of the *INK4b-ARF-INK4a* locus is tightly controlled and the Polycomb group (PcG) complexes are required to initiate and maintain silencing of this locus [176,177]. PcG complexes are targeted to the locus by *ANRIL* [178]. Depletion of *ANRIL* disrupts binding of the PRC2 component SUZ12 to the locus, increases the expression of p15INK4b and inhibits cellular proliferation. *ANRIL*, as a pol II nascent transcript, also controls cellular lifespan by targeting the PRC1 component CBX7 to the INK4 locus [27].

Genome-wide association studies revealed that *ANRIL* is located in a genetic susceptibility locus (9p21) associated with several diseases, including coronary artery disease (CAD), atherosclerosis, intracranial aneurysm, type 2 diabetes, and several cancers, such as glioma, basal cell carcinoma, nasopharyngeal carcinoma, and breast cancer [179]. Several single nucleotide polymorphisms (SNP) in this locus alter *ANRIL* structure [180] and *ANRIL* gene expression [181,182], mediating susceptibility to disease. There is a statistically significant association between an *ANRIL* polymorphism and Philadelphia positive Acute Lymphoblastic Leukemia (Ph+ ALL) [183]. Furthermore, 69% of samples (n = 16) from patients with ALL and AML showed relatively increased expression of *ANRIL* and downregulated p15 compared to controls [130]. The expression of *ANRIL*, CBX7, and EZH2 is coordinated and elevated in preneoplastic and neoplastic tissues, leading to decreased p16INK4a expression and decreased senescence [27]. In fact, the *INK4b-ARF-INK4a* locus is subject to frequent deletion or hypermethylation in cancers, including leukemia, melanoma, lung and bladder cancers [177].

3.3. LncRNAs with Tumor Suppressor Properties

3.3.1. *MEG3*

The maternally expressed gene 3 (*MEG3*) was the first lncRNA proposed to function as a tumor suppressor (Figure 2C). *MEG3* is a paternally imprinted polyadenylated RNA, expressed in many normal human tissues as several alternative splicing variants [184,185]. *MEG3* expression was decreased compared to healthy tissues in various brain cancers (pituitary adenomas, glioma and the majority of meningiomas and meningioma cell lines) [149,154], bladder, lung, renal, breast, cervix, colon and prostate cancers and haematological malignancies such as MM, AML or myelodysplastic syndromes. Surprisingly *MEG3* is overexpressed in Wilms tumor and may be increased or decreased in different hepatocellular carcinomas *versus* healthy livers [186].

The last intron of *MEG3* lncRNA encodes the evolutionarily conserved *miR-770* [187] and *MEG3* isoforms can contain several small open reading frames that are not required for *MEG3* function [152,153]. Instead, the *MEG3* secondary structure, rather than primary sequence, is critical to maintaining function [152]. *MEG3* lncRNA localizes to the nucleus, although some cytoplasmic *MEG3* transcripts have been detected [184,188,189]. In the nucleus, *MEG3* binds to PRC2 to control the imprinting of the *DLK1* locus, where *MEG3* belongs. Furthermore, *MEG3* stimulates both p53-dependent and p53-independent tumor suppressor pathways [149,150,152–155]. *MEG3* activates the tumor suppressor protein p53 at different levels. On one hand *MEG3* down-regulates MDM2 expression, therefore decreasing p53 MDM2-mediated degradation [150]. On the other hand, *MEG3* significantly increases p53 protein levels and stimulates p53-dependent transcription [155]. Finally, *MEG3* enhances p53 binding to some target promoters such as *GDF15* [152,153]. Ectopic expression of *MEG3* RNA leads to p53 accumulation and inhibition of cellular proliferation [153,185]. Inactivation of *MEG3* in the brain increases the expression of genes involved in angiogenesis, suggesting that the tumour suppressor function of *MEG3* works, in part, by inhibiting angiogenesis [190]. In bladder cancer a negative correlation has been shown between *MEG3* expression and autophagy [191].

3.3.2. *DLEU1* and *DLEU2*

Deleted in leukemia 1 (*DLEU1*) and 2 (*DLEU2*) are two genes transcribed head to head in a 30-kb region located in the long arm of chrormosome 13 (13q14), which is lost in more than 50% of patients with CLL and that predicts a poor prognosis [192]. The homozygous loss of this region has great effects on the regulation and control of normal CD5+ B lymphocytes and their homeostasis. Recent studies show that *DLEU1* and *DLEU2* control transcription of their neighbouring candidate tumour suppressor genes, which may act as positive regulators of NF-kB activity [156]. As binding of *DLEU1* and *DLEU2* to chromatin has not been detected, it has been proposed that they regulate neighbouring gene expression by divergent transcription. In addition, the intron 4 of *DLEU2* encodes the miRNAs *hsa-miR-16-1* and *hsa-miR-15a.* This miRNA cluster exerts a crucial role in the tumorigenesis of CLL, in part, regulating the oncogene *BCL2* [193]. Knocking out *hsa-miR-16-1* and *hsa-miR-15a* in mice leads to a lymphoproliferative disease [194]. However the knockout model of *DLEU2*, which includes deletion of *hsa-miR-16-1* and *hsa-miR-15a* as well, shows a more aggressive phenotype than the *hsa-miR-16-1/hsa-miR-15a 6* knockout model alone, suggesting that *DLEU2* can participate in CLL development on its own. In fact, increased expression of *DLEU2* leads to reduced proliferation and clonogenicity [195].

3.3.3. *GAS5*

Growth arrest specific 5 (*GAS5*) is induced under starvation conditions and is highly expressed in cells that have arrested growth [196,197]. *GAS5* modulates cell survival and metabolism by antagonizing the glucocorticoid receptor (GR) [87] (Figure 2D). *GAS5* binds the DNA binding domain of GRs directly, preventing GRs from binding to DNA, from functioning as transcription activators and from reducing cell metabolism [87]. *GAS5* could regulate other receptors (androgen, mineralocorticoid and progesterone but not estrogen receptors) by the same means [87]. Expression of *GAS5* is sufficient to repress GR-induced genes, such as the cellular inhibitor of apoptosis 2 (*cIAP2*) and sensitizes cells to apoptosis [87]. Thus, *GAS5* behaves as a tumor suppressor. *GAS5* expression is decreased in breast cancer and is almost undetectable in growing leukemia cells and increases after density-induced cell cycle arrest [87,196,197]. At the same time, *GAS5* has been shown to be regulated by the mammalian target of rapamycin (mTOR) pathway and to mediate the effect of rapamycin on the cell cycle in T cells [198]. Downregulation of *GAS5* by RNA interference protects leukemic and primary human T cells from the anti-proliferative effect of rapamycin [199].

3.4. LncRNAs with Dual Functions

3.4.1. *H19*

H19 is an imprinted lncRNA located close to the *IGF2* gene. *H19* is expressed form the maternal allele and *IGF2* from the paternal allele [59,200]. A key feature of cancer is the loss of this imprinting, which results in the well documented overexpression of *H19* in cancers of the colon, liver, breast and bladder and in hepatic metastases, compared to healthy tissues [200–204]. Loss of *H19* imprinting has been described in adult T-cell leukaemia/lymphoma (ATL) [157] and decreased *H19* expression was found in the bone marrow of patients with clinically untreated chronic myeloproliferative disorders, including chronic myeloid leukemia (CML), polycythemia vera (PV), essential thrombocythemia (ET), primary myelofibrosis (PMF) and chronic myelomonocytic leukaemia (CMML) [205,206] and AML [207].

H19 can behave as an oncogene or as a tumour suppressor [59]. *H19* expression can be activated by the oncogene c-Myc [200] and downregulated by the tumour suppressor p53 [208,209]. Downregulation of *H19* by RNAi blocks cell growth and clonogenicity of lung cancer cell lines [200] and decreases xenograft tumour growth of a hepatocellular carcinoma cell line [203]. Furthermore, *H19* is the precursor of miR-675, which downregulates the tumor suppressor retinoblastoma in human colorectal cancer [210]. All these results indicate that *H19* is an oncogene [210]. However, depletion of *H19* caused increased polyp count in a mouse model for colorectal cancer [211], larger tumor growth in a mouse teratocarcinoma model and an earlier development of tumours in a mouse hepatocarcinoma model [212]. This dual role as oncogene or tumour suppressor may depend on the cellular environment of the tumour type.

3.4.2. T-UCRs

The expression of many T-UCRs has been described to be significantly altered in tumours such as CLL, colorectal and hepatocellular carcinomas and neuroblastomas [46,162,213,214]. Certain SNPs in T-UCR genes were associated with increased familial breast cancer risk [163]. Moreover, T-UCR transcription profiles can be used to differentiate types of human cancers and predict patient outcome [213]. Some T-UCRs seem tumour specific, such as *UC.73A* and *UC.338*, which are decreased in colon cancer [215]. In fact, some T-UCRs differentially expressed in a particular human cancer locate in fragile sites or cancer-associated genomic regions specifically associated with that type of cancer [216]. This is the case of *UC.349A* and *UC.352*, differentially expressed between normal and leukemic CD5-positive cells [46] and located within a chromosomal region linked to susceptibility to familial CLL [217]. Moreover, a profile of 19 T-UCRs (8 up- and 11 down-regulated) was able to differentiate between normal, CLL, colorectal, and hepatocarcinoma samples. Expression of five T-UCRs was able to divide a CLL cohort into two prognostic groups [46]. Expression of these diagnostic T-UCRs negatively correlated with a previously defined CLL miRNA signature, suggesting a mechanism for miRNA regulation of these T-UCRs [218].

3.5. LncRNAs Poorly Characterized in Haematological Malignancies

***LincRNA-p21*:** is a p53 activated lncRNA identified in mouse that binds to and guides hnRNP K to target genes. *LincRNA-p21* bound hnRNP K acts as a transcriptional repressor that leads to the induction of apoptosis [23]. As BCR-ABL1 stimulates hnRNP-K expression and stability and promotes tumor progression, it has been suggested that *lincRNA-p21* could play a relevant role in acute or chronic leukemia [219,220]. Furthermore, *lincRNA-p21* can inhibit the translation of target mRNAs [104]. In the absence of HuR, *lincRNA-p21* is stable and interacts with the mRNAs *CTNNB1*, *JUNB* and translational repressor Rck, repressing the translation of the targeted mRNAs [104].

TCL6: T cell Leukemia/Lymphoma 6 (*TCL6*) is transcribed from a locus involved in translocations and inversions with T cell receptor (*TCR*) [221]. These rearrangements in TCR commonly lead to activation of *TCL6* lncRNA and other oncogenes related to T cell leukemogenesis [151].

***WT1-AS*:** is an antisense lncRNA to WT-1, a well-characterized developmental gene that is mutated in Wilms' tumor (WT) and AML. *WT1-AS* has been shown to regulate WT1 protein levels. *WT1-AS* binds the exon 1 of WT1 mRNA in the cytoplasm. It has been suggested that the abnormal splicing of *WT1-AS* in AML could play a role in the development of this malignancy [159].

***CRNDE*:** is overexpressed, compared to healthy tissue, in more than 90% of colorectal adenomas tested, but also in hepatocellular, prostate, brain, kidney and pancreas carcinomas and different haematological neoplasia such as AML, MM and T cell leukemia [160]. *CRNDE* has been described as downregulated in ovarian cancer and tends to be overexpressed in non-differentiated tissues *versus* differentiated controls [160]. *CRNDE* binds PRC2 and the downregulation of *CRNDE* causes upregulation of PRC2 regulated genes, decreases growth and increases apoptosis [66].

RMRP: Ribonuclease mitochondrial RNA processing (*RMRP)* is a lncRNA mutated in Cartilage-Hair Hypoplasia (CHH), an autosomal recessive chondrodysplasia with short stature, which entails a high risk of developing Non-Hodgkin lymphoma disease [161,222].

SNHG5: is a precurssor of snoRNAs, similar to *GAS5*, located at the breakpoint of the chromosomal translocation t(3;6)(q27;q15), involved in diffuse large B-cell lymphoma [223].

***HOXA-AS2: HOXA** Cluster Antisense RNA 2* (*HOXA-AS2*) lncRNA is antisense to *HOX3* and *HOX4* coding genes. In an acute promyelocytic leukemia (APL) cell line, *HOXA-AS2* upregulation correlated with inhibition of apoptosis. Treatment with all-*trans* retinoic acid (ATRA) blocked the expression of *HOXA-AS2* and increased apoptosis of the APL cell line [224].

4. LncRNAs Involved in Hematopoiesis

The best studied lncRNA in hematopoiesis is *HOTAIRM1* (HOX antisense intergenic RNA myeloid 1). *HOTAIRM1* is as an essential regulator of myeloid cell differentiation that locates at the 3' end of the *HOXA* cluster and controls *HOXA1* expression [164]. HOXA genes are important transcriptional regulators in normal and malignant hematopoiesis and are known to be important for many cancers including leukemias harbouring MLL rearrangements. *HOTAIRM1* is expressed specifically in the myeloid lineage and is induced during the retinoic acid-driven granulocytic differentiation of the NB4 promyelocytic leukaemia cell line and normal human hematopoietic cells. Knockdown of *HOTAIRM1* affects retinoic acid-induced expression of *HOXA1* and *HOXA4* (but not distal *HOXA* genes) and attenuates induction of myeloid differentiation genes [164].

Other lncRNAs involved in hematopoiesis have also been described. *EGO* (or *EGOT* in human) lncRNA was identified in mouse eosinophil differentiation of CD34+HSCs where it stimulated major basic protein and eosinophil-derived neurotoxin mRNA expression [165]. The lncRNA *PU.1-AS* is an antisense transcript of *PU.1* that negatively regulates *PU.1* mRNA translation by a mechanism similar to miRNAs [166]. *PU.1* is a master hematopoietic transcriptional regulator essential for normal hematopoietic development and suppression of leukaemia development. LincRNA erythroid prosurvival (*EPS*) is one of the about 400 lncRNAs whose expression is modulated during red blood cell formation and is required for differentiation during hematopoiesis in mouse [164,165,167]. *EPS* is an erythroid-specific lncRNA that represses expression of *PYCARD*, a proapoptotic gene, and therefore inhibits apoptosis [167,225]. EPS is not well conserved among mammals. It is presently unclear whether a human version of EPS exists. Finally, *THY-ncR1* is a thymus-specific lncRNA expressed in cell lines derived from stage III immature T cells in which the neighbouring *CD1* gene cluster is also specifically activated [168].

5. Regulation of the Expression of lncRNAs Involved in Haematological Malignancies

Altered expression of lncRNAs, similar to that of coding genes, can be the result of genomic alterations, epigenetic regulation or a change in response to transcription factors or stability effectors such as miRNAs.

The presence of mutations in the lncRNA primary sequence correlates highly with human diseases. In fact, most mutations in the genome occur in noncoding regions [226]. Mutations can be large or small. Large-scale mutations are deletions and amplifications of hundreds of nucleotides and chromosomal translocations occurring at fragile sites. Genome-wide analyses looking for fragile sites in lncRNA genes have not yet been performed. However, it is expected that lncRNAs will have a clear association with common chromosomal aberrations similar to that found for miRNAs in human haematological malignancies and carcinomas [46]. In fact, several studies have described lncRNAs affected by large scale mutations. One of the best examples is *ANRIL*, affected by a large germline deletion that includes the complete INK4/ARF locus. This deletion is associated with hereditary cutaneous malignant melanoma and neural system tumors syndrome [179]. *DLEU1* and *DLEU2* lncRNAs also locate in a region commonly deleted in CLL (see above).

Small scale mutations are deletions or insertions of a few nucleotides. The relevance of small scale mutations for lncRNAs is obscured by the fact that little is known about the relevance of the primary sequence in lncRNA functionality and expression. It is expected that small mutations can lead to disease if they affect relevant linear sequences or they alter the structure of domains important in lncRNA functionality or accumulation. In fact, several disease-associated SNPs have been described as affecting the structure of the 5' and 3' non-translated regions of coding genes [226]. Furthermore, GWAS studies have shown that SNPs in noncoding regions are associated with higher susceptibility to diverse diseases. Germline and somatic mutations in lncRNA genes have been identified in haematological malignancies and colorectal cancers [227]. SNPs that may affect *ANRIL* have been associated with increased risk of type 2 diabetes and increased susceptibility to coronary artery disease and atherosclerosis [228,229]. Some of these mutations did not affect *ANRIL* transcription or stability. Instead, they disrupt *ANRIL* splicing, resulting in a circular transcript, affecting normal *ANRIL* function and influencing *INK4/ARF* locus expression [180]. Moreover, genetic aberrations of the *GAS5* locus have been found in melanoma, breast and prostate cancers [230–232].

Several lncRNAs are regulated at the transcriptional level. Thus, lncRNAs, such as *lincRNA-P21*, are activated in response to DNA damage by the direct binding of the tumour-suppressor protein p53 to the promoter [23]. Similarly, the expression of several lincRNAs responds to pluripotency factors or oncogenes.

Epigenetic modifications are key regulators of lncRNA expression. This has been well described for *MEG3* and *DLEU1/DLEU2*. Expression of the *MEG3* locus is regulated by two regions, which are hypermethylated in several solid tumours leading to downregulation of *MEG3* expression [185,233,234]. AML patients with aberrant hypermethylation of the *MEG3* promoter showed decreased overall survival [235,236]. Thus, *MEG3* methylation status may serve as a useful biomarker in this leukemia. A similar *MEG3* hypermethylation was observed in 35% of the patients with myelodysplastic syndrome, but in this case there was no statistically significant correlation between *MEG3* hypermethylation and prognosis [235]. Similarly, conserved CpG islands at the transcriptional start sites of *DLEU1* and *DLEU2* were found to be significantly demethylated in a cohort of 143 patients with CLL [156]. Demethylation correlated with transcriptional deregulation of the neighbouring candidate tumour suppressor genes. T-UCRs expression has also been shown to be repressed by CpG island hypermethylation [47,213].

Finally, the expression of lncRNAs can be regulated by miRNAs. Several miRNAs have been described as regulating T-URC expression. This has been best described for *miR-155*, which is overexpressed in CLL compared to healthy cells. *miR-155* targets T-UCRs both *in vitro* and in CLL patient samples [46]. Interestingly, *miR-29a* has also been shown to regulate *MEG3* expression in hepatocarcinoma cell lines [186].

6. Concluding Remarks

The identification of lncRNAs and the functional relevance of the lncRNAs studied so far has changed the view about genomes, transcriptomes and gene expression regulation. As the lncRNA field is in its infancy, surprising results are still expected, but a tremendous amount of work needs to be done. Firstly, a systematic identification and annotation of lncRNAs and their expression patterns should be performed and made publically available. As most lncRNAs are tissue specific, all tissues should be profiled. Also, as there is poor sequence conservation between lncRNAs of different species, efforts should be devoted to describing a collection of lncRNAs in different species, including human, mouse, rat, zebra fish, fly, *Arabidopsis* and yeast. As the regulation of expression of lncRNAs is tightly controlled, lncRNAs should also be described in cells responding to different stimuli and in diseased cells. These studies will be complicated further by the fact that lncRNA genes may be transcribed to different transcripts by alternative splicing, polyadenylation and the use of different promoters. It is also necessary to develop a new universal nomenclature that would facilitate routine work with these non coding RNA molecules.

Secondly, functional studies should be performed. Gain and loss of function studies could be carried out to analyze the impact of the lncRNA on the cell phenotype. Transcriptome analysis coupled with gain and loss of function studies could provide clues regarding the cellular pathways affected by the lncRNA, especially if the lncRNA of interest is a regulator of the expression of specific genes. Analysis of lncRNA subcellular localization can also give clues to lncRNA functionality. This can be done with Fish-like techniques that use several labelled oligos at a time. This is essential to detect the expression of lncRNAs, which are generally very structured and not very abundant. The functional domains of lncRNAs should be identified and it should be ruled out that lncRNAs function through the translation of short peptides. Furthermore, it would be desirable to determine the structure of key domains in lncRNAs similarly to what has been done with proteins. This is a major task as there are no reliable methods to determine the secondary structures of lncRNAs with bioinformatic tools. Chemical probing and point mutation studies have been used to determine the structure of many RNAs, but these techniques are very time consuming. Faster results could be obtained by parallel analysis of RNA structure (PARS-Seq) or Frag-Seq, which uses deep sequencing of RNA fragments obtained from RNAs treated with specific RNases that cleave RNA at highly selective structural positions [237]. Furthermore, it would be interesting to identify the factors that bind to relevant lncRNAs. Ideally, specific RNAs should be immunoprecipitated and subjected to mass spectrometry to identify RNA binding proteins. This is not easy, but has been done successfully with pools of cellular RNAs purified by binding to oligodT beads [238]. Theoretically, a lncRNA of interest could be labelled with a domain targeted by a specific protein and the complex could be purified with antibodies specific to the protein. Alternatively, the lncRNA could be immunoprecipitated from cell extracts using biotinylated tiling oligos and streptavidin. Then, lncRNA bound DNA or RNA can be sequenced from the immunoprecipitates. When looking for DNA interactors, this technique has been named Chromatin Isolation by RNA Purification (ChIRP) and has allowed the identification of the natural regions of chromatin that interact with a given lncRNA [58]. Finally, the lncRNA can be transcribed and labelled *in vitro*, incubated with cell extracts and immunoprecipitated with label binding factors.

Finally, detailed analysis of functional lncRNAs will most probably reveal interesting cellular pathways and help to design the architecture of biological tools that may be of interest for biotechnological development. Domains of lncRNAs that function as decoys for miRNAs or transcription factors, mimicking *GAS5* function [239], could be expressed to obtain therapeutic effects. Several lncRNA domains with a specific tertiary structure and a given function could probably be combined to generate lncRNAs with novel functions that could be of therapeutic interest. For instance, an RNA domain involved in the binding to a specific region of the chromatin could be fused to an RNA domain that interacts with factors that silence or activate gene expression or that induce chromosome bendings or genome reorganizations at the specific position. This could be used for silencing of oncogenes or reactivation of tumour suppressor genes. Thus, analysis of the function of lncRNAs is expected to have a tremendous impact on the management of human disease.

Furthermore, strong associations between some lncRNAs and some human diseases have been described. The number of lncRNAs relevant to human diseases is expected to increase as a result of the systematic identification of lncRNAs whose expression is altered in healthy and diseased cells and by genome-wide association studies. In fact GWAS analysis has identified *ANRIL* as a lncRNA involved in atherosclerosis, coronary artery disease, and type 2 diabetes [179]. In the case of cancer and specifically of haematological malignancies, GWAS results at lncRNA loci may identify patient populations at risk of cancer, may classify patients into aggressive or mild cancer groups and may predict a patient's response to a given therapy [240,241]. Once lncRNAs related to a disease are described, the issue should be addressed whether they are useful signatures for early disease detection, for prognosis or can be used as candidate drug targets for disease intervention [242].

lncRNAs may have specific advantages when used as diagnostic biomarkers, as some show tissue-specific and cancer-specific expression patterns [243]. This is the case of *HULC*, a liver-specific lncRNA highly expressed in primary liver tumours and hepatic metastases of colorectal carcinoma, but not found in primary colon cancers or in non-liver metastases [244,245]. Thus, the expression of *HULC* and other lncRNAs can be used to differentiate between subtypes of the same cancer or to identify unknown primary tumours. Similarly, *PCGEM1*, *PCA3* or *PRNCR1* are three lncRNAs exclusively associated with prostate cancer [123,246,247]. Also, as in the case with miRNAs, some lncRNAs can be detected in body fluids by quantitative reverse transcriptase polymerase chain reaction and therefore enable non-invasive diagnoses. In fact, *HULC* can be detected in the blood of hepatocellular carcinoma patients using qRT-PCR [245]. The ProgensaTM PCA3 urine test, a kit to detect *PCA3* in urine samples from patients with prostate cancer is already being clinically used [248,249]. This specific test can help patients who had a first negative prostate biopsy to avoid unnecessary repeated biopsies [250]. In spite of this fast clinical translation for *PCA3* analysis in prostate cancer, the biological function of *PCA3* is unknown.

LncRNAs can also be used as predictive markers, as lncRNA expression can correlate with patient outcome or response to chemotherapy. Thus, the expression of *HOTAIR* correlates with metastasis and poor outcome in primary breast tumors, gastrointestinal, hepatocellular and colorectal cancers and the expression of *MALAT* correlates with survival in early-stage lung adenocarcinoma [68,124,251–253]. Also, the expression of *XIST* correlates with disease-free survival of Taxol-treated cancer patients [254].

Finally, lncRNAs could be used therapeutically. In cancer, expression of tumour suppressor lncRNAs, such as *GAS5* or *MEG3*, should decrease tumour growth. When the downregulation of tumour suppressor lncRNAs results from aberrant epigenetic mechanisms such as DNA hypermethylation or loss of histone acetylation, demethylating agents or histone deacetylase inhibitors could help to reestablish expression. Otherwise, expression of lncRNAs may require gene therapy delivery systems with viral vectors, which are not efficient in targeting all cells within a tumour. Furthermore, RNA interference can be used to decrease the expression of lncRNAs with oncogenic properties. While many lncRNAs have been silenced using siRNAs, it is generally believed that the secondary structure of lncRNAs hinders siRNA functionality. Instead, expression of lncRNAs with oncogenic or tumour suppressor molecules could be altered with small molecules that affect their promoters. Small molecules, aptamers or stable antisense oligonucleotides could also be identified that target essential structures for oncogenic lncRNA functionality. Thus, preventing the interactions of *HOTAIR* with PRC2, for example, may limit the metastatic potential of breast cancer cells [255]. Even if all these strategies are possible, much investment in this field will be required to transfer lncRNA research to clinical oncology.

Acknowledgments

Supported in part by grants from Ministerio de Ciencia e Innovacion and Instituto de Salud Carlos III (ISCIII) BIO2009/09295, SAF2012-40003, PI10/01691, PI10/02983, RD12/0036/0068, CP07/00215, FEDER funding, FP7 Marie Curie Program PIOF-GA-2012-330598 LincMHeM, funds from the "UTE project CIMA" and by the project RNAREG [CSD2009-00080], funded by the Ministry of Science and Innovation under the programme CONSOLIDER INGENIO 2010.

References

1. Mattick, J.S. Non-coding RNAs: The architects of eukaryotic complexity. *EMBO Rep.* **2001**, *2*, 986–991.
2. Dunham, I.; Kundaje, A.; Aldred, S.F.; Collins, P.J.; Davis, C.A.; Doyle, F.; Epstein, C.B.; Frietze, S.; Harrow, J.; Kaul, R.; *et al.* An integrated encyclopedia of DNA elements in the human genome. *Nature* **2012**, *489*, 57–74.
3. Lin, M.F.; Carlson, J.W.; Crosby, M.A.; Matthews, B.B.; Yu, C.; Park, S.; Wan, K.H.; Schroeder, A.J.; Gramates, L.S.; St Pierre, S.E.; *et al.* Revisiting the protein-coding gene catalog of *Drosophila melanogaster* using 12 fly genomes. *Genome Res.* **2007**, *17*, 1823–1836.
4. Lin, M.F.; Deoras, A.N.; Rasmussen, M.D.; Kellis, M. Performance and scalability of discriminative metrics for comparative gene identification in 12 *Drosophila genomes. PLoS Comput. Biol.* **2008**, *4*, e1000067.
5. Pueyo, J.I.; Couso, J.P. Tarsal-less peptides control Notch signalling through the Shavenbaby transcription factor. *Dev. Biol.* **2011**, *355*, 183–193.
6. Ingolia, N.T.; Lareau, L.F.; Weissman, J.S. Ribosome profiling of mouse embryonic stem cells reveals the complexity and dynamics of mammalian proteomes. *Cell* **2011**, *147*, 789–802.
7. Banfai, B.; Jia, H.; Khatun, J.; Wood, E.; Risk, B.; Gundling, W.E., Jr.; Kundaje, A.; Gunawardena, H.P.; Yu, Y.; Xie, L.; *et al.* Long noncoding RNAs are rarely translated in two human cell lines. *Genome Res.* **2012**, *22*, 1646–1657.
8. Derrien, T.; Johnson, R.; Bussotti, G.; Tanzer, A.; Djebali, S.; Tilgner, H.; Guernec, G.; Martin, D.; Merkel, A.; Knowles, D.G.; *et al.* The GENCODE v7 catalog of human long noncoding RNAs: Analysis of their gene structure, evolution, and expression. *Genome Res.* **2012**, *22*, 1775–1789.
9. Kloc, M.; Wilk, K.; Vargas, D.; Shirato, Y.; Bilinski, S.; Etkin, L.D. Potential structural role of non-coding and coding RNAs in the organization of the cytoskeleton at the vegetal cortex of Xenopus oocytes. *Development* **2005**, *132*, 3445–3457.
10. Wadler, C.S.; Vanderpool, C.K. A dual function for a bacterial small RNA: SgrS performs base pairing-dependent regulation and encodes a functional polypeptide. *Proc. Natl. Acad. Sci. USA* **2007**, *104*, 20454–20459.
11. Dinger, M.E.; Pang, K.C.; Mercer, T.R.; Mattick, J.S. Differentiating protein-coding and noncoding RNA: Challenges and ambiguities. *PLoS Comput. Biol.* **2008**, *4*, e1000176.
12. Leygue, E. Steroid receptor RNA activator (SRA1): Unusual bifaceted gene products with suspected relevance to breast cancer. *Nucl. Recept. Signal.* **2007**, *5*, e006.

13. Jenny, A.; Hachet, O.; Zavorszky, P.; Cyrklaff, A.; Weston, M.D.; Johnston, D.S.; Erdelyi, M.; Ephrussi, A. A translation-independent role of oskar RNA in early *Drosophila oogenesis. Development* **2006**, *133*, 2827–2833.
14. Candeias, M.M.; Malbert-Colas, L.; Powell, D.J.; Daskalogianni, C.; Maslon, M.M.; Naski, N.; Bourougaa, K.; Calvo, F.; Fahraeus, R. P53 mRNA controls p53 activity by managing Mdm2 functions. *Nat. Cell Biol.* **2008**, *10*, 1098–1105.
15. Guttman, M.; Amit, I.; Garber, M.; French, C.; Lin, M.F.; Feldser, D.; Huarte, M.; Zuk, O.; Carey, B.W.; Cassady, J.P.; *et al.* Chromatin signature reveals over a thousand highly conserved large non-coding RNAs in mammals. *Nature* **2009**, *458*, 223–227.
16. Tilgner, H.; Knowles, D.G.; Johnson, R.; Davis, C.A.; Chakrabortty, S.; Djebali, S.; Curado, J.; Snyder, M.; Gingeras, T.R.; Guigo, R. Deep sequencing of subcellular RNA fractions shows splicing to be predominantly co-transcriptional in the human genome but inefficient for lncRNAs. *Genome Res.* **2012**, *22*, 1616–1625.
17. Gibb, E.A.; Vucic, E.A.; Enfield, K.S.; Stewart, G.L.; Lonergan, K.M.; Kennett, J.Y.; Becker-Santos, D.D.; MacAulay, C.E.; Lam, S.; Brown, C.J.; *et al.* Human cancer long non-coding RNA transcriptomes. *PLoS One* **2011**, *6*, e25915.
18. Mercer, T.R.; Dinger, M.E.; Sunkin, S.M.; Mehler, M.F.; Mattick, J.S. Specific expression of long noncoding RNAs in the mouse brain. *Proc. Natl. Acad. Sci. USA* **2008**, *105*, 716–721.
19. Djebali, S.; Davis, C.A.; Merkel, A.; Dobin, A.; Lassmann, T.; Mortazavi, A.; Tanzer, A.; Lagarde, J.; Lin, W.; Schlesinger, F.; *et al.* Landscape of transcription in human cells. *Nature* **2012**, *489*, 101–108.
20. Hauptman, N.; Glavac, D. Long non-coding RNA in cancer. *Int. J. Mol. Sci.* **2013**, *14*, 4655–4669.
21. Agirre, X.; Jimenez-Velasco, A.; San Jose-Eneriz, E.; Garate, L.; Bandres, E.; Cordeu, L.; Aparicio, O.; Saez, B.; Navarro, G.; Vilas-Zornoza, A.; *et al.* Down-regulation of hsa-miR-10a in chronic myeloid leukemia CD34+ cells increases USF2-mediated cell growth. *Mol. Cancer Res.* **2008**, *6*, 1830–1840.
22. Agirre, X.; Vilas-Zornoza, A.; Jimenez-Velasco, A.; Martin-Subero, J.I.; Cordeu, L.; Garate, L.; San Jose-Eneriz, E.; Abizanda, G.; Rodriguez-Otero, P.; Fortes, P.; *et al.* Epigenetic silencing of the tumor suppressor microRNA Hsa-miR-124a regulates CDK6 expression and confers a poor prognosis in acute lymphoblastic leukemia. *Cancer Res.* **2009**, *69*, 4443–4453.
23. Huarte, M.; Guttman, M.; Feldser, D.; Garber, M.; Koziol, M.J.; Kenzelmann-Broz, D.; Khalil, A.M.; Zuk, O.; Amit, I.; Rabani, M.; *et al.* A large intergenic noncoding RNA induced by p53 mediates global gene repression in the p53 response. *Cell* **2010**, *142*, 409–419.
24. Katayama, S.; Tomaru, Y.; Kasukawa, T.; Waki, K.; Nakanishi, M.; Nakamura, M.; Nishida, H.; Yap, C.C.; Suzuki, M.; Kawai, J.; *et al.* Antisense transcription in the mammalian transcriptome. *Science* **2005**, *309*, 1564–1566.
25. Carninci, P.; Kasukawa, T.; Katayama, S.; Gough, J.; Frith, M.C.; Maeda, N.; Oyama, R.; Ravasi, T.; Lenhard, B.; Wells, C.; *et al.* The transcriptional landscape of the mammalian genome. *Science* **2005**, *309*, 1559–1563.
26. Galante, P.A.; Vidal, D.O.; de Souza, J.E.; Camargo, A.A.; de Souza, S.J. Sense-antisense pairs in mammals: Functional and evolutionary considerations. *Genome Biol.* **2007**, *8*, R40.

27. Yap, K.L.; Li, S.; Munoz-Cabello, A.M.; Raguz, S.; Zeng, L.; Mujtaba, S.; Gil, J.; Walsh, M.J.; Zhou, M.M. Molecular interplay of the noncoding RNA ANRIL and methylated histone H3 lysine 27 by polycomb CBX7 in transcriptional silencing of INK4a. *Mol. Cell* **2010**, *38*, 662–674.
28. Krystal, G.W.; Armstrong, B.C.; Battey, J.F. *N*-myc mRNA forms an RNA-RNA duplex with endogenous antisense transcripts. *Mol. Cell Biol.* **1990**, *10*, 4180–4191.
29. Hastings, M.L.; Milcarek, C.; Martincic, K.; Peterson, M.L.; Munroe, S.H. Expression of the thyroid hormone receptor gene, erbAalpha, in B lymphocytes: Alternative mRNA processing is independent of differentiation but correlates with antisense RNA levels. *Nucleic Acids Res.* **1997**, *25*, 4296–4300.
30. Yan, M.D.; Hong, C.C.; Lai, G.M.; Cheng, A.L.; Lin, Y.W.; Chuang, S.E. Identification and characterization of a novel gene Saf transcribed from the opposite strand of Fas. *Hum. Mol. Genet.* **2005**, *14*, 1465–1474.
31. Annilo, T.; Kepp, K.; Laan, M. Natural antisense transcript of natriuretic peptide precursor A (NPPA): Structural organization and modulation of NPPA expression. *BMC Mol. Biol.* **2009**, *10*, 81.
32. Allo, M.; Buggiano, V.; Fededa, J.P.; Petrillo, E.; Schor, I.; de la Mata, M.; Agirre, E.; Plass, M.; Eyras, E.; Elela, S.A.; *et al.* Control of alternative splicing through siRNA-mediated transcriptional gene silencing. *Nat. Struct. Mol. Biol.* **2009**, *16*, 717–724.
33. Faghihi, M.A.; Modarresi, F.; Khalil, A.M.; Wood, D.E.; Sahagan, B.G.; Morgan, T.E.; Finch, C.E.; St Laurent, G., 3rd; Kenny, P.J.; Wahlestedt, C. Expression of a noncoding RNA is elevated in Alzheimer's disease and drives rapid feed-forward regulation of beta-secretase. *Nat. Med.* **2008**, *14*, 723–730.
34. Guil, S.; Esteller, M. cis-acting noncoding RNAs: Friends and foes. *Nat. Struct. Mol. Biol.* **2012**, *19*, 1068–1075.
35. Heintzman, N.D.; Stuart, R.K.; Hon, G.; Fu, Y.; Ching, C.W.; Hawkins, R.D.; Barrera, L.O.; Van Calcar, S.; Qu, C.; Ching, K.A.; *et al.* Distinct and predictive chromatin signatures of transcriptional promoters and enhancers in the human genome. *Nat. Genet.* **2007**, *39*, 311–318.
36. Heintzman, N.D.; Hon, G.C.; Hawkins, R.D.; Kheradpour, P.; Stark, A.; Harp, L.F.; Ye, Z.; Lee, L.K.; Stuart, R.K.; Ching, C.W.; *et al.* Histone modifications at human enhancers reflect global cell-type-specific gene expression. *Nature* **2009**, *459*, 108–112.
37. Visel, A.; Blow, M.J.; Li, Z.; Zhang, T.; Akiyama, J.A.; Holt, A.; Plajzer-Frick, I.; Shoukry, M.; Wright, C.; Chen, F.; *et al.* ChIP-seq accurately predicts tissue-specific activity of enhancers. *Nature* **2009**, *457*, 854–858.
38. Kim, T.K.; Hemberg, M.; Gray, J.M.; Costa, A.M.; Bear, D.M.; Wu, J.; Harmin, D.A.; Laptewicz, M.; Barbara-Haley, K.; Kuersten, S.; *et al.* Widespread transcription at neuronal activity-regulated enhancers. *Nature* **2010**, *465*, 182–187.
39. Wang, D.; Garcia-Bassets, I.; Benner, C.; Li, W.; Su, X.; Zhou, Y.; Qiu, J.; Liu, W.; Kaikkonen, M.U.; Ohgi, K.A.; *et al.* Reprogramming transcription by distinct classes of enhancers functionally defined by eRNA. *Nature* **2011**, *474*, 390–394.
40. De Santa, F.; Barozzi, I.; Mietton, F.; Ghisletti, S.; Polletti, S.; Tusi, B.K.; Muller, H.; Ragoussis, J.; Wei, C.L.; Natoli, G. A large fraction of extragenic RNA pol II transcription sites overlap enhancers. *PLoS Biol.* **2010**, *8*, e1000384.

41. Orom, U.A.; Derrien, T.; Beringer, M.; Gumireddy, K.; Gardini, A.; Bussotti, G.; Lai, F.; Zytnicki, M.; Notredame, C.; Huang, Q.; *et al.* Long noncoding RNAs with enhancer-like function in human cells. *Cell* **2010**, *143*, 46–58.
42. Melo, C.A.; Drost, J.; Wijchers, P.J.; van de Werken, H.; de Wit, E.; Oude Vrielink, J.A.; Elkon, R.; Melo, S.A.; Leveille, N.; Kalluri, R.; *et al.* eRNAs are required for p53-dependent enhancer activity and gene transcription. *Mol. Cell.* **2013**, *49*, 524–535.
43. Lai, F.; Orom, U.A.; Cesaroni, M.; Beringer, M.; Taatjes, D.J.; Blobel, G.A.; Shiekhattar, R. Activating RNAs associate with Mediator to enhance chromatin architecture and transcription. *Nature* **2013**, *494*, 497–501.
44. Bejerano, G.; Pheasant, M.; Makunin, I.; Stephen, S.; Kent, W.J.; Mattick, J.S.; Haussler, D. Ultraconserved elements in the human genome. *Science* **2004**, *304*, 1321–1325.
45. Ahituv, N.; Zhu, Y.; Visel, A.; Holt, A.; Afzal, V.; Pennacchio, L.A.; Rubin, E.M. Deletion of ultraconserved elements yields viable mice. *PLoS Biol.* **2007**, *5*, e234.
46. Calin, G.A.; Liu, C.G.; Ferracin, M.; Hyslop, T.; Spizzo, R.; Sevignani, C.; Fabbri, M.; Cimmino, A.; Lee, E.J.; Wojcik, S.E.; *et al.* Ultraconserved regions encoding ncRNAs are altered in human leukemias and carcinomas. *Cancer Cell.* **2007**, *12*, 215–229.
47. Lujambio, A.; Portela, A.; Liz, J.; Melo, S.A.; Rossi, S.; Spizzo, R.; Croce, C.M.; Calin, G.A.; Esteller, M. CpG island hypermethylation-associated silencing of non-coding RNAs transcribed from ultraconserved regions in human cancer. *Oncogene* **2010**, *29*, 6390–6401.
48. Hirotsune, S.; Yoshida, N.; Chen, A.; Garrett, L.; Sugiyama, F.; Takahashi, S.; Yagami, K.; Wynshaw-Boris, A.; Yoshiki, A. An expressed pseudogene regulates the messenger-RNA stability of its homologous coding gene. *Nature* **2003**, *423*, 91–96.
49. Hawkins, P.G.; Morris, K.V. Transcriptional regulation of Oct4 by a long non-coding RNA antisense to Oct4-pseudogene 5. *Transcription* **2010**, *1*, 165–175.
50. Harrison, P.M.; Zheng, D.; Zhang, Z.; Carriero, N.; Gerstein, M. Transcribed processed pseudogenes in the human genome: An intermediate form of expressed retrosequence lacking protein-coding ability. *Nucleic Acids Res.* **2005**, *33*, 2374–2383.
51. Pink, R.C.; Wicks, K.; Caley, D.P.; Punch, E.K.; Jacobs, L.; Carter, D.R. Pseudogenes: Pseudo-functional or key regulators in health and disease? *RNA* **2011**, *17*, 792–798.
52. He, L. Posttranscriptional regulation of PTEN dosage by noncoding RNAs. *Sci. Signal.* **2010**, *3*, pe39.
53. Salmena, L.; Poliseno, L.; Tay, Y.; Kats, L.; Pandolfi, P.P. A ceRNA hypothesis: The Rosetta Stone of a hidden RNA language? *Cell* **2011**, *146*, 353–358.
54. Ledford, H. Circular RNAs throw genetics for a loop. *Nature* **2013**, *494*, 415.
55. Kosik, K.S. Molecular biology: Circles reshape the RNA world. *Nature* **2013**, *495*, 322–324.
56. Memczak, S.; Jens, M.; Elefsinioti, A.; Torti, F.; Krueger, J.; Rybak, A.; Maier, L.; Mackowiak, S.D.; Gregersen, L.H.; Munschauer, M.; *et al.* Circular RNAs are a large class of animal RNAs with regulatory potency. *Nature* **2013**, *495*, 333–338.
57. Hansen, T.B.; Jensen, T.I.; Clausen, B.H.; Bramsen, J.B.; Finsen, B.; Damgaard, C.K.; Kjems, J. Natural RNA circles function as efficient microRNA sponges. *Nature* **2013**, *495*, 384–388.
58. Chu, C.; Qu, K.; Zhong, F.L.; Artandi, S.E.; Chang, H.Y. Genomic maps of long noncoding RNA occupancy reveal principles of RNA-chromatin interactions. *Mol. Cell* **2011**, *44*, 667–678.

59. Gabory, A.; Jammes, H.; Dandolo, L. The H19 locus: Role of an imprinted non-coding RNA in growth and development. *Bioessays* **2010**, *32*, 473–480.
60. Mancini-Dinardo, D.; Steele, S.J.; Levorse, J.M.; Ingram, R.S.; Tilghman, S.M. Elongation of the Kcnq1ot1 transcript is required for genomic imprinting of neighboring genes. *Genes Dev.* **2006**, *20*, 1268–1282.
61. Pauler, F.M.; Koerner, M.V.; Barlow, D.P. Silencing by imprinted noncoding RNAs: Is transcription the answer? *Trends Genet.* **2007**, *23*, 284–292.
62. Engstrom, P.G.; Suzuki, H.; Ninomiya, N.; Akalin, A.; Sessa, L.; Lavorgna, G.; Brozzi, A.; Luzi, L.; Tan, S.L.; Yang, L.; *et al.* Complex Loci in human and mouse genomes. *PLoS Genet.* **2006**, *2*, e47.
63. Li, Y.; Sasaki, H. Genomic imprinting in mammals: Its life cycle, molecular mechanisms and reprogramming. *Cell Res.* **2011**, *21*, 466–473.
64. Mohammad, F.; Mondal, T.; Kanduri, C. Epigenetics of imprinted long noncoding RNAs. *Epigenetics* **2009**, *4*, 277–286.
65. Nagano, T.; Mitchell, J.A.; Sanz, L.A.; Pauler, F.M.; Ferguson-Smith, A.C.; Feil, R.; Fraser, P. The Air noncoding RNA epigenetically silences transcription by targeting G9a to chromatin. *Science* **2008**, *322*, 1717–1720.
66. Khalil, A.M.; Guttman, M.; Huarte, M.; Garber, M.; Raj, A.; Rivea Morales, D.; Thomas, K.; Presser, A.; Bernstein, B.E.; van Oudenaarden, A.; *et al.* Many human large intergenic noncoding RNAs associate with chromatin-modifying complexes and affect gene expression. *Proc. Natl. Acad. Sci. USA* **2009**, *106*, 11667–11672.
67. Rinn, J.L.; Kertesz, M.; Wang, J.K.; Squazzo, S.L.; Xu, X.; Brugmann, S.A.; Goodnough, L.H.; Helms, J.A.; Farnham, P.J.; Segal, E.; *et al.* Functional demarcation of active and silent chromatin domains in human HOX loci by noncoding RNAs. *Cell* **2007**, *129*, 1311–1323.
68. Gupta, R.A.; Shah, N.; Wang, K.C.; Kim, J.; Horlings, H.M.; Wong, D.J.; Tsai, M.C.; Hung, T.; Argani, P.; Rinn, J.L.; *et al.* Long non-coding RNA HOTAIR reprograms chromatin state to promote cancer metastasis. *Nature* **2010**, *464*, 1071–1076.
69. Tsai, M.C.; Manor, O.; Wan, Y.; Mosammaparast, N.; Wang, J.K.; Lan, F.; Shi, Y.; Segal, E.; and Chang, H.Y. Long noncoding RNA as modular scaffold of histone modification complexes. *Science* **2010**, *329*, 689–693.
70. Schmitz, K.M.; Mayer, C.; Postepska, A.; Grummt, I. Interaction of noncoding RNA with the rDNA promoter mediates recruitment of DNMT3b and silencing of rRNA genes. *Genes Dev.* **2010**, *24*, 2264–2269.
71. Buske, F.A.; Mattick, J.S.; Bailey, T.L. Potential *in vivo* roles of nucleic acid triple-helices. *RNA Biol.* **2011**, *8*, 427–439.
72. Aguilera, A.; Garcia-Muse, T. R loops: From transcription byproducts to threats to genome stability. *Mol. Cell* **2012**, *46*, 115–124.
73. Jeon, Y.; Lee, J.T. YY1 tethers Xist RNA to the inactive X nucleation center. *Cell* **2011**, *146*, 119–133.
74. Hung, T.; Chang, H.Y. Long noncoding RNA in genome regulation: Prospects and mechanisms. *RNA Biol.* **2010**, *7*, 582–585.
75. Bonasio, R.; Tu, S.; Reinberg, D. Molecular signals of epigenetic states. *Science* **2010**, *330*, 612–616.

76. Sun, Q.; Csorba, T.; Skourti-Stathaki, K.; Proudfoot, N.J.; Dean, C. R-loop stabilization represses antisense transcription at the *Arabidopsis* FLC locus. *Science* **2013**, *340*, 619–621.
77. Bertani, S.; Sauer, S.; Bolotin, E.; Sauer, F. The noncoding RNA Mistral activates Hoxa6 and Hoxa7 expression and stem cell differentiation by recruiting MLL1 to chromatin. *Mol. Cell* **2011**, *43*, 1040–1046.
78. Dinger, M.E.; Amaral, P.P.; Mercer, T.R.; Pang, K.C.; Bruce, S.J.; Gardiner, B.B.; Askarian-Amiri, M.E.; Ru, K.; Solda, G.; Simons, C.; *et al.* Long noncoding RNAs in mouse embryonic stem cell pluripotency and differentiation. *Genome Res.* **2008**, *18*, 1433–1445.
79. Pandey, R.R.; Mondal, T.; Mohammad, F.; Enroth, S.; Redrup, L.; Komorowski, J.; Nagano, T.; Mancini-Dinardo, D.; Kanduri, C. Kcnq1ot1 antisense noncoding RNA mediates lineage-specific transcriptional silencing through chromatin-level regulation. *Mol. Cell* **2008**, *32*, 232–246.
80. Yao, H.; Brick, K.; Evrard, Y.; Xiao, T.; Camerini-Otero, R.D.; Felsenfeld, G. Mediation of CTCF transcriptional insulation by DEAD-box RNA-binding protein p68 and steroid receptor RNA activator SRA. *Genes Dev.* **2010**, *24*, 2543–2555.
81. Wang, K.C.; Yang, Y.W.; Liu, B.; Sanyal, A.; Corces-Zimmerman, R.; Chen, Y.; Lajoie, B.R.; Protacio, A.; Flynn, R.A.; Gupta, R.A.; *et al.* A long noncoding RNA maintains active chromatin to coordinate homeotic gene expression. *Nature* **2011**, *472*, 120–124.
82. Feng, J.; Bi, C.; Clark, B.S.; Mady, R.; Shah, P.; Kohtz, J.D. The Evf-2 noncoding RNA is transcribed from the Dlx-5/6 ultraconserved region and functions as a Dlx-2 transcriptional coactivator. *Genes Dev.* **2006**, *20*, 1470–1484.
83. Lanz, R.B.; McKenna, N.J.; Onate, S.A.; Albrecht, U.; Wong, J.; Tsai, S.Y.; Tsai, M.J.; O'Malley, B.W. A steroid receptor coactivator, SRA, functions as an RNA and is present in an SRC-1 complex. *Cell* **1999**, *97*, 17–27.
84. Caretti, G.; Schiltz, R.L.; Dilworth, F.J.; di Padova, M.; Zhao, P.; Ogryzko, V.; Fuller-Pace, F.V.; Hoffman, E.P.; Tapscott, S.J.; Sartorelli, V. The RNA helicases p68/p72 and the noncoding RNA SRA are coregulators of MyoD and skeletal muscle differentiation. *Dev. Cell.* **2006**, *11*, 547–560.
85. Watanabe, M.; Yanagisawa, J.; Kitagawa, H.; Takeyama, K.; Ogawa, S.; Arao, Y.; Suzawa, M.; Kobayashi, Y.; Yano, T.; Yoshikawa, H.; *et al.* A subfamily of RNA-binding DEAD-box proteins acts as an estrogen receptor alpha coactivator through the N-terminal activation domain (AF-1) with an RNA coactivator, SRA. *EMBO J.* **2001**, *20*, 1341–1352.
86. Lanz, R.B.; Chua, S.S.; Barron, N.; Soder, B.M.; DeMayo, F.; O'Malley, B.W. Steroid receptor RNA activator stimulates proliferation as well as apoptosis *in vivo*. *Mol. Cell Biol.* **2003**, *23*, 7163–7176.
87. Kino, T.; Hurt, D.E.; Ichijo, T.; Nader, N.; Chrousos, G.P. Noncoding RNA gas5 is a growth arrest- and starvation-associated repressor of the glucocorticoid receptor. *Sci. Signal.* **2010**, *3*, ra8.
88. Willingham, A.T.; Orth, A.P.; Batalov, S.; Peters, E.C.; Wen, B.G.; Aza-Blanc, P.; Hogenesch, J.B.; Schultz, P.G. A strategy for probing the function of noncoding RNAs finds a repressor of NFAT. *Science* **2005**, *309*, 1570–1573.
89. Hung, T.; Wang, Y.; Lin, M.F.; Koegel, A.K.; Kotake, Y.; Grant, G.D.; Horlings, H.M.; Shah, N.; Umbricht, C.; Wang, P.; *et al.* Extensive and coordinated transcription of noncoding RNAs within cell-cycle promoters. *Nat. Genet.* **2011**, *43*, 621–629.

90. Martianov, I.; Ramadass, A.; Serra Barros, A.; Chow, N.; Akoulitchev, A. Repression of the human dihydrofolate reductase gene by a non-coding interfering transcript. *Nature* **2007**, *445*, 666–670.
91. Mazo, A.; Hodgson, J.W.; Petruk, S.; Sedkov, Y.; Brock, H.W. Transcriptional interference: An unexpected layer of complexity in gene regulation. *J. Cell Sci.* **2007**, *120*, 2755–2761.
92. Tripathi, V.; Ellis, J.D.; Shen, Z.; Song, D.Y.; Pan, Q.; Watt, A.T.; Freier, S.M.; Bennett, C.F.; Sharma, A.; Bubulya, P.A.; *et al.* The nuclear-retained noncoding RNA MALAT1 regulates alternative splicing by modulating SR splicing factor phosphorylation. *Mol. Cell* **2010**, *39*, 925–938.
93. Mao, Y.S.; Sunwoo, H.; Zhang, B.; Spector, D.L. Direct visualization of the co-transcriptional assembly of a nuclear body by noncoding RNAs. *Nat. Cell Biol.* **2011**, *13*, 95–101.
94. Bond, C.S.; Fox, A.H. Paraspeckles: Nuclear bodies built on long noncoding RNA. *J. Cell Biol.* **2009**, *186*, 637–644.
95. Clemson, C.M.; Hutchinson, J.N.; Sara, S.A.; Ensminger, A.W.; Fox, A.H.; Chess, A.; Lawrence, J.B. An architectural role for a nuclear noncoding RNA: NEAT1 RNA is essential for the structure of paraspeckles. *Mol. Cell* **2009**, *33*, 717–726.
96. Chen, L.L.; Carmichael, G.G. Altered nuclear retention of mRNAs containing inverted repeats in human embryonic stem cells: Functional role of a nuclear noncoding RNA. *Mol. Cell* **2009**, *35*, 467–478.
97. Sunwoo, H.; Dinger, M.E.; Wilusz, J.E.; Amaral, P.P.; Mattick, J.S.; Spector, D.L. MEN epsilon/beta nuclear-retained non-coding RNAs are up-regulated upon muscle differentiation and are essential components of paraspeckles. *Genome Res.* **2009**, *19*, 347–359.
98. Eissmann, M.; Gutschner, T.; Hammerle, M.; Gunther, S.; Caudron-Herger, M.; Gross, M.; Schirmacher, P.; Rippe, K.; Braun, T.; Zornig, M.; *et al.* Loss of the abundant nuclear non-coding RNA MALAT1 is compatible with life and development. *RNA Biol.* **2012**, *9*, 1076–1087.
99. Ip, J.Y.; Nakagawa, S. Long non-coding RNAs in nuclear bodies. *Dev Growth Differ.* **2012**, *54*, 44–54.
100. Nakagawa, S.; Naganuma, T.; Shioi, G.; Hirose, T. Paraspeckles are subpopulation-specific nuclear bodies that are not essential in mice. *J. Cell Biol.* **2011**, *193*, 31–39.
101. Zhang, B.; Arun, G.; Mao, Y.S.; Lazar, Z.; Hung, G.; Bhattacharjee, G.; Xiao, X.; Booth, C.J.; Wu, J.; Zhang, C.; *et al.* The lncRNA Malat1 is dispensable for mouse development but its transcription plays a *cis*-regulatory role in the adult. *Cell Rep.* **2012**, *2*, 111–123.
102. Yin, Q.F.; Yang, L.; Zhang, Y.; Xiang, J.F.; Wu, Y.W.; Carmichael, G.G.; Chen, L.L. Long noncoding RNAs with snoRNA ends. *Mol. Cell.* **2012**, *48*, 219–230.
103. Carrieri, C.; Cimatti, L.; Biagioli, M.; Beugnet, A.; Zucchelli, S.; Fedele, S.; Pesce, E.; Ferrer, I.; Collavin, L.; Santoro, C.; *et al.* Long non-coding antisense RNA controls Uchl1 translation through an embedded SINEB2 repeat. *Nature* **2012**, *491*, 454–457.
104. Yoon, J.H.; Abdelmohsen, K.; Srikantan, S.; Yang, X.; Martindale, J.L.; De, S.; Huarte, M.; Zhan, M.; Becker, K.G.; Gorospe, M. LincRNA-p21 suppresses target mRNA translation. *Mol. Cell* **2012**, *47*, 648–655.

105. Kretz, M.; Siprashvili, Z.; Chu, C.; Webster, D.E.; Zehnder, A.; Qu, K.; Lee, C.S.; Flockhart, R.J.; Groff, A.F.; Chow, J.; *et al.* Control of somatic tissue differentiation by the long non-coding RNA TINCR. *Nature* **2013**, *493*, 231–235.
106. Gong, C.; Maquat, L.E. lncRNAs transactivate STAU1-mediated mRNA decay by duplexing with 3' UTRs via Alu elements. *Nature* **2011**, *470*, 284–288.
107. Cesana, M.; Cacchiarelli, D.; Legnini, I.; Santini, T.; Sthandier, O.; Chinappi, M.; Tramontano, A.; Bozzoni, I. A long noncoding RNA controls muscle differentiation by functioning as a competing endogenous RNA. *Cell* **2011**, *147*, 358–369.
108. Cai, X.; Cullen, B.R. The imprinted H19 noncoding RNA is a primary microRNA precursor. *RNA* **2007**, *13*, 313–316.
109. Wang, K.C.; Chang, H.Y. Molecular mechanisms of long noncoding RNAs. *Mol. Cell* **2011**, *43*, 904–914.
110. Spitale, R.C.; Tsai, M.C.; Chang, H.Y. RNA templating the epigenome: Long noncoding RNAs as molecular scaffolds. *Epigenetics* **2011**, *6*, 539–543.
111. Zappulla, D.C.; Cech, T.R. RNA as a flexible scaffold for proteins: Yeast telomerase and beyond. *Cold Spring Harb. Symp. Quant. Biol.* **2006**, *71*, 217–224.
112. Kertesz, M.; Wan, Y.; Mazor, E.; Rinn, J.L.; Nutter, R.C.; Chang, H.Y.; Segal, E. Genome-wide measurement of RNA secondary structure in yeast. *Nature* **2010**, *467*, 103–107.
113. Castello, A.; Fischer, B.; Eichelbaum, K.; Horos, R.; Beckmann, B.M.; Strein, C.; Davey, N.E.; Humphreys, D.T.; Preiss, T.; Steinmetz, L.M.; *et al.* Insights into RNA biology from an atlas of mammalian mRNA-binding proteins. *Cell* **2012**, *149*, 1393–1406.
114. Guttman, M.; Donaghey, J.; Carey, B.W.; Garber, M.; Grenier, J.K.; Munson, G.; Young, G.; Lucas, A.B.; Ach, R.; Bruhn, L.; *et al.* lincRNAs act in the circuitry controlling pluripotency and differentiation. *Nature* **2011**, *477*, 295–300.
115. Sheik Mohamed, J.; Gaughwin, P.M.; Lim, B.; Robson, P.; Lipovich, L. Conserved long noncoding RNAs transcriptionally regulated by Oct4 and Nanog modulate pluripotency in mouse embryonic stem cells. *RNA* **2010**, *16*, 324–337.
116. Taft, R.J.; Pang, K.C.; Mercer, T.R.; Dinger, M.; Mattick, J.S. Non-coding RNAs: Regulators of disease. *J. Pathol.* **2010**, *220*, 126–139.
117. Ishii, N.; Ozaki, K.; Sato, H.; Mizuno, H.; Saito, S.; Takahashi, A.; Miyamoto, Y.; Ikegawa, S.; Kamatani, N.; Hori, M.; *et al.* Identification of a novel non-coding RNA, MIAT, that confers risk of myocardial infarction. *J. Hum. Genet.* **2006**, *51*, 1087–1099.
118. Pasmant, E.; Laurendeau, I.; Heron, D.; Vidaud, M.; Vidaud, D.; Bieche, I. Characterization of a germ-line deletion, including the entire INK4/ARF locus, in a melanoma-neural system tumor family: Identification of ANRIL, an antisense noncoding RNA whose expression coclusters with ARF. *Cancer Res.* **2007**, *67*, 3963–3969.
119. Sonkoly, E.; Bata-Csorgo, Z.; Pivarcsi, A.; Polyanka, H.; Kenderessy-Szabo, A.; Molnar, G.; Szentpali, K.; Bari, L.; Megyeri, K.; Mandi, Y.; *et al.* Identification and characterization of a novel, psoriasis susceptibility-related noncoding RNA gene, PRINS. *J. Biol. Chem.* **2005**, *280*, 24159–24167.
120. Daughters, R.S.; Tuttle, D.L.; Gao, W.; Ikeda, Y.; Moseley, M.L.; Ebner, T.J.; Swanson, M.S.; Ranum, L.P. RNA gain-of-function in spinocerebellar ataxia type 8. *PLoS Genet.* **2009**, *5*, e1000600.

121. Khalil, A.M.; Faghihi, M.A.; Modarresi, F.; Brothers, S.P.; Wahlestedt, C. A novel RNA transcript with antiapoptotic function is silenced in fragile X syndrome. *PLoS One* **2008**, *3*, e1486.
122. Kogo, R.; Shimamura, T.; Mimori, K.; Kawahara, K.; Imoto, S.; Sudo, T.; Tanaka, F.; Shibata, K.; Suzuki, A.; Komune, S.; *et al.* Long noncoding RNA HOTAIR regulates polycomb-dependent chromatin modification and is associated with poor prognosis in colorectal cancers. *Cancer Res.* **2011**, *71*, 6320–6326.
123. Chung, S.; Nakagawa, H.; Uemura, M.; Piao, L.; Ashikawa, K.; Hosono, N.; Takata, R.; Akamatsu, S.; Kawaguchi, T.; Morizono, T.; *et al.* Association of a novel long non-coding RNA in 8q24 with prostate cancer susceptibility. *Cancer Sci.* **2011**, *102*, 245–252.
124. Yang, Z.; Zhou, L.; Wu, L.M.; Lai, M.C.; Xie, H.Y.; Zhang, F.; Zheng, S.S. Overexpression of long non-coding RNA HOTAIR predicts tumor recurrence in hepatocellular carcinoma patients following liver transplantation. *Ann. Surg. Oncol.* **2011**, *18*, 1243–1250.
125. Lai, M.C.; Yang, Z.; Zhou, L.; Zhu, Q.Q.; Xie, H.Y.; Zhang, F.; Wu, L.M.; Chen, L.M.; Zheng, S.S. Long non-coding RNA MALAT-1 overexpression predicts tumor recurrence of hepatocellular carcinoma after liver transplantation. *Med. Oncol.* **2012**, *29*, 1810–1816.
126. Calin, G.A.; Pekarsky, Y.; Croce, C.M. The role of microRNA and other non-coding RNA in the pathogenesis of chronic lymphocytic leukemia. *Best. Pract. Res. Clin. Haematol.* **2007**, *20*, 425–437.
127. Khaitan, D.; Dinger, M.E.; Mazar, J.; Crawford, J.; Smith, M.A.; Mattick, J.S.; Perera, R.J. The melanoma-upregulated long noncoding RNA SPRY4-IT1 modulates apoptosis and invasion. *Cancer Res.* **2011**, *71*, 3852–3862.
128. Huarte, M.; Rinn, J.L. Large non-coding RNAs: Missing links in cancer? *Hum. Mol. Genet.* **2010**, *19*, R152–R161.
129. Li, L.; Feng, T.; Lian, Y.; Zhang, G.; Garen, A.; Song, X. Role of human noncoding RNAs in the control of tumorigenesis. *Proc. Natl. Acad. Sci. USA* **2009**, *106*, 12956–12961.
130. Yu, W.; Gius, D.; Onyango, P.; Muldoon-Jacobs, K.; Karp, J.; Feinberg, A.P.; Cui, H. Epigenetic silencing of tumour suppressor gene p15 by its antisense RNA. *Nature* **2008**, *451*, 202–206.
131. Lawrie, C.H. MicroRNAs and lymphomagenesis: A functional review. *Br. J. Haematol.* **2013**, *160*, 571–581.
132. Agirre, X.; Martinez-Climent, J.A.; Odero, M.D.; Prosper, F. Epigenetic regulation of miRNA genes in acute leukemia. *Leukemia* **2012**, *26*, 395–403.
133. Elton, T.S.; Selemon, H.; Elton, S.M.; Parinandi, N.L. Regulation of the MIR155 host gene in physiological and pathological processes. *Gene* **2012**, doi:10.1016/j.gene.2012.12.009.
134. Vargova, K.; Curik, N.; Burda, P.; Basova, P.; Kulvait, V.; Pospisil, V.; Savvulidi, F.; Kokavec, J.; Necas, E.; Berkova, A.; *et al.* MYB transcriptionally regulates the miR-155 host gene in chronic lymphocytic leukemia. *Blood* **2011**, *117*, 3816–3825.
135. Nielsen, C.B.; Shomron, N.; Sandberg, R.; Hornstein, E.; Kitzman, J.; Burge, C.B. Determinants of targeting by endogenous and exogenous microRNAs and siRNAs. *RNA* **2007**, *13*, 1894–1910.
136. Ji, M.; Rao, E.; Ramachandrareddy, H.; Shen, Y.; Jiang, C.; Chen, J.; Hu, Y.; Rizzino, A.; Chan, W.C.; Fu, K.; *et al.* The miR-17–92 microRNA cluster is regulated by multiple mechanisms in B-cell malignancies. *Am. J. Pathol.* **2011**, *179*, 1645–1656.

137. Rinaldi, A.; Poretti, G.; Kwee, I.; Zucca, E.; Catapano, C.V.; Tibiletti, M.G.; Bertoni, F. Concomitant MYC and microRNA cluster miR-17–92 (C13orf25) amplification in human mantle cell lymphoma. *Leuk Lymphoma* **2007**, *48*, 410–412.
138. Hayashita, Y.; Osada, H.; Tatematsu, Y.; Yamada, H.; Yanagisawa, K.; Tomida, S.; Yatabe, Y.; Kawahara, K.; Sekido, Y.; Takahashi, T. A polycistronic microRNA cluster, miR-17–92, is overexpressed in human lung cancers and enhances cell proliferation. *Cancer Res.* **2005**, *65*, 9628–9632.
139. Humphreys, K.J.; Cobiac, L.; Le Leu, R.K.; Van der Hoek, M.B.; Michael, M.Z. Histone deacetylase inhibition in colorectal cancer cells reveals competing roles for members of the oncogenic miR-17–92 cluster. *Mol. Carcinog.* **2013**, *52*, 459–474.
140. Kunkeaw, N.; Jeon, S.H.; Lee, K.; Johnson, B.H.; Tanasanvimon, S.; Javle, M.; Pairojkul, C.; Chamgramol, Y.; Wongfieng, W.; Gong, B.; *et al.* Cell death/proliferation roles for nc886, a non-coding RNA, in the protein kinase R pathway in cholangiocarcinoma. *Oncogene* **2012**, doi:10.1038/onc.2012.382.
141. Treppendahl, M.B.; Qiu, X.; Sogaard, A.; Yang, X.; Nandrup-Bus, C.; Hother, C.; Andersen, M.K.; Kjeldsen, L.; Mollgaard, L.; Hellstrom-Lindberg, E.; *et al.* Allelic methylation levels of the noncoding VTRNA2–1 located on chromosome 5q31.1 predict outcome in AML. *Blood* **2012**, *119*, 206–216.
142. He, L.; Thomson, J.M.; Hemann, M.T.; Hernando-Monge, E.; Mu, D.; Goodson, S.; Powers, S.; Cordon-Cardo, C.; Lowe, S.W.; Hannon, G.J.; *et al.* A microRNA polycistron as a potential human oncogene. *Nature* **2005**, *435*, 828–833.
143. Nagoshi, H.; Taki, T.; Hanamura, I.; Nitta, M.; Otsuki, T.; Nishida, K.; Okuda, K.; Sakamoto, N.; Kobayashi, S.; Yamamoto-Sugitani, M.; *et al.* Frequent PVT1 rearrangement and novel chimeric genes PVT1-NBEA and PVT1-WWOX occur in multiple myeloma with 8q24 abnormality. *Cancer Res.* **2012**, *72*, 4954–4962.
144. Zeidler, R.; Joos, S.; Delecluse, H.J.; Klobeck, G.; Vuillaume, M.; Lenoir, G.M.; Bornkamm, G.W.; Lipp, M. Breakpoints of Burkitt's lymphoma t(8;22) translocations map within a distance of 300 kb downstream of MYC. *Genes Chromosomes Cancer* **1994**, *9*, 282–287.
145. Enciso-Mora, V.; Broderick, P.; Ma, Y.; Jarrett, R.F.; Hjalgrim, H.; Hemminki, K.; van den Berg, A.; Olver, B.; Lloyd, A.; Dobbins, S.E.; *et al.* A genome-wide association study of Hodgkin's lymphoma identifies new susceptibility loci at 2p16.1 (REL), 8q24.21 and 10p14 (GATA3). *Nat. Genet.* **2010**, *42*, 1126–1130.
146. Huppi, K.; Volfovsky, N.; Runfola, T.; Jones, T.L.; Mackiewicz, M.; Martin, S.E.; Mushinski, J.F.; Stephens, R.; Caplen, N.J. The identification of microRNAs in a genomically unstable region of human chromosome 8q24. *Mol. Cancer Res.* **2008**, *6*, 212–221.
147. Beck-Engeser, G.B.; Lum, A.M.; Huppi, K.; Caplen, N.J.; Wang, B.B.; Wabl, M. Pvt1-encoded microRNAs in oncogenesis. *Retrovirology* **2008**, *5*, 4.
148. Barsotti, A.M.; Beckerman, R.; Laptenko, O.; Huppi, K.; Caplen, N.J.; Prives, C. p53-Dependent induction of PVT1 and miR-1204. *J. Biol. Chem.* **2011**, *287*, 2509–2519.
149. Zhang, X.; Gejman, R.; Mahta, A.; Zhong, Y.; Rice, K.A.; Zhou, Y.; Cheunsuchon, P.; Louis, D.N.; Klibanski, A. Maternally expressed gene 3, an imprinted noncoding RNA gene, is associated with meningioma pathogenesis and progression. *Cancer Res.* **2010**, *70*, 2350–2358.

150. Zhou, Y.; Zhang, X.; Klibanski, A. MEG3 noncoding RNA: A tumor suppressor. *J. Mol. Endocrinol.* **2012**, *48*, R45–R53.
151. Saitou, M.; Sugimoto, J.; Hatakeyama, T.; Russo, G.; Isobe, M. Identification of the TCL6 genes within the breakpoint cluster region on chromosome 14q32 in T-cell leukemia. *Oncogene* **2000**, *19*, 2796–2802.
152. Zhang, X.; Rice, K.; Wang, Y.; Chen, W.; Zhong, Y.; Nakayama, Y.; Zhou, Y.; Klibanski, A. Maternally expressed gene 3 (MEG3) noncoding ribonucleic acid: Isoform structure, expression, and functions. *Endocrinology* **2010**, *151*, 939–947.
153. Zhou, Y.; Zhong, Y.; Wang, Y.; Zhang, X.; Batista, D.L.; Gejman, R.; Ansell, P.J.; Zhao, J.; Weng, C.; Klibanski, A. Activation of p53 by MEG3 non-coding RNA. *J. Biol. Chem.* **2007**, *282*, 24731–24742.
154. Gejman, R.; Batista, D.L.; Zhong, Y.; Zhou, Y.; Zhang, X.; Swearingen, B.; Stratakis, C.A.; Hedley-Whyte, E.T.; Klibanski, A. Selective loss of MEG3 expression and intergenic differentially methylated region hypermethylation in the MEG3/DLK1 locus in human clinically nonfunctioning pituitary adenomas. *J. Clin. Endocrinol. Metab.* **2008**, *93*, 4119–4125.
155. Benetatos, L.; Vartholomatos, G.; Hatzimichael, E. MEG3 imprinted gene contribution in tumorigenesis. *Int. J. Cancer.* **2011**, *129*, 773–779.
156. Garding, A.; Bhattacharya, N.; Claus, R.; Ruppel, M.; Tschuch, C.; Filarsky, K.; Idler, I.; Zucknick, M.; Caudron-Herger, M.; Oakes, C.; *et al.* Epigenetic upregulation of lncRNAs at 13q14.3 in leukemia is linked to the in cis downregulation of a gene cluster that targets NF-kB. *PLoS Genet.* **2013**, *9*, e1003373.
157. Takeuchi, S.; Hofmann, W.K.; Tsukasaki, K.; Takeuchi, N.; Ikezoe, T.; Matsushita, M.; Uehara, Y.; Phillip Koeffler, H. Loss of H19 imprinting in adult T-cell leukaemia/lymphoma. *Br. J. Haematol.* **2007**, *137*, 380–381.
158. Li, J.; Rhodes, J.C.; Askew, D.S. Evolutionary conservation of putative functional domains in the human homolog of the murine His-1 gene. *Gene* **1997**, *184*, 169–176.
159. Dallosso, A.R.; Hancock, A.L.; Malik, S.; Salpekar, A.; King-Underwood, L.; Pritchard-Jones, K.; Peters, J.; Moorwood, K.; Ward, A.; Malik, K.T.; *et al.* Alternately spliced WT1 antisense transcripts interact with WT1 sense RNA and show epigenetic and splicing defects in cancer. *RNA* **2007**, *13*, 2287–2299.
160. Ellis, B.C.; Molloy, P.L.; Graham, L.D. CRNDE: A long non-coding RNA involved in CanceR, neurobiology, and development. *Front. Genet.* **2012**, *3*, 270.
161. Taskinen, M.; Ranki, A.; Pukkala, E.; Jeskanen, L.; Kaitila, I.; Makitie, O. Extended follow-up of the Finnish cartilage-hair hypoplasia cohort confirms high incidence of non-Hodgkin lymphoma and basal cell carcinoma. *Am. J. Med. Genet A.* **2008**, *146A*, 2370–2375.
162. Braconi, C.; Valeri, N.; Kogure, T.; Gasparini, P.; Huang, N.; Nuovo, G.J.; Terracciano, L.; Croce, C.M.; Patel, T. Expression and functional role of a transcribed noncoding RNA with an ultraconserved element in hepatocellular carcinoma. *Proc. Natl. Acad. Sci. USA* **2011**, *108*, 786–791.
163. Yang, R.; Frank, B.; Hemminki, K.; Bartram, C.R.; Wappenschmidt, B.; Sutter, C.; Kiechle, M.; Bugert, P.; Schmutzler, R.K.; Arnold, N.; *et al.* SNPs in ultraconserved elements and familial breast cancer risk. *Carcinogenesis* **2008**, *29*, 351–355.

164. Zhang, X.; Lian, Z.; Padden, C.; Gerstein, M.B.; Rozowsky, J.; Snyder, M.; Gingeras, T.R.; Kapranov, P.; Weissman, S.M.; Newburger, P.E. A myelopoiesis-associated regulatory intergenic noncoding RNA transcript within the human HOXA cluster. *Blood* **2009**, *113*, 2526–2534.

165. Wagner, L.A.; Christensen, C.J.; Dunn, D.M.; Spangrude, G.J.; Georgelas, A.; Kelley, L.; Esplin, M.S.; Weiss, R.B.; Gleich, G.J. EGO, a novel, noncoding RNA gene, regulates eosinophil granule protein transcript expression. *Blood* **2007**, *109*, 5191–5198.

166. Ebralidze, A.K.; Guibal, F.C.; Steidl, U.; Zhang, P.; Lee, S.; Bartholdy, B.; Jorda, M.A.; Petkova, V.; Rosenbauer, F.; Huang, G.; *et al.* PU.1 expression is modulated by the balance of functional sense and antisense RNAs regulated by a shared cis-regulatory element. *Genes Dev.* **2008**, *22*, 2085–2092.

167. Hu, W.; Yuan, B.; Flygare, J.; Lodish, H.F. Long noncoding RNA-mediated anti-apoptotic activity in murine erythroid terminal differentiation. *Genes Dev.* **2011**, *25*, 2573–2578.

168. Aoki, K.; Harashima, A.; Sano, M.; Yokoi, T.; Nakamura, S.; Kibata, M.; Hirose, T. A thymus-specific noncoding RNA, Thy-ncR1, is a cytoplasmic riboregulator of MFAP4 mRNA in immature T-cell lines. *BMC Mol. Biol.* **2010**, *11*, 99.

169. Marasa, B.S.; Srikantan, S.; Martindale, J.L.; Kim, M.M.; Lee, E.K.; Gorospe, M.; Abdelmohsen, K. MicroRNA profiling in human diploid fibroblasts uncovers miR-519 role in replicative senescence. *Aging (Albany NY)* **2010**, *2*, 333–343.

170. Meyer, K.B.; Maia, A.T.; O'Reilly, M.; Ghoussaini, M.; Prathalingam, R.; Porter-Gill, P.; Ambs, S.; Prokunina-Olsson, L.; Carroll, J.; Ponder, B.A. A functional variant at a prostate cancer predisposition locus at 8q24 is associated with PVT1 expression. *PLoS Genet.* **2011**, *7*, e1002165.

171. Guan, Y.; Kuo, W.L.; Stilwell, J.L.; Takano, H.; Lapuk, A.V.; Fridlyand, J.; Mao, J.H.; Yu, M.; Miller, M.A.; Santos, J.L.; *et al.* Amplification of PVT1 contributes to the pathophysiology of ovarian and breast cancer. *Clin. Cancer Res.* **2007**, *13*, 5745–5755.

172. Palumbo, A.P.; Boccadoro, M.; Battaglio, S.; Corradini, P.; Tsichlis, P.N.; Huebner, K.; Pileri, A.; Croce, C.M. Human homologue of Moloney leukemia virus integration-4 locus (MLVI-4), located 20 kilobases 3' of the myc gene, is rearranged in multiple myelomas. *Cancer Res.* **1990**, *50*, 6478–6482.

173. Borg, A.; Baldetorp, B.; Ferno, M.; Olsson, H.; Sigurdsson, H. c-myc amplification is an independent prognostic factor in postmenopausal breast cancer. *Int. J. Cancer.* **1992**, *51*, 687–691.

174. You, L.; Chang, D.; Du, H.Z.; Zhao, Y.P. Genome-wide screen identifies PVT1 as a regulator of Gemcitabine sensitivity in human pancreatic cancer cells. *Biochem. Biophys. Res. Commun.* **2011**, *407*, 1–6.

175. Carramusa, L.; Contino, F.; Ferro, A.; Minafra, L.; Perconti, G.; Giallongo, A.; Feo, S. The PVT-1 oncogene is a Myc protein target that is overexpressed in transformed cells. *J. Cell. Physiol.* **2007**, *213*, 511–518.

176. Gil, J.; Peters, G. Regulation of the INK4b-ARF-INK4a tumour suppressor locus: All for one or one for all. *Nat. Rev. Mol. Cell Biol.* **2006**, *7*, 667–677.

177. Popov, N.; Gil, J. Epigenetic regulation of the INK4b-ARF-INK4a locus: In sickness and in health. *Epigenetics* **2010**, *5*, 685–690.
178. Kotake, Y.; Nakagawa, T.; Kitagawa, K.; Suzuki, S.; Liu, N.; Kitagawa, M.; Xiong, Y. Long non-coding RNA ANRIL is required for the PRC2 recruitment to and silencing of p15(INK4B) tumor suppressor gene. *Oncogene* **2011**, *30*, 1956–1962.
179. Pasmant, E.; Sabbagh, A.; Vidaud, M.; Bieche, I. ANRIL, a long, noncoding RNA, is an unexpected major hotspot in GWAS. *FASEB J.* **2011**, *25*, 444–448.
180. Burd, C.E.; Jeck, W.R.; Liu, Y.; Sanoff, H.K.; Wang, Z.; Sharpless, N.E. Expression of linear and novel circular forms of an INK4/ARF-associated non-coding RNA correlates with atherosclerosis risk. *PLoS Genet.* **2010**, *6*, e1001233.
181. Liu, Y.; Sanoff, H.K.; Cho, H.; Burd, C.E.; Torrice, C.; Mohlke, K.L.; Ibrahim, J.G.; Thomas, N.E.; Sharpless, N.E. INK4/ARF transcript expression is associated with chromosome 9p21 variants linked to atherosclerosis. *PLoS One* **2009**, *4*, e5027.
182. Cunnington, M.S.; Santibanez Koref, M.; Mayosi, B.M.; Burn, J.; Keavney, B. Chromosome 9p21 SNPs associated with multiple disease phenotypes correlate with anril expression. *PLoS Genet.* **2010**, *6*, e1000899.
183. Iacobucci, I.; Sazzini, M.; Garagnani, P.; Ferrari, A.; Boattini, A.; Lonetti, A.; Papayannidis, C.; Mantovani, V.; Marasco, E.; Ottaviani, E.; *et al.* A polymorphism in the chromosome 9p21 ANRIL locus is associated to Philadelphia positive acute lymphoblastic leukemia. *Leuk Res.* **2011**, *35*, 1052–1059.
184. Miyoshi, N.; Wagatsuma, H.; Wakana, S.; Shiroishi, T.; Nomura, M.; Aisaka, K.; Kohda, T.; Surani, M.A.; Kaneko-Ishino, T.; Ishino, F. Identification of an imprinted gene, Meg3/Gtl2 and its human homologue MEG3, first mapped on mouse distal chromosome 12 and human chromosome 14q. *Genes Cells* **2000**, *5*, 211–220.
185. Zhang, X.; Zhou, Y.; Mehta, K.R.; Danila, D.C.; Scolavino, S.; Johnson, S.R.; Klibanski, A. A pituitary-derived MEG3 isoform functions as a growth suppressor in tumor cells. *J. Clin. Endocrinol. Metab.* **2003**, *88*, 5119–5126.
186. Braconi, C.; Kogure, T.; Valeri, N.; Huang, N.; Nuovo, G.; Costinean, S.; Negrini, M.; Miotto, E.; Croce, C.M.; Patel, T. microRNA-29 can regulate expression of the long non-coding RNA gene MEG3 in hepatocellular cancer. *Oncogene* **2011**, *30*, 4750–4756.
187. Hagan, J.P.; O'Neill, B.L.; Stewart, C.L.; Kozlov, S.V.; Croce, C.M. At least ten genes define the imprinted Dlk1-Dio3 cluster on mouse chromosome 12qF1. *PLoS One* **2009**, *4*, e4352.
188. Schuster-Gossler, K.; Bilinski, P.; Sado, T.; Ferguson-Smith, A.; Gossler, A. The mouse Gtl2 gene is differentially expressed during embryonic development, encodes multiple alternatively spliced transcripts, and may act as an RNA. *Dev. Dyn.* **1998**, *212*, 214–228.
189. Mondal, T.; Rasmussen, M.; Pandey, G.K.; Isaksson, A.; Kanduri, C. Characterization of the RNA content of chromatin. *Genome Res.* **2010**, *20*, 899–907.
190. Gordon, F.E.; Nutt, C.L.; Cheunsuchon, P.; Nakayama, Y.; Provencher, K.A.; Rice, K.A.; Zhou, Y.; Zhang, X.; Klibanski, A. Increased expression of angiogenic genes in the brains of mouse meg3-null embryos. *Endocrinology* **2010**, *151*, 2443–2452.
191. Ying, L.; Huang, Y.; Chen, H.; Wang, Y., Xia, L.; Chen, Y.; Liu, Y.; Qiu, F. Downregulated MEG3 activates autophagy and increases cell proliferation in bladder cancer. *Mol. Biosyst.* **2013**, *9*, 407–411.

192. Dal Bo, M.; Rossi, F.M.; Rossi, D.; Deambrogi, C.; Bertoni, F.; Del Giudice, I.; Palumbo, G.; Nanni, M.; Rinaldi, A.; Kwee, I.; *et al.* 13q14 deletion size and number of deleted cells both influence prognosis in chronic lymphocytic leukemia. *Genes Chromosomes Cancer* **2011**, *50*, 633–643.
193. Cimmino, A.; Calin, G.A.; Fabbri, M.; Iorio, M.V.; Ferracin, M.; Shimizu, M.; Wojcik, S.E.; Aqeilan, R.I.; Zupo, S.; Dono, M.; *et al.* miR-15 and miR-16 induce apoptosis by targeting BCL2. *Proc. Natl. Acad. Sci. USA* **2005**, *102*, 13944–13949.
194. Klein, U.; Lia, M.; Crespo, M.; Siegel, R.; Shen, Q.; Mo, T.; Ambesi-Impiombato, A.; Califano, A.; Migliazza, A.; Bhagat, G.; *et al.* The DLEU2/miR-15a/16–1 cluster controls B cell proliferation and its deletion leads to chronic lymphocytic leukemia. *Cancer Cell* **2010**, *17*, 28–40.
195. Lerner, M.; Harada, M.; Loven, J.; Castro, J.; Davis, Z.; Oscier, D.; Henriksson, M.; Sangfelt, O.; Grander, D.; Corcoran, M.M. DLEU2, frequently deleted in malignancy, functions as a critical host gene of the cell cycle inhibitory microRNAs miR-15a and miR-16–1. *Exp. Cell Res.* **2009**, *315*, 2941–2952.
196. Coccia, E.M.; Cicala, C.; Charlesworth, A.; Ciccarelli, C.; Rossi, G.B.; Philipson, L.; Sorrentino, V. Regulation and expression of a growth arrest-specific gene (gas5) during growth, differentiation, and development. *Mol. Cell Biol.* **1992**, *12*, 3514–3521.
197. Mourtada-Maarabouni, M.; Pickard, M.R.; Hedge, V.L.; Farzaneh, F.; Williams, G.T. GAS5, a non-protein-coding RNA, controls apoptosis and is downregulated in breast cancer. *Oncogene* **2009**, *28*, 195–208.
198. Williams, G.T.; Mourtada-Maarabouni, M.; Farzaneh, F. A critical role for non-coding RNA GAS5 in growth arrest and rapamycin inhibition in human T-lymphocytes. *Biochem. Soc. Trans.* **2011**, *39*, 482–486.
199. Mourtada-Maarabouni, M.; Hasan, A.M.; Farzaneh, F.; Williams, G.T. Inhibition of human T-cell proliferation by mammalian target of rapamycin (mTOR) antagonists requires noncoding RNA growth-arrest-specific transcript 5 (GAS5). *Mol. Pharmacol.* **2010**, *78*, 19–28.
200. Barsyte-Lovejoy, D.; Lau, S.K.; Boutros, P.C.; Khosravi, F.; Jurisica, I.; Andrulis, I.L.; Tsao, M.S.; Penn, L.Z. The c-Myc oncogene directly induces the H19 noncoding RNA by allele-specific binding to potentiate tumorigenesis. *Cancer Res.* **2006**, *66*, 5330–5337.
201. Hibi, K.; Nakamura, H.; Hirai, A.; Fujikake, Y.; Kasai, Y.; Akiyama, S.; Ito, K.; Takagi, H. Loss of H19 imprinting in esophageal cancer. *Cancer Res.* **1996**, *56*, 480–482.
202. Fellig, Y.; Ariel, I.; Ohana, P.; Schachter, P.; Sinelnikov, I.; Birman, T.; Ayesh, S.; Schneider, T.; de Groot, N.; Czerniak, A.; *et al.* H19 expression in hepatic metastases from a range of human carcinomas. *J. Clin. Pathol.* **2005**, *58*, 1064–1068.
203. Matouk, I.J.; DeGroot, N.; Mezan, S.; Ayesh, S.; Abu-lail, R.; Hochberg, A.; Galun, E. The H19 non-coding RNA is essential for human tumor growth. *PLoS One* **2007**, *2*, e845.
204. Berteaux, N.; Lottin, S.; Monte, D.; Pinte, S.; Quatannens, B.; Coll, J.; Hondermarck, H.; Curgy, J.J.; Dugimont, T.; Adriaenssens, E. H19 mRNA-like noncoding RNA promotes breast cancer cell proliferation through positive control by E2F1. *J. Biol. Chem.* **2005**, *280*, 29625–29636.
205. Nunez, C.; Bashein, A.M.; Brunet, C.L.; Hoyland, J.A.; Freemont, A.J.; Buckle, A.M.; Murphy, C.; Cross, M.A.; Lucas, G.; Bostock, V.J.; *et al.* Expression of the imprinted tumour-suppressor gene H19 is tightly regulated during normal haematopoiesis and is reduced in haematopoietic precursors of patients with the myeloproliferative disease polycythaemia vera. *J. Pathol.* **2000**, *190*, 61–68.

206. Bock, O.; Schlue, J.; Kreipe, H. Reduced expression of H19 in bone marrow cells from chronic myeloproliferative disorders. *Leukemia* **2003**, *17*, 815–816.
207. Tessema, M.; Langer, F.; Bock, O.; Seltsam, A.; Metzig, K.; Hasemeier, B.; Kreipe, H.; Lehmann, U. Down-regulation of the IGF-2/H19 locus during normal and malignant hematopoiesis is independent of the imprinting pattern. *Int. J. Oncol.* **2005**, *26*, 499–507.
208. Dugimont, T.; Montpellier, C.; Adriaenssens, E.; Lottin, S.; Dumont, L.; Iotsova, V.; Lagrou, C.; Stehelin, D.; Coll, J.; Curgy, J.J. The H19 TATA-less promoter is efficiently repressed by wild-type tumor suppressor gene product p53. *Oncogene* **1998**, *16*, 2395–2401.
209. Farnebo, M.; Bykov, V.J.; Wiman, K.G. The p53 tumor suppressor: A master regulator of diverse cellular processes and therapeutic target in cancer. *Biochem. Biophys. Res. Commun.* **2010**, *396*, 85–89.
210. Tsang, W.P.; Ng, E.K.; Ng, S.S.; Jin, H.; Yu, J.; Sung, J.J.; Kwok, T.T. Oncofetal H19-derived miR-675 regulates tumor suppressor RB in human colorectal cancer. *Carcinogenesis* **2010**, *31*, 350–358.
211. Yoshimizu, T.; Miroglio, A.; Ripoche, M.A.; Gabory, A.; Vernucci, M.; Riccio, A.; Colnot, S.; Godard, C.; Terris, B.; Jammes, H.; *et al.* The H19 locus acts *in vivo* as a tumor suppressor. *Proc. Natl. Acad. Sci. USA* **2008**, *105*, 12417–12422.
212. Leighton, P.A.; Saam, J.R.; Ingram, R.S.; Stewart, C.L.; Tilghman, S.M. An enhancer deletion affects both H19 and Igf2 expression. *Genes Dev.* **1995**, *9*, 2079–2089.
213. Scaruffi, P.; Stigliani, S.; Moretti, S.; Coco, S.; De Vecchi, C.; Valdora, F.; Garaventa, A.; Bonassi, S.; Tonini, G.P. Transcribed-Ultra Conserved Region expression is associated with outcome in high-risk neuroblastoma. *BMC Cancer* **2009**, *9*, 441.
214. Mestdagh, P.; Fredlund, E.; Pattyn, F.; Rihani, A.; Van Maerken, T.; Vermeulen, J.; Kumps, C.; Menten, B.; de Preter, K.; Schramm, A.; *et al.* An integrative genomics screen uncovers ncRNA T-UCR functions in neuroblastoma tumours. *Oncogene* **2010**, *29*, 3583–3592.
215. Sana, J.; Hankeova, S.; Svoboda, M.; Kiss, I.; Vyzula, R.; Slaby, O. Expression levels of transcribed ultraconserved regions uc.73 and uc.388 are altered in colorectal cancer. *Oncology* **2012**, *82*, 114–118.
216. Rossi, S.; Sevignani, C.; Nnadi, S.C.; Siracusa, L.D.; Calin, G.A. Cancer-associated genomic regions (CAGRs) and noncoding RNAs: Bioinformatics and therapeutic implications. *Mamm. Genome*. **2008**, *19*, 526–540.
217. Ng, D.; Toure, O.; Wei, M.H.; Arthur, D.C.; Abbasi, F.; Fontaine, L.; Marti, G.E.; Fraumeni, J.F., Jr.; Goldin, L.R.; Caporaso, N.; *et al.* Identification of a novel chromosome region, 13q21.33-q22.2, for susceptibility genes in familial chronic lymphocytic leukemia. *Blood* **2007**, *109*, 916–925.
218. Calin, G.A.; Ferracin, M.; Cimmino, A.; Di Leva, G.; Shimizu, M.; Wojcik, S.E.; Iorio, M.V.; Visone, R.; Sever, N.I.; Fabbri, M.; *et al.* A MicroRNA signature associated with prognosis and progression in chronic lymphocytic leukemia. *N. Engl. J. Med.* **2005**, *353*, 1793–1801.
219. Notari, M.; Neviani, P.; Santhanam, R.; Blaser, B.W.; Chang, J.S.; Galietta, A.; Willis, A.E.; Roy, D.C.; Caligiuri, M.A.; Marcucci, G.; *et al.* A MAPK/HNRPK pathway controls BCR/ABL oncogenic potential by regulating MYC mRNA translation. *Blood* **2006**, *107*, 2507–2516.

220. Du, Q.; Wang, L.; Zhu, H.; Zhang, S.; Xu, L.; Zheng, W.; Liu, X. The role of heterogeneous nuclear ribonucleoprotein K in the progression of chronic myeloid leukemia. *Med. Oncol.* **2010**, *27*, 673–679.
221. Tsujimoto, Y.; Gorham, J.; Cossman, J.; Jaffe, E.; Croce, C.M. The t(14;18) chromosome translocations involved in B-cell neoplasms result from mistakes in VDJ joining. *Science* **1985**, *229*, 1390–1393.
222. Ridanpaa, M.; van Eenennaam, H.; Pelin, K.; Chadwick, R.; Johnson, C.; Yuan, B.; vanVenrooij, W.; Pruijn, G.; Salmela, R.; Rockas, S.; *et al.* Mutations in the RNA component of RNase MRP cause a pleiotropic human disease, cartilage-hair hypoplasia. *Cell* **2001**, *104*, 195–203.
223. Tanaka, R.; Satoh, H.; Moriyama, M.; Satoh, K.; Morishita, Y.; Yoshida, S.; Watanabe, T.; Nakamura, Y.; Mori, S. Intronic U50 small-nucleolar-RNA (snoRNA) host gene of no protein-coding potential is mapped at the chromosome breakpoint t(3;6)(q27;q15) of human B-cell lymphoma. *Genes Cells* **2000**, *5*, 277–287.
224. Zhao, H.; Zhang, X.; Frazao, J.B.; Condino-Neto, A.; Newburger, P.E. HOX antisense lincRNA HOXA-AS2 is an apoptosis repressor in all trans retinoic acid treated NB4 promyelocytic leukemia cells. *J. Cell Biochem.* **2013**, doi:10.1002/jcb.24586.
225. Paralkar, V.R.; Weiss, M.J. A new 'Linc' between noncoding RNAs and blood development. *Genes Dev.* **2011**, *25*, 2555–2558.
226. Halvorsen, M.; Martin, J.S.; Broadaway, S.; Laederach, A. Disease-associated mutations that alter the RNA structural ensemble. *PLoS Genet.* **2010**, *6*, e1001074.
227. Wojcik, S.E.; Rossi, S.; Shimizu, M.; Nicoloso, M.S.; Cimmino, A.; Alder, H.; Herlea, V.; Rassenti, L.Z.; Rai, K.R.; Kipps, T.J.; *et al.* Non-codingRNA sequence variations in human chronic lymphocytic leukemia and colorectal cancer. *Carcinogenesis* **2010**, *31*, 208–215.
228. Scott, L.J.; Mohlke, K.L.; Bonnycastle, L.L.; Willer, C.J.; Li, Y.; Duren, W.L.; Erdos, M.R.; Stringham, H.M.; Chines, P.S.; Jackson, A.U.; *et al.* A genome-wide association study of type 2 diabetes in Finns detects multiple susceptibility variants. *Science* **2007**, *316*, 1341–1345.
229. Broadbent, H.M.; Peden, J.F.; Lorkowski, S.; Goel, A.; Ongen, H.; Green, F.; Clarke, R.; Collins, R.; Franzosi, M.G.; Tognoni, G.; *et al.* Susceptibility to coronary artery disease and diabetes is encoded by distinct, tightly linked SNPs in the ANRIL locus on chromosome 9p. *Hum. Mol. Genet.* **2008**, *17*, 806–814.
230. Morrison, L.E.; Jewell, S.S.; Usha, L.; Blondin, B.A.; Rao, R.D.; Tabesh, B.; Kemper, M.; Batus, M.; Coon, J.S. Effects of ERBB2 amplicon size and genomic alterations of chromosomes 1, 3, and 10 on patient response to trastuzumab in metastatic breast cancer. *Genes Chromosomes Cancer* **2007**, *46*, 397–405.
231. Nupponen, N.N.; Carpten, J.D. Prostate cancer susceptibility genes: Many studies, many results, no answers. *Cancer Metastasis Rev.* **2001**, *20*, 155–164.
232. Smedley, D.; Sidhar, S.; Birdsall, S.; Bennett, D.; Herlyn, M.; Cooper, C.; Shipley, J. Characterization of chromosome 1 abnormalities in malignant melanomas. *Genes Chromosomes Cancer* **2000**, *28*, 121–125.
233. Kagami, M.; O'Sullivan, M.J.; Green, A.J.; Watabe, Y.; Arisaka, O.; Masawa, N.; Matsuoka, K.; Fukami, M.; Matsubara, K.; Kato, F.; *et al.* The IG-DMR and the MEG3-DMR at human chromosome 14q32.2: Hierarchical interaction and distinct functional properties as imprinting control centers. *PLoS Genet.* **2010**, *6*, e1000992.

234. Astuti, D.; Latif, F.; Wagner, K.; Gentle, D.; Cooper, W.N.; Catchpoole, D.; Grundy, R.; Ferguson-Smith, A.C.; Maher, E.R. Epigenetic alteration at the DLK1-GTL2 imprinted domain in human neoplasia: Analysis of neuroblastoma, phaeochromocytoma and Wilms' tumour. *Br. J. Cancer*. **2005**, *92*, 1574–1580.

235. Benetatos, L.; Hatzimichael, E.; Dasoula, A.; Dranitsaris, G.; Tsiara, S.; Syrrou, M.; Georgiou, I.; Bourantas, K.L. CpG methylation analysis of the MEG3 and SNRPN imprinted genes in acute myeloid leukemia and myelodysplastic syndromes. *Leuk Res*. **2010**, *34*, 148–153.

236. Khoury, H.; Suarez-Saiz, F.; Wu, S.; Minden, M.D. An upstream insulator regulates DLK1 imprinting in AML. *Blood* **2010**, *115*, 2260–2263.

237. Rojas, J.J.; Guedan, S.; Searle, P.F.; Martinez-Quintanilla, J.; Gil-Hoyos, R.; Alcayaga-Miranda, F.; Cascallo, M.; Alemany, R. Minimal RB-responsive E1A promoter modification to attain potency, selectivity, and transgene-arming capacity in oncolytic adenoviruses. *Mol. Ther*. **2010**, *18*, 1960–1971.

238. Castello, A.; Horos, R.; Strein, C.; Fischer, B.; Eichelbaum, K.; Steinmetz, L.M.; Krijgsveld, J.; Hentze, M.W. System-wide identification of RNA-binding proteins by interactome capture. *Nat. Protoc*. **2013**, *8*, 491–500.

239. Fan, M.; Li, X.; Jiang, W.; Huang, Y.; Li, J.; Wang, Z. A long non-coding RNA, PTCSC3, as a tumor suppressor and a target of miRNAs in thyroid cancer cells. *Exp. Ther. Med*. **2013**, *5*, 1143–1146.

240. Chung, C.C.; Chanock, S.J. Current status of genome-wide association studies in cancer. *Hum. Genet*. **2011**, *130*, 59–78.

241. Giacomini, K.M.; Brett, C.M.; Altman, R.B.; Benowitz, N.L.; Dolan, M.E.; Flockhart, D.A.; Johnson, J.A.; Hayes, D.F.; Klein, T.; Krauss, R.M.; *et al.* The pharmacogenetics research network: From SNP discovery to clinical drug response. *Clin. Pharmacol. Ther*. **2007**, *81*, 328–345.

242. Mattick, J.S. The genetic signatures of noncoding RNAs. *PLoS Genet*. **2009**, *5*, e1000459.

243. Prensner, J.R.; Iyer, M.K.; Balbin, O.A.; Dhanasekaran, S.M.; Cao, Q.; Brenner, J.C.; Laxman, B.; Asangani, I.A.; Grasso, C.S.; Kominsky, H.D.; *et al.* Transcriptome sequencing across a prostate cancer cohort identifies PCAT-1, an unannotated lincRNA implicated in disease progression. *Nat. Biotechnol*. **2011**, *29*, 742–749.

244. Matouk, I.J.; Abbasi, I.; Hochberg, A.; Galun, E.; Dweik, H.; Akkawi, M. Highly upregulated in liver cancer noncoding RNA is overexpressed in hepatic colorectal metastasis. *Eur. J. Gastroenterol. Hepatol*. **2009**, *21*, 688–692.

245. Panzitt, K.; Tschernatsch, M.M.; Guelly, C.; Moustafa, T.; Stradner, M.; Strohmaier, H.M.; Buck, C.R.; Denk, H.; Schroeder, R.; Trauner, M.; *et al.* Characterization of HULC, a novel gene with striking up-regulation in hepatocellular carcinoma, as noncoding RNA. *Gastroenterology* **2007**, *132*, 330–342.

246. De Kok, J.B.; Verhaegh, G.W.; Roelofs, R.W.; Hessels, D.; Kiemeney, L.A.; Aalders, T.W.; Swinkels, D.W.; Schalken, J.A. DD3(PCA3), a very sensitive and specific marker to detect prostate tumors. *Cancer Res*. **2002**, *62*, 2695–2698.

247. Ifere, G.O.;Ananaba, G.A. Prostate cancer gene expression marker 1 (PCGEM1): A patented prostate- specific non-coding gene and regulator of prostate cancer progression. *Recent Pat. DNA Gene Seq*. **2009**, *3*, 151–163.

248. Lee, G.L.; Dobi, A.; Srivastava, S. Prostate cancer: Diagnostic performance of the PCA3 urine test. *Nat. Rev. Urol.* **2011**, *8*, 123–124.
249. Tomlins, S.A.; Aubin, S.M.; Siddiqui, J.; Lonigro, R.J.; Sefton-Miller, L.; Miick, S.; Williamsen, S.; Hodge, P.; Meinke, J.; Blase, A.; *et al.* Urine TMPRSS2:ERG fusion transcript stratifies prostate cancer risk in men with elevated serum PSA. *Sci. Transl. Med.* **2011**, *3*, 94ra72.
250. Durand, X.; Moutereau, S.; Xylinas, E.; de la Taille, A. Progensa PCA3 test for prostate cancer. *Expert Rev. Mol. Diagn.* **2011**, *11*, 137–144.
251. Niinuma, T.; Suzuki, H.; Nojima, M.; Nosho, K.; Yamamoto, H.; Takamaru, H.; Yamamoto, E.; Maruyama, R.; Nobuoka, T.; Miyazaki, Y.; *et al.* Upregulation of miR-196a and HOTAIR drive malignant character in gastrointestinal stromal tumors. *Cancer Res.* **2012**, *72*, 1126–1136.
252. Geng, Y.J.; Xie, S.L.; Li, Q.; Ma, J.; Wang, G.Y. Large intervening non-coding RNA HOTAIR is associated with hepatocellular carcinoma progression. *J. Int. Med. Res.* **2011**, *39*, 2119–2128.
253. Ji, P.; Diederichs, S.; Wang, W.; Boing, S.; Metzger, R.; Schneider, P.M.; Tidow, N.; Brandt, B.; Buerger, H.; Bulk, E.; *et al.* MALAT-1, a novel noncoding RNA, and thymosin beta4 predict metastasis and survival in early-stage non-small cell lung cancer. *Oncogene* **2003**, *22*, 8031–8041.
254. Huang, K.C.; Rao, P.H.; Lau, C.C.; Heard, E.; Ng, S.K.; Brown, C.; Mok, S.C.; Berkowitz, R.S.; Ng, S.W. Relationship of XIST expression and responses of ovarian cancer to chemotherapy. *Mol. Cancer Ther.* **2002**, *1*, 769–776.
255. Tsai, M.C.; Spitale, R.C.; Chang, H.Y. Long intergenic noncoding RNAs: New links in cancer progression. *Cancer Res.* **2011**, *71*, 3–7.

10

MicroRNA-143 Downregulates Interleukin-13 Receptor Alpha1 in Human Mast Cells

Shaoqing Yu [1,2], Ruxin Zhang [1,*], Chunshen Zhu [3], Jianqiu Cheng [3], Hong Wang [1] and Jing Wu [1]

[1] Department of Otolaryngology, Huadong Hospital, Fudan University, Shanghai 200040, China; E-Mails: yu_shaoqing@163.com (S.Y.); whhb1984@163.com (H.W.); dushixingniu@163.com (J.W.)

[2] Department of Otolaryngology, Tongji Hospital, Tongji University, Shanghai 200065, China

[3] Department of Otolaryngology, Jinan General Hospital of PLA, Jinan 250031, Shandong, China; E-Mails: zhucs1@sina.cn (C.Z.); showing100@163.com (J.C.)

* Author to whom correspondence should be addressed; E-Mail: nosezhang@163.com;

Abstract: MicroRNA-143 (miR-143) was found to be downregulated in allergic rhinitis, and bioinformatics analysis predicted that IL-13Rα1 was a target gene of miR-143. To understand the molecular mechanisms of miR-143 involved in the pathogenesis of allergic inflammation, recombinant miR-143 plasmid vectors were constructed, and human mast cell-1(HMC-1) cells which play a central role in the allergic response were used for study. The plasmids were transfected into HMC-1 cells using a lentiviral vector. Expression of IL-13Rα1 mRNA was then detected by reverse transcriptase polymerase chain reaction (RT-PCR) and Western Blotting. The miR-143 lentiviral vector was successfully stably transfected in HMC-1 cells for target gene expression. Compared to the control, the target gene IL-13Rα1 was less expressed in HMC-1 transfected with miR-143 as determined by RT-PCR and Western Blotting ($p < 0.05$); this difference in expression was statistically significant and the inhibition efficiency was 71%. It indicates that miR-143 directly targets IL-13Rα1 and suppresses IL-13Rα1 expression in HMC-1 cells. Therefore, miR-143 may be associated with allergic reaction in human mast cells.

Key words: microRNA-143; IL-13Rα1; mast cell; lentiviral vector; allergy

1. Introduction

Allergy is a chronic inflammatory condition and many inflammatory factors, such as IL-4, IL-13, IL-8, *etc.* have been related to this disorder [1,2]; however, how these factors regulate mechanisms responsible for allergic reactions has not yet been thoroughly elucidated. Recently, it has been found that microRNAs (miRNAs) inhibit protein translation to regulate gene expression; miRAs are small RNAs produced by cells through a unique process, involving RNA polymerase II, microprocessor protein complex, and the RNAase III/Dicer endonuclease complex. miRNA ribonucleoprotein complex is formed at the end of this process [3]. The biological functions of miRNAs depend on their ability to silence gene expression, primarily through degradation of the target mRNA and/or translational suppression mediated by the RNA-induced silencing complex (RISC) [3]. The function of miRNAs in gene regulation has been investigated in a variety of diseases including allergic disease.

In our previous study, miRNA profiles and RT-PCR were used to reveal the difference between nasal mucosa biopsies of patients with allergic rhinitis (AR) and those of healthy volunteers. When compared with normal tissue, microRNA-143 (miR-143) was the most significantly downregulated miRNA in nasal mucosa of tissues exhibiting AR. It was found that expression of miR-143 in smooth muscle cells of airways triggers allergic conditions such as asthma [4]. However, we have not yet been able to understand how miR-143 regulates allergic inflammation in upper airways such as AR.

To study the mechanism of miR-143 regulation in allergic inflammation, we investigated the target genes of miR-143 through the TargetScan procedure (http://genes.mit.edu/tscan/targetscanS.html and http://pictar.mdc-berlin.de/) to predict miR-143 target genes, and found that the IL-13Rα1 gene 3' UTR had 15 sequential pairing bases with miR-143 (Table 1), which indicates that IL-13Rα1, known to play an important role in allergy, may be a potential target of miR-143. To investigate this, we used mast cells, which are an important AR target cell and built the miR-143 overexpression system. This system was transfected into mast cells to observe changes induced through target gene expression. The aim of this study was thus to identify the effects of miR-143 on IL-13Rα1 in mast cells to further an understanding of the molecular mechanisms of allergic inflammation.

Table 1. Target prediction of IL-13Aa1 for Has-miR-143 based on conservation.

Nuclei mapped to alignments	Nuclei mapped to sequence	Structure of predicted duplex	Probabilities	Free energies kcal/mol
1397	1338	_GAG__GCAG__G________CAUCUCA_: _CUC__UGUC__C________GUAGAGU_	0.30	−24.1

2. Results

2.1. Construction and Identification of miR-143 pLenO-RIP Plasmid

We connected the target gene DNA fragment into pLenO-RIP empty vector (Figure 1a), and transformed the plasmid ligation product into *E. coli* DH5. Subsequently, we extracted the plasmids from amplified DH5. An approximately 600 bp DNA fragment was identified from the recombinant plasmid (Figure 1b). Finally, DNA sequencing confirmed the recombinant plasmid DNA sequence.

Figure 1. Images of gel of plasmid digest of pLenO-RIP, PUC-miR-143 and recombinant plasmid. Lanes 1–4 (from left to right) were products of DNA marker (DL15000), pLenO-RIP, PUC-miR-143 and DNA marker (DL2000), respectively (**a**). From left to right were positive clones, negative control and marker (DL2000) respectively. The approximate 600bp size of the product can be found in positive clones (**b**).

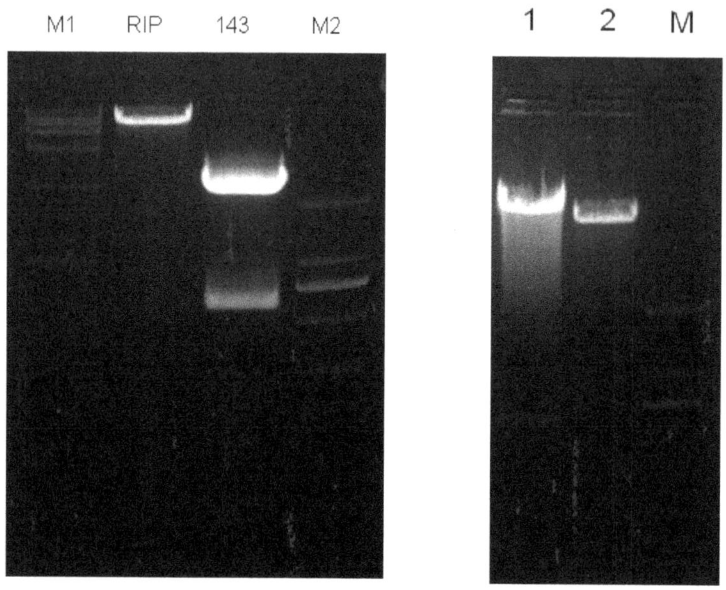

(a) **(b)**

2.2. Lentiviral Vector Packaging and Identification

The lentiviral packaging systems, comprising four kinds of plasmid DNA solutions, were co-transfected into 293T cells. Cells grew well and strong fluorescence intensity was observed under the fluorescent microscope, indicating that the virus packaging was successful (Figure 2). Forty-eight hours after transfection, supernatant viral material was collected and concentrated, then a 10-fold dilution was transformed into a $100–10^{-5}$ concentration gradient to infect 293T cells, and the supernatant viral material was collected. 48 h later, the virus titer was determined via the virus titer formula (the Tu/mL) = (GFP − positive cell count × virus supernatant dilution factor)/0.1. The production of virus droplets was 2.9×10^8 of TU/mL, and they were then prepared for transfection of HMC-1 cells.

Forty-eight hours after developing HMC-1 cell culture, the lentiviral vector of miR-143 was transfected into cell culture, and RFP blank lentivirus was used as the control. Approximately 48 h–72 h after infection, the cell culture was found in good condition under an inverted fluorescence microscope, wherein the infection efficiency was up to 80% (Figure 3). The results indicate that the miR-143 lentiviral vector was successfully transfected into HMC-1 cells, and that it expressed the relevant target genes in a stable manner.

Figure 2. Different concentrations of virus were co-transfected into 293T cells: 0.1 μL (**a**), 0.1 μL (**b**) and 0.01 μL (**c**). Cells grew well and strong fluorescence intensity was observed under the fluorescent microscope, suggesting that the virus packaging was successful.

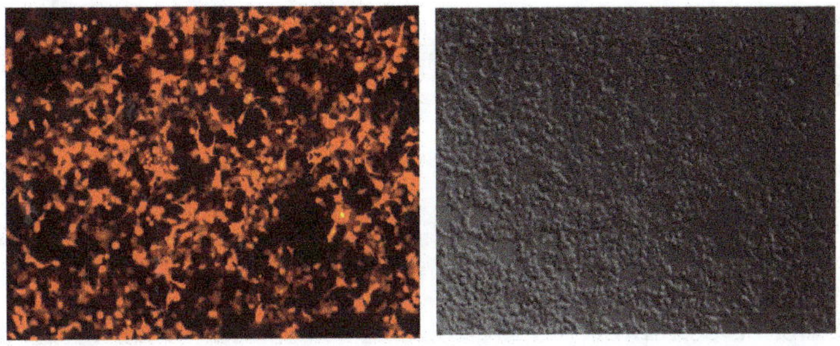

(**a**)

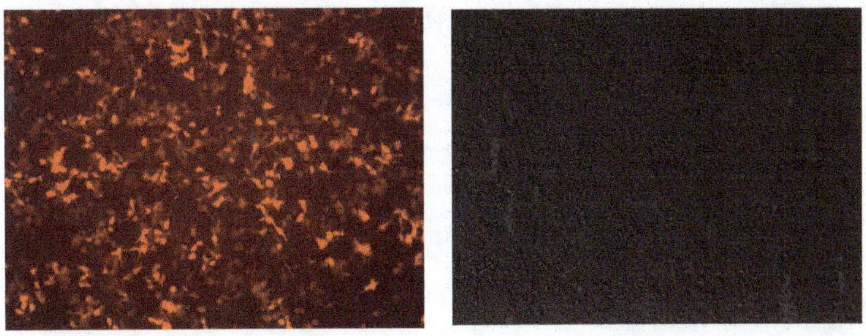

(**b**)

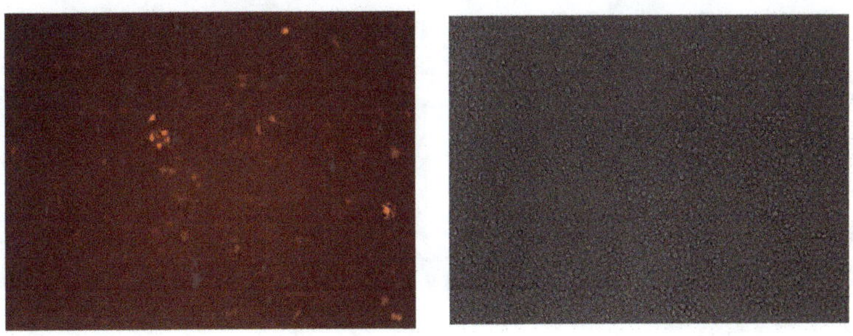

(**c**)

Figure 3. miR-143 transfected HMC-1 cells (**a**) fluorescent vision and (**b**) normal vision. 36 h after transfection, the RFP red fluorescent protein gene were seen expressed in up to 80% cells under fluorescent vision, it showed that the target genes were successfully expressed in the cells. (original magnification ×40)

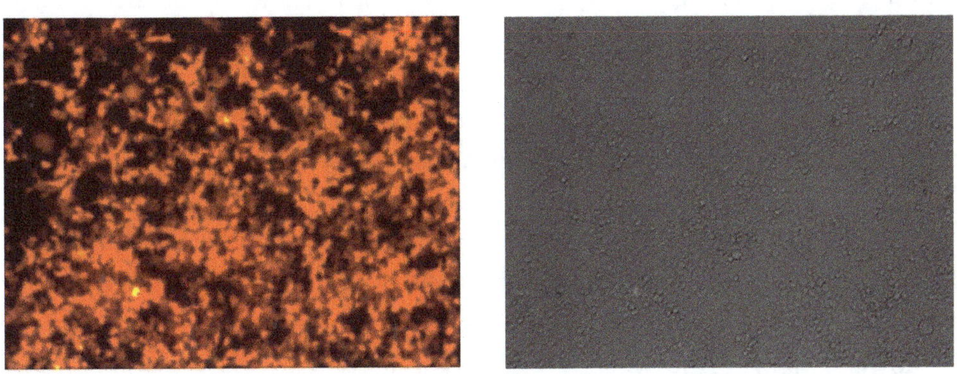

(**a**) (**b**)

2.3. miR-143 Suppresses IL-13Rα1 Expression in HMC-1 Cells

The cumulative data for mRNA expression of IL-13Rα1 are presented in Figure 4. Compared with the control, IL-13Rα1 mRNA expression was significantly downregulated in miR-143 transfected cells ($p > 0.05$). No significant changes of IL-13Rα1 mRNA levels were observed in HMC-1 cells and empty vector transfected cells ($p > 0.05$).

Figure 4. Expression of target gene IL-13Rα1 in different groups by RT-PCR. For the three groups: miR-143 mimics transfected cells, negative control group (empty vector transfected cells) and blank control group (HMC-1 cells), and every experiment was conducted in triplicate. The vertical axis represents the relative expression level of IL-13Rα1 control to 18sRNA (ΔCt). With the same cycle number, among them, miR-143 transfected cells had less amplification of IL-13Rα1 control to 18sRNA (* $p < 0.05$ for each).

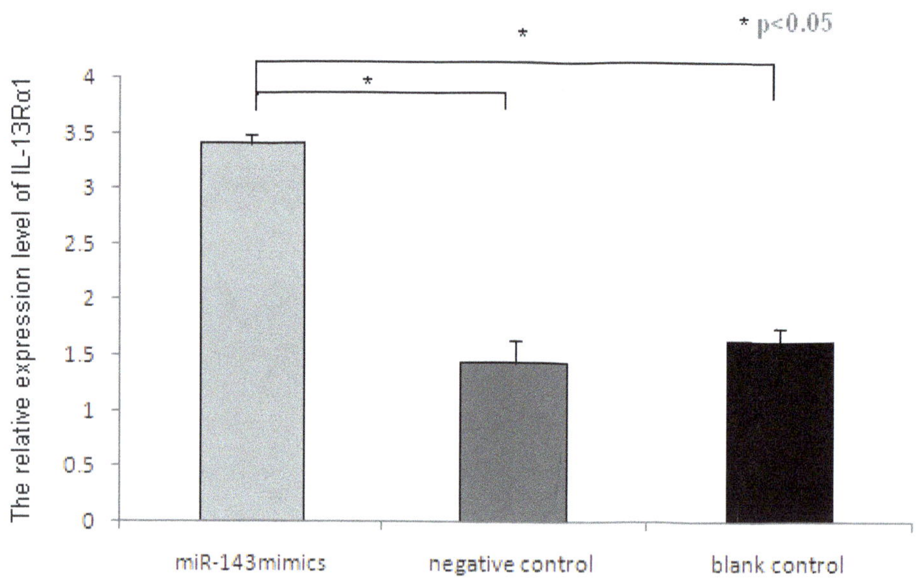

As shown in Figure 4, the expression of IL-13Rα1 in the miR-143 transfected HMC-1 cells group was reported to be 3.41; this value is relative to the internal reference 18sRNA. Furthermore, the expression of IL-13Rα1 in untreated controls HMC-1 and empty vector transfected control group were 1.63 and 1.43, respectively. The results indicated there was significantly reduced expression of IL-13Rα1 when miR-143 is overexpressed in HMC-1 cells, with an inhibition efficiency of 71%. This indicates that IL-13Rα1 is a target gene of miR-143 in HMC-1 cells, and that miR-143 can significantly inhibit the target gene IL-13Rα1 expression in HMC-1 cells.

2.4. Western Blotting Results of IL-13Rα1 Downregulation in miR-143 Transfected HMC-1 Cells

Western blot analysis showed that the expression of IL-13Rα1 protein in the miR-143 group was much lower than in the negative and blank control groups ($p < 0.05$ for each). There was no significant difference in the expression of IL-13Rα1 protein between the negative control and blank control group ($p > 0.05$). As shown in Figure 5, the over-expression of miR-143 caused reduction in IL-13Rα1 protein expression in HMC-1 cells.

Figure 5. Western blot analysis of IL-13Rα1 protein 48 h after transfection. For the three groups: miR-143 mimics, negative control group and blank control group. Experiments were done in duplicate. Western blot analysis showed that the relative expression of IL-13Rα1 protein in the miR-143 group was much lower than in negative and blank control groups (* $p < 0.05$ for each).

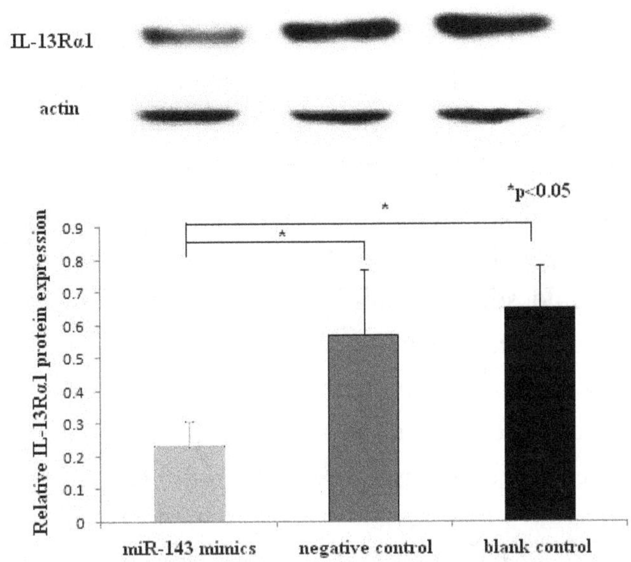

3. Discussion

MiRNAs are short (20–24 nt), non-coding RNAs that are involved in post-transcriptional regulation of gene expression in multicellular organisms, because they affect both the stability and translation of mRNAs [5]. In mammals, thousands of miRNAs that perform diverse functions have been identified. Among them, miR-143 was found to be associated with allergic rhinitis in our research [6], but the mechanism of its action was unclear. In order to study this mechanism, we determined the target genes of miR-143 and investigated related mechanisms. The target genes of miR-143 were screened using miRanda (http://www.microrna.org) software. It was found that miR-143 may target multiple genes, including NOVA1, ZIC3 and MARCKS etc. Among these target genes, IL-13α1 has been the focus of allergic inflammation-associated research in recent years [7]. miR-143 caused allergic inflammation, which might be associated with the target gene IL-13α1. Therefore, we carried out this microRNA intervention project.

To determine the gene regulation of miRNA, methods of overexpression and suppression of miRNAs are often used in research studies focused on this subject [8]. In this study, the overexpression of miR-143 was chosen for exploring the regulation of miR-143 on genes of inflammatory cytokines in allergic inflammation. The precursor sequence of miR-143 was synthesized directly, and then this sequence amplified genes using the primer extension method. The product was digested and inserted into the lentivirus expression vector; thus lentiviral packaging plasmids were used to transfect 293T cells. The results indicated that the miR-143 lentiviral expression vector was successfully built which could be perfomed for the next step.

Lentiviral transduction vectora were used in this study because they have the ability to infect different types of cells. In addition, lentiviral transduction vectors can also carry exogenous genes and integrate them into the genome of the host cell for the purpose of achieving long-term stable expression [9]. It will not encode viral proteins, so it can be used for miRNA overexpression transfection experiments [10]. The viral vector carrying a red fluorescent protein gene RFP can be transfected into HMC-1 by lentiviral vectors [11]. The high expression of RFP indicates the high efficiency of miR-143 transfection. Moreover, the sustained expression of RFP in HMC-1 also showed that there was long-term overexpression of miR-143 in cells.

It has been reported that miR-143 plays an important role in regulating smooth muscle cell (SMC) fate and plasticity. Thomas Boettger's study revealed that the expression of miR-143, which formed a small cluster on mouse chromosome 18, strongly correlated with the number of SMCs. Vascular smooth muscle cells (VSMCs) from miR-143-deficient mice were locked in the synthetic state, which incapacitated their contractile abilities and favored neo-intimal lesion development, thereby revealing an unanticipated role of miR-143 in the regulation of VSMC phenotype [12]. Pleiotropic cytokines, IFN-beta and IFN-gamma, could stimulate miR-143 expression of smooth muscle cell in airways, contributing to airway allergic diseases such as asthma [13]. Some studies also demonstrated that miR-143 deficiency is associated not only with altered vasocontraction but also with impaired vasodilation [14]. Furthermore, some authors showed that miR-143 is a critical regulator of cell cycle activity in stem cells [15]. Recently it was discovered that miR-143 has tumor suppressor activity [16], but the role of miR-143 in allergic inflammation, especially in functional cells, such as mast cell, EOS, and T cell has been little investigated. This is the first study to illustrate that the allergic reaction in mast cells was attributable to inhibition of IL-13Rα1 by miR-143.

IL-13 is an immunoregulatory cytokine predominantly secreted by activated Th2 cells. Over the past several years, it has become evident that IL-13 is a key mediator in the pathogenesis of allergic inflammation. Like IL-4, IL-13 responds by signaling through the T-cell antigen receptor and mast cells. IL-13 responds through basophils when there is a cross-linkage of the high-affinity receptor for IgE. It is also produced by activated eosinophils. IL-13 plays a pivotal role in IgE-dependent inflammatory reactions, and it acts on B cells to produce IgE. IL-13 has been implicated in a variety of allergic responses, including airway hypersensitivity, mucus hypersecretion, AR, and asthma. IL-13 induces many of the same responses as IL-4 and shares a receptor subunit with IL-4 [17,18].

IL-13 mediates its effects by interacting with a complex receptor system comprised of IL-4Rα and two IL-13 binding proteins: IL-13Rα1 and IL-13Rα2. The expression of IL-13 receptors has been detected on human B cells, basophils, eosinophils, mast cells, endothelial cells, fibroblasts, monocytes, macrophages, respiratory epithelial cells, and smooth muscle cells. Human IL-13Rα1 is a single gene on chromosome Xq13 [19]. It binds IL-13 with low affinity, and also combines with IL-4Rα to form a high-affinity IL-13 binding complex. A variety of cells, including B cells, basophils, eosinophils, mast cells, endothelial cells, fibroblasts, monocytes, epithelial cells, and smooth muscle cells, respond to IL-13 owing to this binding complex.

Some studies have shown that human mast cell can express IL-13Rα1 and can be activated by IL-13 [20]. In this study, the overexpression of miR-143 can suppress the expression of IL-13Rα1 in mast cells, which may be stimulated by combined IL-13/IL-4. This ultimately translates into minimizing the allergic response. Thus, the role of miR-143 in allergic response may be associated with regulation of IL-13 pathway. Nevertheless, further investigation is needed to validate these inferences.

4. Experimental Section

4.1. Construction of miR-143 Target Sequence-Luciferase Reporter Plasmid

Using the miR-143 target sequence obtained from Ensembl, we designed its pre-miRNA sequence (GCGCAGCGCCCUGUCUCCCAGCCUGAGGUGCAGUGCUGCAUCUCUGGUCAGUUGGGAG UCUGAGAUGAAGCACUGUAGCUCAGGAAGAGAGAAGUUGUUCUGCAGC). We extended both sides of the sequence of pre-miRNA about 200 bp as the pri-miRNA. The sequence with an over-hanging Mlu I site and a Not I site was synthesized by Genscript (Shanghai, China) and digested to obtain the target gene fragment. The miR 143-coding genomic DNA fragment was cloned down stream from the RFP gene of the lentiviral vector pLenO-RIP by digesting with Not I and Mlu I (Figure 6). The constructs were confirmed by DNA sequencing.

Figure 6. Construction of miR-143 target sequence-luciferase reporter plasmid.

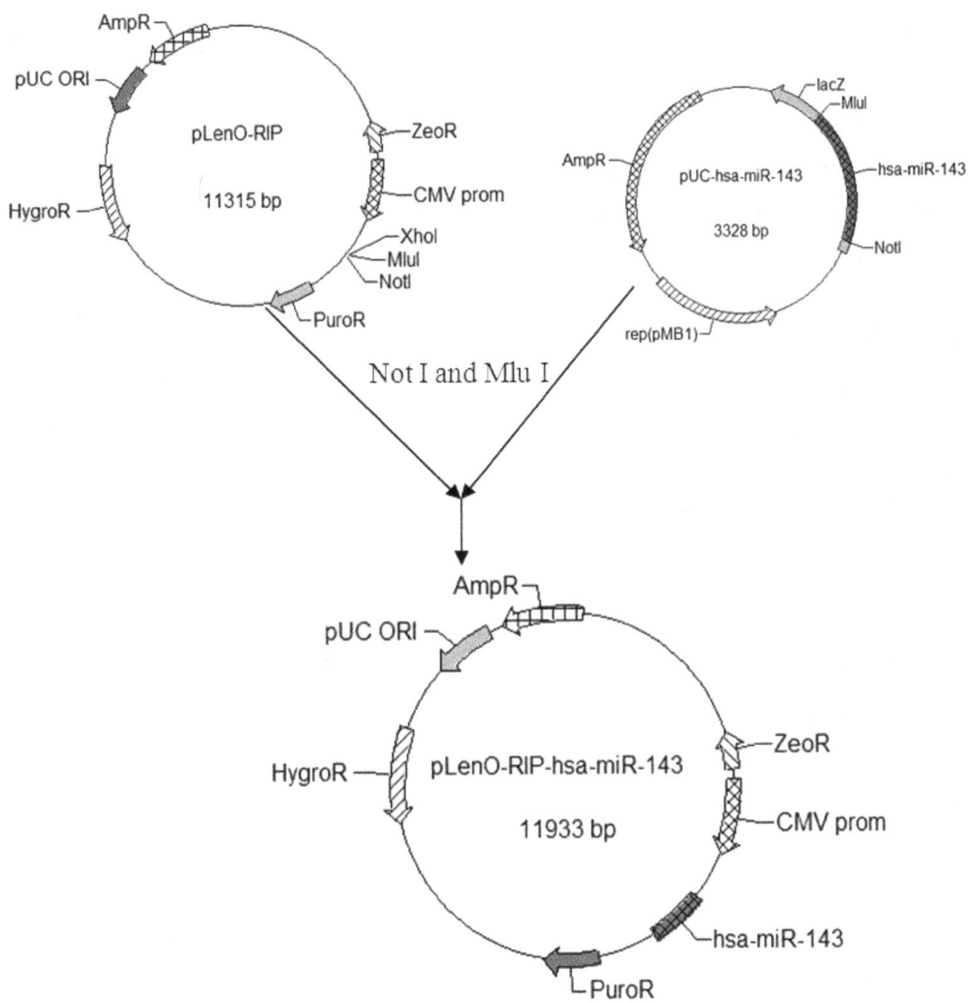

4.2. Lentivirus Production and Transfection

Lentivirus vector was used to transfect genes. This virus packaging system was composed of pRsv-REV, pMDlg-pRRE pMD2G, and interference plasmid. With the help of Lectivirus vector, we could determine the gene expression plasmid, the (pLenO-RIP + miR-143), which also containing the red fluorescent protein (RFP). The production of lentiviral vector pLenO-RIP + miR-143 was performed by simultaneously delivering lentiviral transfer vectors and packaging plasmids (pRsv-REV, pMDlg-pRRE and pMD2.G) into 293 T cells.

Pseudo-viral particles that were generated by 293 T cells within 48 h were centrifuged at 100,000× *g* for 2 h and frozen at −70 °C for future experiments. 293 T cells were seeded into 24-well plates, and lentivirus was used to transduce for 48 h with a multiplicity of infection (MOI) of 10.

4.3. Cell Culture

The HMC-1 cells were cultured at 37 °C in an atmosphere of 5% CO_2. In this case, HMC-1 cells were cultured in DMEM media supplemented with 10% FBS, and the media were replaced routinely. The cells were subjected to transfection when the cell density was about 60%–70%.

4.4. Transfection and Luciferase Assays

To eliminate the interference of transfection reagent, cell lines were seeded into 24-well plates and set up as follows: (a) miR-143 mimics group, in which the cells were transfected with lentivirus at a multiplicity of infection (MOI) of 10 for 48 h; (b) negative control group, in which the cells were subjected to transfection with lentiviral without miR-143; and (c) blank control group, in which the HMC-1 cells were cultured routinely without any treatment. All cells of groups were cultured in normal conditions. The cells were subjected to transfection when the cell density was about 60%–70%. All transfections were performed in quadruplicate and repeated in at least three independent experiments.

After transfection, cell count was determined at 24 h, 48 h and 72 h, respectively for the purpose of determining cell proliferation. The expression of RFP was observed as red fluorescent protein under fluorescent microscope. Moreover, RNA and protein were extracted after transfection at 48 h or 72 h, respectively.

4.5. RNA Isolation and Reverse Transcription and Polymerase Chain Reaction (RT-PCR) for miR-143-Target Gene Expression

Total cellular RNA was extracted from 1×10^6 cells using TrizolTM reagent (Invitrogen Inc., Carlsbad, California, MD, USA) according to the manufacturer's protocol. The concentration and purity of RNA were determined spectrophotometrically. Then, the synthesis of cDNA was performed according to a cDNA cycle kit K1621 (Fermentas Inc., Burlington Canada). To determine the expression of the target gene of miR-143, IL-13Rα1, we performed fluorescent quantitative real time RT-PCR assay. The sequences of the primers (TaKaRa Inc., Dalian, China) specific for IL-13Rα1 were performed with sense (ACCCGAGGGAGCCAGCTCAA) and antisense (GGTGCTACACTGGGACCCCACT) primers, wherein the expected size of the amplified sequence was 111 bp. 18sRNA was used as control. Then, the incubation of cDNA and primer was performed at 95 °C for 15 s, and the PCR reaction proceeded for 45 cycles as per the following conditions: 95 °C for 15 s and 60 °C for 30 s, in a programmable thermal cycler (Eppendorf realplex.2s, Eppendorf Co., Ltd., Hamburger, Germany) using a thermostable Taq DNA polymerase (SYBR premix ex taq, TaKaRa Inc., Dalian, China). Experiments were done in triplicate. For each sample, the amounts of the target and an endogenous control (18sRNA) were determined. To obtain a normalized target value, the amount of the target molecule was divided by the amount of the endogenous reference.

4.6. Western Blotting Analysis

Forty-eight hours after transfection, HMC-1 cells were harvested and centrifuged, and total protein was extracted. Protein concentrations were determined using the BCA protein assay. After heated for 10 min at 100 °C, 20 g of denatured protein was subjected to 10% SDS-PAGE. Then proteins were transferred electrophoretically for 1 h at 200 mA at 4 °C onto PVDF membranes. Membranes were blocked for 1 h at room temperature in TBS containing 5% non-fat dry milk. Blots were washed 3 times for 10 min each with 0.1% TBS-T and subsequently treated overnight at 4 °C with primary antibodies against IL-13Rα1 and actin (1:500). After washing 3 times for 10 min each with 0.1% TBS-T, the blots were incubated with anti-mouse antibody (1:5000) conjugated with horseradish peroxidase for 1 h at room temperature. Bands were visualized by using EZ-ECL detection reagents. The scanned images were quantified using Quantity One software. Actin used as an endogenous protein for normalization. Experiments were done in duplicate. The ratio of IL-13Rα1/actin was used for semi-quantification and comparison between different groups.

4.7. Statistical Analysis

All data were analyzed by SPSS10.0 software (SPSS, Chicago, IL, USA, 2000). Microarray data were analyzed by Cluster Analysis. All data were represented by mean values ± standard deviation. P values less than 0.05 were considered statistically significant. Independent simple T test was used to compare the results of different groups.

5. Conclusions

We successfully constructed the lentiviral miR-143 overexpression vector and transfected it into human mast cell. In this study, we also identified IL-13Rα1 as a target gene of miR-143, and that it was inhibited by overexpression of miR-143 in mast cells. All these findings indicate that miR-143 may play important roles in triggering allergic inflammation.

Acknowledgments

This study was supported by a grant from the National Science Foundation of China (No. 81170893) and Postdoctoral Science Foundation of China (No. 20100470108).

References

1. Meltzer, E.O.; Nathan, R.; Derebery, J.; Stang, P.E.; Campbell, U.B.; Yeh, W.S.; Corrao, M.; Stanford, R. Sleep, quality of life, and productivity impact of nasal symptoms in the United States: Findings from the burden of rhinitis in America survey. *Allergy Asthma Proc.* **2009**, *30*, 244–254.
2. Blaiss, M.S. Allergic rhinitis: Direct and indirect costs. *Allergy Asthma Proc.* **2010**, *31*, 375–380.
3. Lee, Y.; Ahn, C.; Han, J.; Choi, H.; Kim, J.; Yim, J.; Lee, J.; Provost, P.; Rådmark, O.; Kim, S.; *et al.* The nuclear RNase III Drosha initiates microRNA processing. *Nature* **2003**, *425*, 415–419.

4. Ji, X.; Li, J.; Xian, X. microRNAs: Potential regulators of airway smooth muscle cell plasticity involved in asthma-induced airway remodeling. *Asian Biomed.* **2013**, *7*, 3–14.
5. Farh, K.K.; Grimson, A.; Jan, C.; Lewis, B.P.; Johnston, W.K.; Lim, L.P.; Burge, C.B.; Bartel, D.P. The widespread impact of mammalian microRNAs on mRNA repression and evolution. *Science* **2005**, *310*, 1817–1821.
6. Yu, S.; Zhang, R.; Liu, G.; Yan, Z.; Hu, H.; Yu, S.; Zhang, J. Microarray analysis of differentially expressed microRNAs in allergic rhinitis. *Am. J. Rhinol. Allergy* **2011**, *25*, e242–e246.
7. Oh, C.K.; Geba, G.P.; Molfino, N. Investigational therapeutics targeting the IL-4/IL-13/STAT-6 pathway for the treatment of asthma. *Eur. Respir. Rev.* **2010**, *19*, 46–54.
8. Zhang, X.; Graves, P.R.; Zeng, Y. Stable Argonaute2 overexpression differentially regulates microRNA production. *Biochim. Biophys. Acta* **2009**, *1789*, 153–159.
9. Pfeifer, A.; Hofmann, A. Lentiviral transgenesis. *Methods Mol. Biol.* **2009**, *530*, 391–405.
10. Loewen, N.; Poeschla, E.M. Lentiviral vectors. *Adv. Biochem. Eng. Biotechnol.* **2005**, *99*, 169–191.
11. Scherr, M.; Battmer, K.; Ganser, A.; Eder, M. Modulation of gene expression by lentiviral-mediated delivery of small interfering RNA. *Cell Cycle* **2003**, *2*, 251–257.
12. Elia, L.; Quintavalle, M.; Zhang, J.; Contu, R.; Cossu, L.; Latronico, M.V.; Peterson, K.L.; Indolfi, C.; Catalucci, D.; Chen, J.; *et al.* The knockout of miR-143 and -145 alters smooth muscle cell maintenance and vascular homeostasis in mice: correlates with human disease. *Cell Death Differ.* **2009,***16*, 1590–1598.
13. Boettger, T.; Beetz, N.; Kostin, S.; Schneider, J.; Krüger, M.; Hein, L.; Braun, T. Acquisition of the contractile phenotype by murine arterial smooth muscle cells depends on the miR143/145 gene cluster. *J. Clin. Invest.* **2009**, *119*, 2634–2647.
14. Norata, G.D.; Pinna, C.; Zappella, F.; Elia, L.; Sala, A.; Condorelli, G.; Catapano, A.L. microRNA 143–145 deficiency impairs vascular function. *Int. J. Immunopathol. Pharmacol.* **2012**, *25*, 467–474.
15. Lai, V.K.; Ashraf, M.; Jiang, S.; Haider, K. microRNA-143 is a critical regulator of cell cycle activity in stem cells with co-overexpression of Akt and angiopoietin-1 via transcriptional regulation of Erk5/cyclin D1 signaling. *Cell Cycle* **2012**, *11*, 767–777.
16. Ugras, S.; Brill, E.; Jacobsen, A.; Hafner, M.; Socci, N.D.; Decarolis, P.L.; Khanin, R.; O'Connor , R.; Mihailovic, A.; Taylor, B.S.; *et al.* Small RNA sequencing and functional characterization reveals microRNA-143 tumor suppressor activity in liposarcoma. *Cancer Res.* **2011**, *71*, 5659–5669.
17. Kelly-Welch, A.; Hanson, E.M.; Keegan, A.D. Interleukin-13 (IL-13) pathway. *Sci. STKE* **2005**, *2005*, cm8.
18. Munitz, A.; Brandt, E.B.; Mingler, M.; Finkelman, F.D.; Rothenberg, M.E. Distinct roles for IL-13 and IL-4 via IL-13 receptor alpha1 and the type II IL-4 receptor in asthma pathogenesis. *Proc. Natl. Acad. Sci. USA* **2008**, *105*, 7240–7245.
19. Guo, J.; Apiou, F.; Mellerin, M.P.; Lebeau, B.; Jacques, Y.; Minvielle, S. Chromosome mapping and expression of the human interleukin-13 receptor. *Genomics* **1997**, *42*, 141–145.
20. Kaur, D.; Hollins, F.; Woodman, L.; Yang, W.; Monk, P.; May, R.; Bradding, P.; Brightling, C.E. Mast cells express IL-13R alpha 1: IL-13 promotes human lung mast cell proliferation and Fc epsilon RI expression. *Allergy* **200**, *61*, 1047–1053.

11

An Emerging Role for Long Non-Coding RNA Dysregulation in Neurological Disorders

Chiara Fenoglio *, Elisa Ridolfi, Daniela Galimberti and Elio Scarpini

Department of Pathophysiology and Transplantation, University of Milan, "Dino Ferrari" Center, IRCCS Cà Granda Foundation Ospedale Maggiore Policlinico, Via F.Sforza 35, Milan 20122, Italy; E-Mails: elisa.ridolfi@unimi.it (E.R.); daniela.galimberti@unimi.it (D.G.); elio.scarpini@unimi.it (E.S.)

* Author to whom correspondence should be addressed; E-Mail: chiara.fenoglio@unimi.it;

Abstract: A novel class of transcripts, long non coding RNAs (lncRNAs), has recently emerged as key players in several biological processes, including dosage compensation, genomic imprinting, chromatin regulation, embryonic development and segmentation, stem cell pluripotency, cell fate determination and potentially many other biological processes, which still are to be elucidated. LncRNAs are pervasively transcribed in the genome and several lines of evidence correlate dysregulation of different lncRNAs to human diseases including neurological disorders. Although their mechanisms of action are yet to be fully elucidated, evidence suggests lncRNA contributions to the pathogenesis of a number of diseases. In this review, the current state of knowledge linking lncRNAs to different neurological disorders is discussed and potential future directions are considered.

Keywords: lncRNA; RNA; neurodegenerative diseases; epigenetics

1. Introduction

For the last few decades of the 20th century, the underlying dogma of molecular biology has been that the purpose of RNA is to direct the assembly of proteins from amino acids. A few exceptions to this paradigm were known (such as ribosomal RNA and transfer RNA, which are functional RNA macromolecules that do not code for protein). Non-coding RNAs (ncRNAs) include the familial "housekeeping" RNAs and the thousands of regulatory RNAs that are the subject of recent intense investigation. NcRNAs come in many different sizes and for this reason have been divided into small and long classes: small ncRNAs (sncRNA) being less than 200 nucleotides (nt), and long ncRNAs (lncRNA) greater than 200 nt to over 100 kb in length [1]. The current cut-off is arbitrary and corresponds to certain biochemical fractionation protocols and excludes most categories of small infrastructural or regulatory RNAs (tRNAs, snRNAs, miRNAs, siRNAs, piRNAs, tiRNAs, spliRNAs, sdRNAs and others, Figure 1).

Figure 1. Non-coding RNAs (ncRNAs) are arbitrarily grouped basing on size. Small ncRNAs being less than 200 nucleotides and long ncRNAs greater than 200 nucleotides.

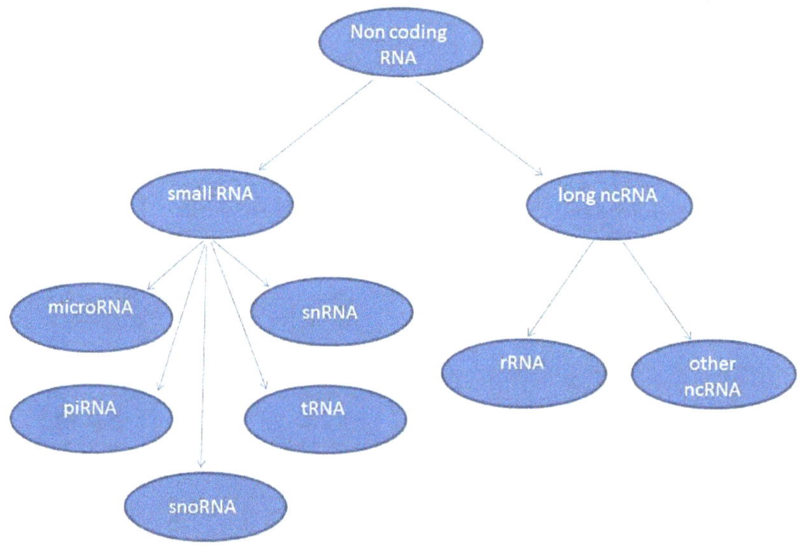

An estimation of the number of lncRNA loci from transcriptional surveys in mammals suggests that they are at least as numerous as protein-coding genes [2] with many of lncRNAs identified in intergenic regions alone [3]. With a few exceptions, it is only within the past few years that the functions and mechanisms of lncRNAs have begun to emerge.

The lncRNAs that have been studied in detail were found to be involved in different biological processes including X chromosome inactivation, nuclear structure, genomic imprinting and development.

In the following paragraphs some of the known functions of lncRNAs will be discussed with particular regard to their role in neurological disorders.

2. Identification of lncRNAs

Data derived from massive cloning and traditional sequencing methods have demonstrated that mammalian genomes produce thousands of RNA transcripts in addition to protein-coding genes [2]. These studies modified our simplistic view of the genome and suggested the presence of thousands of previously unknown transcripts. Recently more than 3000 intergenic ncRNAs have been found in the mammalian genome, by using information from chromatin modifications [3,4]. Evidence gained from ChIP-Seq analyses demonstrated that transcribed protein-coding genes have unusual chromatin modifications; these modifications are trimethylation of histone H3 at lysine 4 (H3K4me3) at the promoter region and trimethylation of histone H3 at lysine 36 (H3K36me3) in the body of the gene [5]. By eliminating these chromatin domains that correspond to protein-coding genes more than 3000 intergenic domains transcribed into lncRNAs have been found. Bioinformatics analyses showed that the majority of these transcripts have no protein-coding capability. Moreover, it was shown that many of these lncRNAs are able to interact with multiple chromatin-modifying complexes in different human and mouse cell types suggesting that lncRNAs may be involved in epigenetic regulation [3].

Recent advances in RNA sequencing technology (RNA-Seq) addressed further interrogation of the total cellular RNA, or, transcriptome, at a much higher resolution [6]. Thousands of transcripts, in addition to protein-coding mRNAs and microRNAs, have been found to be expressed in a wide range of tissues and cell types [7].

3. Epigenetics

Many of the lncRNAs seem to be involved in epigenetic mechanisms of gene regulation. Although the field of epigenetics earned its name over 50 years ago, just in this past decade the significance of epigenetics has been recognized in human health and disease [8]. The term epigenetics refers to changes in gene expression and/or phenotype that can be heritable without a change in the underlying DNA sequence. Several factors contribute to epigenetic mechanisms of gene regulation including DNA methylation, histone modifications, and ncRNAs. DNA methylation in the promoter region of genes is typically associated with transcriptional repression. Several mammalian enzymes are responsible for establishing and maintaining DNA methylation in the genome [9].

The modification of histone proteins has also been found in epigenetic regulation [10]. Histones are highly conserved proteins that package DNA in the nucleus and modulate the accessibility of transcription factors and RNA polymerases to DNA. Histone modifications typically take place at amino acids located in the *N*-termini of histones such as lysine, arginine and serine residues that can be acetylated or methylated (lysine and arginine) and phosphorylated (serine). Histone modifications are placed on and removed from histone residues by numerous enzymes that usually work as part of multi-protein complexes [10].

A relevant question in biology is how chromatin modifying complexes are targeted to specific genomic loci since many of these enzymes lack DNA binding capacity. Recent studies suggest a potential role for ncRNA in driving chromatin-modifying complexes to genomic loci [11–13], however, the exact mechanism by which ncRNAs drive complexes is not known and is currently under investigation.

4. Functions of lncRNAs

Basing on our current knowledge of lncRNAs, it appears that such molecules are involved in many different aspects of cellular functions (reviewed in [14,15]). The roles of lncRNAs in the regulation of gene expression and organismal development are different and just beginning to be discovered. Biological processes dependent upon lncRNAs include imprinting and gene dosage regulation, stem cell pluripotency, embryonic development and segmentation, hematopoiesis, and neural cell fate determination (see for review ref [16]). LncRNAs may employ a number of mechanisms to impact gene expression via *cis* and *trans* processes.

4.1. Gene Imprinting

While the function of parental gene imprinting is still unclear, lncRNAs have been found to participate in imprinting processes. These are referred to as the events that influence the monoallelic expression of a gene according to its parents of origin. Imprinting control regions (ICRs) are DNA regions that are differentially methylated depending on their parental origins. Unmethylated ICRs cause specific expression of nearby lncRNAs, which then suppress neighboring genes in *cis*. Several imprinted clusters contain protein-coding genes and lncRNAs that are reciprocally expressed, such as *IGf2r/Air* [17], *Dlk1/Gtl2* [18,19], and *Nesp/Nespas/Gnas* [20]. Some of these lncRNAs can exert their function by recruiting epigenetic factors, such as PRC2 and G9a, in order to control the imprinted expression of neighboring coding genes [21,22].

Airn and Kcnq1ot1 are examples of lncRNAs that cause suppression of paternally inherited genes. Kcnq1ot1, in particular, is involved in the repression of several protein-coding genes in *cis* through interaction with repressive chromatin modifying complexes [23,24].

H19 was one of the first mammalian lncRNAs to be identified and is highly expressed in the embryo [25,26]. Though it does not seem to act as an lncRNA [27], H19 likely functions as an miRNA precursor [28,29] and is mutually imprinted with the protein-coding gene *Igf2*.

4.2. Gene Dosage and X Chromosome Inactivation

The X chromosome inactivation (XCI) indicates the mechanisms by which the difference in X-linked gene dosage between XX females and XY males is exerted in therian mammals in which one of the two X chromosomes in females is silenced (the inactive X or Xi) so that only one X remains active and is expressed in each female cell (the active X, or Xa) [30]. It is well-known that XCI in placental mammals is largely controlled by a cluster of lncRNA loci known as the X-inactivation center (*Xic*) [31]. The X (inactive)-specific transcript (*Xist*) is highly expressed from Xi during the onset of *XCI* but not from *Xa*, contributing a defining moment for the realization that lncRNAs can have profound roles in the control of gene expression even though the exact mechanism of action is still not completely understood. Some evidence suggests that Xist mediates the chromosome X silencing effects by interacting with repressive chromatin-modifying complexes such as PRC2 [22].

Xist itself is also regulated by other lncRNAs. Initially, *Xist* and *Tsix*, its antisense transcript, transcribed from a promoter downstream of *Xist*, are expressed on both X chromosomes. However, *Tsix* expression continues on the X that will remain active (*Xa*) and this activity recruits dnmt3a to suppress *Xist* from being transcribed on *Xa* [32]. Conversely on *Xi*, it is Tsix that is suppressed, potentially through another lncRNA that is part of the X inactivation center, Jpx [33]. With *Tsix* suppressed, the protein PRC2 is recruited to induce histone modification marks at the 5' end of *Xist*. This upregulates *Xist* expression on *Xi* and causes further propagation of these silencing marks throughout Xi, which are maintained across the lifetime of the organism [33].

4.3. Embryonic Development and Segmentation

LncRNAs are likely implicated in processes involving animal development. The *Hox* genes encode homeodomain TFs that are crucial for anterior-posterior pattern formation in bilateral metazoans [34].

Hox genes are structured in linear clusters along the chromosome, and mammals have four paralogous clusters, *HoxA*, *-B*, -C, and -D. Several lncRNAs are encoded within these clusters, including HOTAIR (Hox antisense intergenic RNA) from *HoxC*, and HOTTIP and Mistral from *HoxA* [35–37]. The expression of *Hox* genes is also regulated by lncRNAs [36]. Some Hox-related lncRNAs operate in *cis*, having either enhancing or repressive effects. However, some like the human HOTAIR works in *trans*, and it is expressed from the *Hox* locus marking a boundary of active and inactive chromatin [35]. Furthermore, HOTAIR, similar to *Xist*, interacts with chromatin-modifying complexes such as PRC2 and the corepressor complex CoREST [35] and may guide these complexes to genomic loci. Overexpression of HOTAIR caused cells to become metastatic when injected into mice compared to control cells with an empty vector. Moreover, HOTAIR may serve as a scaffold for targeting chromatin-modifying complexes to chromatin [35].

Recently HOTAIR was found to be crucial for cell growth and viability and that its knockdown induced apoptosis in breast cancer cells Moreover it was found that HOTAIR is transcriptionally induced by estradiol [38]. It is possible that other lncRNAs that interact with chromatin-modifying complexes also function in a manner similar to HOTAIR.

4.4. Stem Cell Pluripotency and Cell Fate Determination

The promoters of more than 100 lncRNAs are bound by stem cell factors. Disruption of these lncRNAs can alter cell differentiation. The human lncRNA-RoR (RoR) is a recently identified lncRNA that is capable of reprogramming differentiated cells to induce pluripotent stem cells [39,40]. RoR is highly expressed both in embryonic stem cells and in induced pluripotent stem cells, due to the regulation of RoR by pluripotency transcription factors such as Oct4, Sox2 and Nanog. Interestingly it was observed that knockdown of RoR leads to a modest increase in apoptosis and activation of p53 pathways [40]. Although the underlying mechanisms still remain to be fully clarified, Zhang *et al.* recently [41] demonstrated that human RoR is a strong negative regulator of p53 influencing the inhibition of p53-mediated cell cycle arrest and apoptosis.

Recently, a refined analysis from Guttman *et al.* [39] performed loss-of-function studies on 226 lncRNAs expressed in mouse embryonic stem cells characterizing the effects on gene expression. The authors identified 26 lncRNAs able to maintain the pluripotent status. In particular, knockdown of these lncRNAs resulted in a loss of pluripotency markers, and reduction of Nanog promoter activity. Simultaneously, expression patterns similar to differentiation into specific lineages were produced, suggesting that lncRNAs repressed differentiation programs in mouse embryonic stem cells. Altogether these findings support the hypothesis that some lncRNAs are integral members of a regulatory network, together with key pluripotent transcription factors, which modulate pluripotency and lineage-specific differentiation pathways in mouse embryonic stem cells.

lncRNAs are implicated in cell fate determination events in multiple cells lineages, including the nervous system. Taurine upregulated gene 1 (*TUG1*) is an lncRNA expressed in the developing retina and brain, as well as in adult tissues. It has been found that in the newborn retina, loss of TUG1 resulted in malformed or non-existent outer segments of transfected photoreceptors, thus suggesting that TUG1 is required for the proper formation of photoreceptors in the developing rodent retina [42]. Evf2 is instead a mouse lncRNA that appears to recruit Dlx and Mecp2 transcription factors to important DNA regulatory elements in the *Dlx5-Dlx6* intergenic regions and controlled Dlx5, *Dlx6* and Gad1 expression through *cis* and *trans* acting mechanisms. Evf2 mouse mutants appeared to have reduced numbers of GABAergic interneurons in early postnatal hippocampus and the dentate gyrus. This situation is restored to normality in Evf2 mutant adult hippocampus although reduced synaptic inhibition still occurred [43,44].

Although many lncRNAs have been shown to regulate gene expression only a few have been shown to have other cellular functions. For example, NEAT1 has been shown to play an important role in paraspeckle formation [45]. Also NRON has a role in nuclear import/export [46]. All together, these studies suggest that the lncRNAs have different cellular functions, many of which are yet to be identified and characterized for the mechanism of their function.

5. LncRNAs in Human Diseases

As the functions and mechanisms of lncRNAs are beginning to emerge, there is an intense interest in identifying any potential role of these molecules in human diseases. Several studies have shown that lncRNAs are dysregulated in human pathologies, however it has yet to be shown that these molecules are enough to drive the disease status.

lncRNAs have been strongly associated with cancer [47]. Recently, the lncRNA PCAT-1, was found to promote cell proliferation and is a target of PRC2 regulation, [48]. Moreover, ANRIL, which is upregulated in prostate cancer, is required for the expression of the tumor suppressors INK4a/p16 and INK4b/p15 [49]. HOTAIR upregulation is associated with poor prognosis in breast cancer [11], liver [50], colorectal [51], gastrointestinal [52] and pancreatic [53] cancers and probably contributes to increase also tumor invasiveness and metastasis [11].

MALAT-1, which is another lncRNA associated with various cancers and metastasis [54] was found to affect the transcriptional and post-transcriptional regulation of cytoskeletal and extracellular matrix genes [55]. Although lncRNAs have been extensively investigated in cancers several lines of evidence suggest a possible role also in different disease conditions such as cardiovascular diseases. For example, two lncRNAs have been found to be dysregulated in heart disease; the expression of the lncRNA MIAT is associated with increased risk of myocardial infarction, whereas the lncRNA ANRIL is associated with increased risk to coronary heart disease [56,57]. Recently, a novel lncRNA has been discovered, named DBE-T, that functions in *cis* and localizes to the Facioscapulohumeral muscular dystrophy (FSHD) locus. FSHD is the third most common myopathy and is predominantly caused by a contraction of specific repeats mapping on chromosome 4q35 [58]. It is suggested that DBE-T likely acts as a locus control element by promoting active chromatin domains and thus FSHD would be caused from lncRNA "promoter mutations" able to perturb DBE-T regulation [58]. Furthermore, a novel lncRNA seems to be involved in the pathogenic mechanisms underneath the HELLP syndrome (hemolysis, elevated liver enzymes, low platelets) that is a recessively inherited life-threatening pregnancy complication [59].

6. Role of lncRNAs in the Central Nervous System

Recent evidence demonstrate that lncRNAs contribute to the complex biological system organization and gene regulatory networks of the central nervous system (CNS), affecting brain patterning, neural stem cell maintenance, neurogenesis and gliogenesis, stress responses, and synaptic and neural plasticity.

Mercer and colleagues identified 849 lncRNAs (among the 1328 examined), that are expressed in the adult mouse brain and found that the majority were associated with specific neuroanatomical regions, cell types, or subcellular compartments [60]. A complementary study showed that over 200 of these lncRNAs are expressed in developing and adult mouse brain and are largely derived from genomic loci located proximal to protein-coding genes with similar expression profiles in the brain [61].

Guttman *et al.* discovered more than 1000 evolutionarily conserved intergenic lncRNAs in mouse by analyzing chromatin signatures from four mouse cell types [5]. A functional analysis of the expression of these lncRNAs revealed the presence of a "brain cluster" of lncRNAs that is associated with biological processes including hippocampal development, oligodendrocyte (OL) myelination, brain aging, CREB and PGC1-alpha transcriptional regulation, and GABAergic neuronal (GABAN), G protein coupled receptor and calcineurin signaling pathways. An additional study demonstrated that 169 lncRNAs are differentially expressed during the sequential processes of mouse ventral forebrain-derived neural precursor cells mediated lineage restriction, GABAN and OL lineage specification, progressive OL lineage maturation, and terminal differentiation including myelination [62].

Detailed analyses of specific lncRNAs, dynamically expressed in the CNS, reveal potential roles in mediating neural cell fate decisions. The Sox2OT lncRNA, which contains the *Sox2* gene within one of its introns and is subsequently transcribed in the same direction [63], is expressed in regions of constitutive adult neurogenesis [60]. Moreover, Sox2OT is dynamically regulated in CNS structures during development, where it may be responsible for modulating Sox2 expression [64]. The lncRNA Nkx2.2AS regulates Nkx2.2, a transcription factor that is critical for OL lineage specification. A recent study reported that forced expression of Nkx2.2AS in NSCs *in vitro* enhances their differentiation along the OL lineage, in part, by inducing an increase in Nkx2.2 mRNA levels [65].

LncRNAs also modulate synaptic plasticity and promote long-term changes in synaptic strength. The rodent-specific *BC1* and primate-specific BC200 lncRNAs, are selectively targeted to postsynaptic dendritic compartments, where they regulate local protein synthesis by repressing the initiation of translation through an eIF4A-dependent mechanism [66–69]. Similarly, NTAB is a lncRNA that is expressed in developing and adult rat brain, where it is also found in neuronal processes [70]. Another lncRNA, MALAT-1, is enriched in hippocampal neurons, where it regulates several serine/arginine splicing factors important for synapse formation, density and maturation [71].

6.1. Dysregulation of lncRNAs in Neurological Disorders

6.1.1. LncRNAs Play a Role in the Pathophysiology of Several Neurological Disorders

Angelman syndrome (AS) is a neurodevelopmental disorder associated with genomic imprinting and characterized by severe neurologic abnormalities [72]. Ube3a-as is a lncRNA transcribed antisense to the maternally expressed *Ube3a* gene, mutated or deleted in AS, suggesting that Ube3a-as may repress paternal Ube3a expression. Other studies have shown that repression of Ube3a is dependent on Ube3a-as [73,74]. However, other data has demonstrated that silencing of paternal *Ube3a* can occur in the absence of Ube3a-as and implies a more complex regulatory relationship underlying the imprinting of *Ube3a* [75].

LncRNAs may influence the pathogenesis of fragile X syndrome (FXS), which is characterized by a triplet nucleotide repeat expansion in the 5'UTR of *FMR1*, the gene encoding the neuronal development protein, FMRP. The lncRNAs ASFMR1 and FMR4 are generated from the *FMR1* gene locus. ASFMR1 has multiple alternative splicing patterns and overlaps the 5' untranslated region (UTR) CGG repeat region of *FMR1* [76]; FMR4 is initiated upstream of the *FMR1* start site. Alternative splicing of ASFMR1 seems to exhibit pre-mutation-specific profiles and is also silenced in FXS patients and upregulated in pre-mutation carriers, suggesting that a common process is responsible for regulating the expression of these transcripts. *FMR4* is also silenced in FXS patients because of a CGG expansion repeat in the 5'UTR of the *FMR1* gene and upregulated in pre-mutation carriers [77]; thus, their absence in the neurons of affected patients might contribute to the pathogenesis of this neurological disorder.

Another microsatellite expansion disease in which lncRNAs are involved is the spinocerebellar ataxia type 8 (SCA8), an autosomal dominant disorder, characterized by bidirectional transcription of this expansion repeat from opposite strands, forming both a protein-coding transcript encoding a polyglutamine expansion, ATXN8, and a lncRNA transcript containing a CUG expansion, ATXN8OS [78]. This suggests that SCA8 pathogenesis involves a toxic gain of function at both the protein and RNA levels. A recent study found that the expanded ATXN8OS transcript accumulates in ribonuclear inclusions in the GABAergic neurons of SCA8 patients [79]. These inclusions co-localized with splicing factor, MBNL1, altering the activity of MBNL alternative splicing proteins [79].

Huntington's disease (HD) is caused by an expansion repeat mutation in the *Htt* gene, which encodes a ubiquitously expressed 3144 amino acid protein of unknown function, leading to a toxic gain of function in the mutant protein [80], which promotes aberrant nuclear-cytoplasmic trafficking of the master neuronal regulator REST. The result is the deregulation of REST target gene expression in tissues from animal models of HD and human HD, which include both protein-coding genes as well as ncRNAs, such as lncRNAs. It is therefore likely that HD tissues are also characterized by dysregulation of lncRNA expression. ChiP-seq data showed that the *HAR1* locus is under control of REST [81]. The *HAR1* region contains lncRNAs, *HAR1F* and *HAR1R*. HAR1F is specifically expressed in the neurons of the marginal zone during development of the cortex and in the frontal cortex, hippocampus, cerebellum, thalamus and hypothalamus in the adult brain [82]. The levels of HAR1F and HAR1R are decreased in HD brains compared with normal brains [83].

6.1.2. LncRNAs in Neurodegenerative Diseases, such as Alzheimer's Disease (AD)

β-site amyloid precursor protein-cleaving enzyme 1 (BACE1) is a crucial enzyme in AD pathophysiology, involved in the cleavage of the amyloid precursor protein (*APP*) and the generation of amyloid peptides which can aggregate and form plaques. Faghihi and colleagues characterized a conserved non-coding antisense transcript for BACE1, called BACE1-AS, which functions as a regulator of BACE1 gene expression. BACE1-AS upregulates BACE1 levels in response to a variety of stresses, including Aβ 1–42 exposure, and is elevated in several brain regions of patients with AD. These findings imply that BACE1-AS is deregulated in AD, which induces feed-forward regulation of BACE1, increases Aβ levels, and thus may promote the pathogenesis of the disease [84].

Mus *et al.* found a link between altered levels of a lncRNA, BC200, and AD [85]. Increased levels of BC200 were found in brain regions that are preferentially affected in AD, such as the hippocampus, which correlated with disease severity. Further, in advanced stages of AD, BC200 was mis-localized and clustered in the perikaryon. These observations suggest that deregulation of this synaptic lncRNA is involved in the synaptic and neural network dysfunction that is found in both early and later stages of AD [85].

Together with neurological diseases, a number of psychiatric disorders have also been associated with lncRNAs (Table 1). The risk of developing schizophrenia (SZ), schizoaffective disorder, bipolar disorder, major depression, and autistic spectrum disorders has been linked to the disruption of the *DISC* genomic locus, which encodes both the *DISC1* protein-coding gene and the *DISC2* lncRNA [86–88]. *DISC2* controls the expression of its partner, *DISC1*, which modulates multiple aspects of CNS structure and function such as embryonic and adult neurogenesis [89]. Another lncRNA, Gomafu, implicated in brain and retinal development [62,90] binds directly to the splicing factors QKI and SRSF1 and the dysregulation of Gomafu induces alternative splicing that resemble those observed in SZ for the archetypal SZ-associated genes *DISC1* and *ERB4*. Moreover, Gomafu is downregulated in post-mortem cortex of SZ subjects, suggesting a role in SZ pathogenesis for this lncRNA [91].

Table 1. Examples of lncRNAs that are dysregulated in neurological disorders.

lncRNA	Disease association	Biological function	Reference
Ube3a-as	Angelman Syndrome	Repress paternal *Ube3a* expression	[72]
ASFMR1 FMR4	Fragile X Syndrome	Regulate the expression of *ASFMR1* and *FMR1* genes	[76,77]
ATXN8OS	SCA8	Alteration of the activity of the splicing factor *MBNL1*	[78,79]
HAR1F HAR1R	Hungtington's Disease	Influence genes promoting aberrant nuclear-cytoplasmatic trafficking of *REST* gene	[83]
BC200 BACE1-AS	Alzheimer's	Involved in the synaptic and neural network dysfunction Regulates *BACE1* gene expression	[84,85]
DISC2 Gomafu	Psychiatric disorders	Controls the expression of *DISC1* Implicated in brain and retinal development	[86–89] [91]

7. Conclusions and Future Perspectives

Several recent studies suggest that lncRNAs play a pivotal role in many key biological processes, although their mechanisms of action are yet to be fully elucidated. Currently there is great interest in identifying the functions of this novel class of transcripts. There is strong evidence that many lncRNAs are biologically relevant, with a large percentage of these molecules functioning through their interactions with chromatin-modifying complexes to alter gene expression [3]. These findings are beginning to shed light on how chromatin-modifying complexes are targeted to specific genomic loci and suggest the interesting idea that lncRNAs are in some way driving chromatin-modifying complexes to genomic loci.

LncRNAs have, in a relatively short period of time, become recognized as a major new class of genes that may potentially comprise a major component of the genome's information content, complementary and comparable in abundance and complexity to the proteome. Furthermore lncRNAs have already been reported in a wide range of human diseases suggesting that their activity is crucial for human health.

In addition, therapeutic strategies that target endogenous mRNA molecules, such as those employing RNA interference (RNAi) and other customized oligonucleotide approaches with the capacity to reprogram disease-associated mRNAs, are now being developed [92]. These approaches could be adapted to target lncRNAs whose expression is dysregulated in CNS disorders. These observations suggest that lncRNAs represent a versatile class of factors that are centrally important to the modulation of different CNS processes and may represent a major layer underlying the genetic programming of brain development that could potentially be utilized for developing novel diagnostic and therapeutic tools for the cure of CNS disorders.

References

1. Mattick, J.S. Non-coding RNAs: The architects of eukaryotic complexity. *EMBO* **2001**, *2*, 986–991.
2. Harrow, J.; Frankish, A.; Gonzalez, J.M.; Tapanari, E.; Diekhans, M.; Kokocinski, F.; Aken, B.L.; Barrell, D.; Zadissa, A.; Searle, S.; *et al.* GENCODE: The reference human genome annotation for The ENCODE Project. *Genome Res.* **2012**, *22*, 1760–1774.
3. Khalil, A.M.; Guttman, M.; Huarte, M.; Garber, M.; Raj, A.; Rivea Morales, D.; Thomas, K.; Presser, A.; Bernstein, B.E.; van Oudenaarden, A.; *et al.* Many human large intergenic noncoding RNAs associate with chromatin-modifying complexes and affect gene expression. *Proc. Natl. Acad. Sci. USA* **2009**, *106*, 11667–11672.
4. Marques, A.C.; Ponting, C.P. Catalogues of mammalian long noncoding RNAs: Modest conservation and incompleteness. *Genome Biol.* **2009**, *10*, R124.
5. Guttman, M.; Amit, I.; Garber, M.; French, C.; Lin, M.F.; Feldser, D.; Huarte, M.; Zuk, O.; Carey, B.W.; Cassady, J.P.; *et al.* Lander Chromatin signature reveals over a thousand highly conserved large non-coding RNAs in mammals. *Nature* **2009**, *458*, 223–227.
6. Nagalakshmi, U.; Waern, K.; Snyder, M. RNA-Seq: A method for comprehensive transcriptome analysis. *Curr. Protoc. Mol. Biol.* **2010**, doi:10.1002/0471142727.
7. Wang, E.T.; Sandberg, R.; Luo, S.; Khrebtukova, I.; Zhang, L.; Mayr, C.; Kingsmore, S.F.; Schroth, G.P.; Burge, C.B. Alternative isoform regulation in human tissue transcriptomes. *Nature* **2008**, *456*, 470–476.
8. Egger, G.; Liang, G.; Aparicio, A.; Jones, P.A. Epigenetics in human disease and prospects for epigenetic therapy. *Nature* **2004**, *429*, 457–463.
9. Kaneda, M.; Okano, M.; Hata, K.; Sado, T.; Tsujimoto, N.; Li, E.; Sasaki, H. Essential role for de novo DNA methyltransferase Dnmt3a in paternal and maternal imprinting. *Nature* **2004**, *429*, 900–903.
10. Bernstein, B.E.; Meissner, A.; Lander, E.S. The mammalian epigenome. *Cell* **2007**, *128*, 669–681.
11. Gupta, R.A.; Shah, N.; Wang, K.C.; Kim, J.; Horlings, H.M.; Wong, D.J.; Tsai, M.C.; Hung, T.; Argani, P.; Rinn, J.L.; *et al.* Long non-coding RNA HOTAIR reprograms chromatin state to promote cancer metastasis. *Nature* **2010**, *464*, 1071–1076.

12. Reis, E.M.; Louro, R.; Nakaya, H.I.; Verjovski-Almeida, S. As antisense RNA gets intronic. *OMICS* **2005**, *9*, 2–12.
13. Tsai, M.C.; Manor, O.; Wan, Y.; Mosammaparast, N.; Wang, J.K.; Lan, F.; Shi, Y.; Segal, E.; Chang, H.Y. Long noncoding RNA as modular scaffold of histone modification complexes. *Science* **2010**, *329*, 689–693.
14. Mattick, J.S. The genetic signatures of noncoding RNAs. *PLoS Genet.* **2009**, *5*, e1000459.
15. Mattick, J.S.; Amaral, P.P.; Dinger, M.E.; Mercer, T.R.; Mehler, M.F. RNA regulation of epigenetic processes. *Bioessays* **2009**, *31*, 51–59.
16. Clark, M.B.; Mattick, J.S. Long noncoding RNAs in cell biology. *Semin. Cell Dev. Biol.* **2011**, *22*, 366–376.
17. Lyle, R.; Watanabe, D.; Vruchte, W.; Lerchner, O.; Smrzka, W. The imprinted antisense RNA at the Igf2r locus overlaps but does not imprint Mas1. *Nat. Genet.* **2000**, *25*, 19–21.
18. Schmidt, J.V.; Matteson, P.G.; Jones, B.K.; Guan, X.J.; Tilghman, S.M. The *Dlk1* and *Gtl2* genes are linked and reciprocally imprinted. *Genes Dev.* **2000**, *14*, 1997–2002.
19. Da Rocha, S.T.; Edwards, C.A.; Ito, M.; Ogata, T.; Ferguson-Smith, A.C. Genomic imprinting at the mammalian Dlk1-Dio3 domain. *Trends Genet.* **2008**, *24*, 306–316.
20. Williamson, C.M.; Turner, M.D.; Ball, S.T.; Nottingham, W.T.; Glenister, P. Identification of an imprinting control region affecting the expression of all transcripts in the Gnas cluster. *Nat. Genet.* **2006**, *38*, 350–355.
21. Nagano, T.; Mitchell, J.A.; Sanz, L.A.; Pauler, F.M.; Ferguson-Smith, A.C.; Feil, R.; Fraser, P. The Air noncoding RNA epigenetically silences transcription by targeting G9a to chromatin. *Science* **2008**, *322*, 1717–1720.
22. Zhao, J.; Sun, B.K.; Erwin, J.A.; Song, J.J.; Lee, J.T. Polycomb proteins targeted by a short repeat RNA to the mouse X chromosome. *Science* **2008**, *322*, 750–756.
23. Pandey, R.R.; Mondal, T.; Mohammad, F.; Enroth, S.; Redrup, L.; Komorowski, J.; Nagano, T.; Mancini-Dinardo, D.; Kanduri, C. Kcnq1ot1 antisense noncoding RNA mediates lineage-specific transcriptional silencing through chromatin-level regulation. *Mol. Cell* **2008**, *32*, 232–234.
24. Kanduri, C.; Thakur, N.; Pandey, R. The length of the transcript encoded from the Kcnq1ot1 antisense promoter determines the degree of silencing. *EMBO J.* **2006**, *25*, 2096–2106.
25. Bartolomei, M.S.; Zemel, S.; Tilghman, S.M. Parental imprinting of the mouse *H19* gene. *Nature* **1991**, *351*, 153–155.
26. Brannan, C.I.; Dees, E.C.; Ingram, R.S.; Tilghman, S.M. The product of the *H19* gene may function as an RNA. *Mol. Cell Biol.* **1990**, *20*, 28–36.
27. Gabory, A.; Jammes, H.; Dandolo, L. The H19 locus: Role of an imprinted non-coding RNA in growth and development. *Bioessays* **2010**, *32*, 473–480.
28. Cai, X.; Cullen, B.R. The imprinted H19 noncoding RNA is a primary microRNA precursor. *RNA* **2007**, *13*, 313–316.
29. Keniry, A.; Oxley, D.; Monnier, P.; Kyba, M.; Dandolo, L. The H19 lincRNA is a developmental reservoir of miR-675 that suppresses growth and Igf1r. *Nat. Cell Biol.* **2012**, *14*, 859–865.
30. Lee, J.T. Gracefully ageing at 50, X-chromosome inactivation becomes a paradigm for RNA and chromatin control. *Nat. Rev. Mol. Cell Biol.* **2011**, *12*, 815–826.

31. Brown, C.J.; Lafrenière, R.G.; Powers, V.E.; Sebastio, G.; Ballabio, A. Localization of the X inactivation centre on the human X chromosome in Xq13. *Nature* **1991**, *349*, 82–84.
32. Do, J.T.; Han, D.W.; Gentile, L.; Sobek-Klocke, I.; Stehling, M.; Schöler, H.R. Enhanced reprogramming of Xist by induced upregulation of Tsix and Dnmt3a. *Stem Cells* **2008**, *26*, 2821–2831.
33. Tian, D.; Sun, S.; Lee, J.T. The long noncoding RNA, Jpx, is a molecular switch for X chromosome inactivation. *Cell* **2010**, *143*, 390–403.
34. Pearson, J.C.; Lemons, D.; McGinnis, W. Modulating *Hox* gene functions during animal body patterning. *Nat. Rev. Genet.* **2005**, *6*, 893–904.
35. Rinn, J.L.; Kertesz, M.J.; Wang, K.; Squazzo, S.L.; Xu, X. Functional demarcation of active and silent chromatin domains in human HOX loci by noncoding RNAs. *Cell* **2007**, *129*, 1311–1323.
36. Bertani, S.; Sauer, S.; Bolotin, E.; Sauer, F. The noncoding RNA Mistral activates Hoxa6 and Hoxa7 expression and stem cell differentiation by recruiting MLL1 to chromatin. *Mol. Cell* **2011**, *43*, 1040–1046.
37. Wang, K.C.; Yang, Y.W.; Liu, B.; Sanyal, A.; Corces-Zimmerman, R. A long noncoding RNA maintains active chromatin to coordinate homeotic gene expression. *Nature* **2011b**, *472*, 120–124.
38. Bhan, A.; Hussain, I.; Ansari, K.I.; Kasiri, S.; Bashyal, A.; Mandal, S.S. Antisense Transcript Long Noncoding RNA (lncRNA) HOTAIR is transcriptionally induced by Estradiol. *J. Mol. Biol.* **2013**, doi:10.1016/j.jmb.2013.01.022.
39. Guttman, M.; Donaghey, J.; Carey, B.W.; Garber, M.; Grenier, J.K.; Munson, G.; Young, G.; Lucas, A.B.; Ach, R.; Bruhn, L.; *et al.* lincRNAs act in the circuitry controlling pluripotency and differentiation. *Nature* **2011**, *477*, 295–300.
40. Loewer, S.; Cabili, M.N.; Guttman, M.; Loh, Y.H.; Thomas, K.; Park, I.H.; Garber, M.; Curran, M.; Onder, T.; Agarwal, S.; *et al.* Large intergenic non-coding RNA-RoR modulates reprogramming of human induced pluripotent stem cells. *Nat. Genet.* **2010**, *42*, 1113–1117.
41. Zhang, A.; Zhou, N.; Huang, J.; Liu, Q.; Fukuda, K.; Ma, D.; Lu, Z.; Bai, C.; Watabe, K.; Mo, Y.Y. The human long non-coding RNA-RoR is a p53 repressor in response to DNA damage. *Cell Res.* **2013**, *23*, 340–350.
42. Young, T.L.; Matsuda, T.; Cepko, C.L. The noncoding RNA taurine upregulated gene 1 is required for differentiation of the murine retina. *Curr. Biol.* **2005**, *15*, 501–512.
43. Feng, J.; Bi, C.; Clark, B.S.; Mady, R.; Shah, P.; Kohtz, J.D. The Evf-2 noncoding RNA is transcribed from the Dlx-5/6 ultraconserved region and functions as a Dlx-2 transcriptional coactivator. *Genes Dev.* **2006**, *20*, 1470–1484.
44. Bond, A.M.; Vangompel, M.J.; Sametsky, E.A.; Clark, M.F.; Savage, J.C.; Disterhoft, J.F. Balanced gene regulation by an embryonic brain ncRNA is critical for adult hippocampal GABA circuitry. *Nat. Neurosci.* **2009**, *12*, 1020–1027.
45. Clemson, C.M.; Hutchinson, J.N.; Sara, S.A.; Ensminger, A.W.; Fox, A.H.; Chess, A.; Lawrence J.B. An architectural role for a nuclear noncoding RNA: NEAT1 RNA is essential for the structure of paraspeckles. *Mol. Cell* **2009**, *33*, 717–726.
46. Willingham, A.T.; Orth, A.P.; Batalov, S.; Peters, E.C.; Wen, B.G.; Aza-Blanc, P.; Hogenesch, J.B.; Schultz, P.G. A strategy for probing the function of noncoding RNAs finds a repressor of NFAT. *Science* **2005**, *309*, 1570–1573.

47. Gutschner, T.; Diederichs, S. The hallmarks of cancer: A long non-coding RNA point of view. *RNA Biol.* **2012**, *9*, 703–719.
48. Prensner, J.R.; Iyer, M.K.; Balbin, O.A.; Dhanasekaran, S.M.; Cao, Q. Transcriptome sequencing across a prostate cancer cohort identifies PCAT-1, an unannotated lincRNA implicated in disease progression. *Nat. Biotechnol.* **2011**, *29*, 742–749.
49. Kotake, Y.; Nakagawa, T.; Kitagawa, K.; Suzuki, S.; Liu, N. Long non-coding RNA ANRIL is required for the PRC2 recruitment to and silencing of p15(INK4B) tumor suppressor gene. *Oncogene* **2011**, *30*, 1956–1962.
50. Yang, Z.; Zhou, L.; Wu, L.M.; Lai, M.C.; Xie, H.Y.; Zhang, F.; Zheng, S.S. Overexpression of long non-coding RNA HOTAIR predicts tumor recurrence in hepatocellular carcinoma patients following liver transplantation. *Ann. Surg. Oncol.* **2011**, *18*, 1243–1250.
51. Kogo, R.; Shimamura, T.; Mimori, K.; Kawahara, K.; Imoto, S.; Sudo, T.; Tanaka, F.; Shibata, K.; Suzuki, A.; Komune, S.; *et al.* Long noncoding RNA HOTAIR regulates polycombdependent chromatin modification and is associated with poor prognosis in colorectal cancers. *Cancer Res.* **2011**, *71*, 6320–6326.
52. Niinuma, T.; Suzuki, H.; Nojima, M.; Nosho, K.; Yamamoto, H. Upregulation of miR-196a and HOTAIR drive malignant character in gastrointestinal stromal tumors. *Cancer Res.* **2012**, *72*, 1126–1136.
53. Kim, K.; Jutooru, I.; Chadalapaka, G.; Johnson, G.; Frank J.; Burghardt, R.; Kim, S.; Safe, S. HOTAIR is a negative prognostic factor and exhibits pro-oncogenic activity in pancreatic cancer. *Oncogene* **2012**, doi:10.1038/onc.2012.193.
54. Lin, R.; Roychowdhury-Saha, M.; Black, C.; Watt, A.T.; Marcusson, E.G. Control of RNA processing by a large non-coding RNA over-expressed in carcinomas. *FEBS Lett.* **2011**, *585*, 671–676.
55. Tano, K.; Mizuno, R.; Okada, T.; Rakwal, R.; Shibato, J. MALAT-1 enhances cell motility of lung adenocarcinoma cells by influencing the expression of motility-related genes. *FEBS Lett.* **2010**, *584*, 4575–4580.
56. Ishii, N.; Ozaki, K.; Sato, H.; Mizuno, H.; Saito, S.; Takahashi, A.; Miyamoto, Y.; Ikegawa, S.; Kamatani, N.; Hori, M.; *et al.* Identification of a novel non-coding RNA, MIAT, that confers risk of myocardial infarction. *J. Hum. Genet.* **2006**, *51*, 1087–1099.
57. Broadbent, H.M.; Peden, J.F.; Lorkowski, S.; Goel, A.; Ongen, H.; Green, F.; Clarke, R.; Collins, R.; Franzosi, M.G.; Tognoni, G.; *et al.* PROCARDIS consortium. Susceptibility to coronary artery disease and diabetes is encoded by distinct, tightly linked SNPs in the ANRIL locus on chromosome 9p. *Hum. Mol. Genet.* **2008**, *17*, 806–814.
58. Cabianca, D.S.; Casa, V.; Bodega, B.; Xynos, A.; Ginelli, E.; Tanaka, Y.; Gabellini, D. Long ncRNA links copy number variation to a polycomb/trithorax epigenetic switch in FSHD muscular dystrophy. *Cell* **2012**, *149*, 819–831.
59. Van Dijk, M.; Thulluru, H.K.; Mulders, J.; Michel, O.J.; Poutsma, A. Windhorst, HELLP babies link a novel lincRNA to the trophoblast cell cycle. *J. Clin. Invest.* **2012**, *122*, 4003–4011.
60. Mercer, T.R.; Dinger, M.E.; Sunkin, S.M.; Mehler, M.F.; Mattick, J.S. Specific expression of long noncoding RNAs in the mouse brain. *Proc. Natl. Acad. Sci. USA* **2008**, *105*, 716–721.

61. Ponjavic, J.; Oliver, P.L.; Lunter, G.; Ponting, C.P. Genomic and transcriptional co-localization of protein-coding and long non-coding RNA pairs in the developing brain. *PLoS Genet.* **2009**, *5*, e1000617.
62. Mercer, T.R.; Qureshi, I.A.; Gokhan, S.; Dinger, M.E.; Li, G.; Mattick, J.S.; Mehler, M.F. Long noncoding RNAs in neuronal-glial fate specification and oligodendrocyte lineage maturation *BMC Neurosci.* **2010**, *11*, 14.
63. Fantes, J.; Ragge, N.K.; Lynch, S.A.; McGill, N.I.; Collin, J.R.; Howard-Peebles, P.N.; Hayward, C.; Vivian, A.J.; Williamson, K.; van Heyningen V.; *et al.* Mutations in *SOX2* cause anophthalmia. *Nat. Genet.* **2003**, *33*, 461–463.
64. Amaral, P.P.; Neyt, C.; Wilkins, S.J.; Askarian-Amiri, M.E.; Sunkin, S.M.; Perkins, A.C.; Mattick, J.S. Complex architecture and regulated expression of the Sox2ot locus during vertebrate development. *RNA* **2009**, *15*, 2013–2027.
65. Tochitani, S.; Hayashizaki, Y. Nkx2.2 antisense RNA overexpression enhanced oligodendrocytic differentiation. *Biochem. Biophys. Res. Commun.* **2008**, *372*, 691–696.
66. Brosius, J. RNAs from all categories generate retrosequences that may be exapted as novel genes or regulatory elements. *Gene* **1999**, *238*, 115–134.
67. Kondrashov, A.V.; Kiefmann, M.; Ebnet, K.; Khanam, T.; Muddashetty, R.S.; Brosius, J. Inhibitory effect of naked neural BC1 RNA or BC200 RNA on eukaryotic *in vitro* translation systems is reversed by poly(A)-binding protein (PABP). *J. Mol. Biol.* **2005**, *353*, 88–103.
68. Lin, D.; Pestova, T.V.; Hellen, C.U.; Tiedge, H. Translational control by a small RNA: Dendritic BC1 RNA targets the eukaryotic initiation factor 4A helicase mechanism. *Mol. Cell. Biol.* **2008**, *28*, 3008–3019.
69. Martignetti, J.A.; Brosius, J. BC200 RNA: A neural RNA polymerase III product encoded by a monomeric Alu element. *Proc. Natl. Acad. Sci. USA* **1993**, *90*, 11563–11567.
70. French, P.J.; Bliss, T.V.; O'Connor, V. Ntab, a novel non-coding RNA abundantly expressed in rat brain. *Neuroscience* **2001**, *108*, 207–215.
71. Bernard, D.; Prasanth, K.V.; Tripathi, V.; Colasse, S.; Nakamura, T.; Xuan, Z.; Zhang, M.Q.; Sedel, F.; Jourdren, L.; Coulpier. F.; *et al.* A long nuclear-retained non-coding RNA regulates synaptogenesis by modulating gene expression. *EMBO J.* **2010**, *29*, 3082–3093.
72. Williams, C.J. Angelman syndrome scientific symposium on the structure and function of UBE3A/E6AP. *Child Neurol.* **2009**, *24*, 904–908.
73. Chamberlain, S.J.; Brannan, C.I. The Prader–Willi syndrome imprinting center activates the paternally expressed murine Ube3a antisense transcript but represses paternal Ube3a. *Genomics* **2001**, *73*, 316–322.
74. Johnstone, K.A.; DuBose, A.J.; Futtner, C.R.; Elmore, M.D.; Brannan, C.I.; Resnick, J.L. A human imprinting centre demonstrates conserved acquisition but diverged maintenance of imprinting in a mouse model for Angelman syndrome imprinting defects. *Hum. Mol. Genet.* **2006**, *15*, 393–404.
75. Yamasaki, K.; Joh, K.; Ohta, T.; Masuzaki, H.; Ishimaru, T.; Mukai, T.; Niikawa, N.; Ogawa, M.; Wagstaff, J.; Kishino, T. Neurons but not glial cells show reciprocal imprinting of sense and antisense transcripts of Ube3a. *Hum. Mol. Genet.* **2003**, *15*, 837–847.

76. Khalil, A.M.; Faghihi, M.A.; Modarresi, F.; Brothers, S.P.; Wahlestedt, C. A novel RNA transcript with antiapoptotic function is silenced in fragile X syndrome. *PLoS One* **2008**, *3*, e1486.
77. Ladd, P.D.; Smith, L.E.; Rabaia, N.A.; Moore, J.M.; Georges, S.A.; Hansen, R.S.; Hagerman, R.J.; Tassone, F.; Tapscott, S.J.; Filippova, G.N. An antisense transcript spanning the CGG repeat region of FMR1 is upregulated in premutation carriers but silenced in full mutation individuals. *Hum. Mol. Genet.* **2007**, *16*, 3174–3187.
78. Moseley, M.L.; Zu, T.; Ikeda, Y.; Gao, W.; Mosemiller, A.K.; Daughters, R.S.; Chen, G.; Weatherspoon, M.R.; Clark, H.B.; *et al.* Ranum Bidirectional expression of CUG and CAG expansion transcripts and intranuclear polyglutamine inclusions in spinocerebellar ataxia type 8. *Nat. Genet.* **2006**, *38*, 758–769.
79. Daughters, R.S.; Tuttle, D.L.; Gao, W.; Ikeda, Y.; Moseley, M.L.; Ebner, T.J.; Swanson, M.S.; Ranum, L.P. RNA gain-of-function in spinocerebellar ataxia type 8. *PLoS Genet.* **2009**, *5*, e1000600.
80. Scahill, R.I.; Wild, E.J.; Tabrizi, S.J. Biomarkers for Huntington's disease: An update. *Expert Opin. Med. Diagn.* **2012**, *6*, 371–375.
81. Johnson, D.S.; Mortazavi, A.; Myers, R.M.; Wold, B. Genome-Wide mapping of *in vivo* protein-DNA interactions. *Science* **2007**, *316*, 1497–1502.
82. Pollard, K.S.; Salama, S.R.; Lambert, N.; Lambot, M.A.; Coppens, S.; Pedersen, J.S.; Katzman, S.; King, B.; Onodera, C.; *et al.* An RNA gene expressed during cortical development evolved rapidly in humans. *Nature* **2006**, *443*, 167–172.
83. Johnson, R.; Richter, N.; Jauch, R.; Gaughwin, P.M.; Zuccato, C.; Cattaneo, E.; Stanton, L.W. The human accelerated region 1 noncoding RNA is repressed by REST in Huntington's disease. *Physiol. Genomics* **2010**, *41*, 269.
84. Faghihi, M.A.; Modarresi, F.; Khalil, A.M.; Wood, D.E.; Sahagan, B.G.; Morgan, T.E.; Finch, C.E.; St Laurent, G.; Kenny, P.J.; Wahlestedt, C. Expression of a noncoding RNA is elevated in Alzheimer's disease and drives rapid feed-forward regulation of beta-secretase. *Nat. Med.* **2008**, *14*, 723–730.
85. Mus, E.; Hof, P.R.; Tiedge, H. Dendritic BC200 RNA in aging and in Alzheimer's disease. *Proc. Natl. Acad. Sci. USA* **2007**, *104*, 10679–10684.
86. Chubb, J.E.; Bradshaw, N.J.; Soares, D.C.; Porteous, D.J.; Millar, J.K. The DISC locus in psychiatric illness. *Mol. Psychiatr.* **2008**, *13*, 36–64.
87. Millar, J.K.; Wilson-Annan, J.C.; Anderson, S.; Christie, S.; Taylor, M.S.; Semple, C.A.; Devon, R.S.; St Clair, D.M.; Muir, W.J.; Blackwood, D.H.; *et al.* Disruption of two novel genes by a translocation co-segregating with schizophrenia. *Hum. Mol. Genet.* **2000**, *9*, 1415–1423.
88. Williams, J.M.; Beck, T.F.; Pearson, D.M.; Proud, M.B.; Cheung, S.W.; Scott, D.A. A 1q42 deletion involving DISC1, DISC2, and TSNAX in an autism spectrum disorder. *Am. J. Med. Genet.* **2009**, *149A*, 1758–1762.
89. Brandon, N.J.; Millar, J.K.; Korth, C.; Sive, H.; Singh, K.K.; Sawa, A. Understanding the role of DISC1 in psychiatric disease and during normal development. *J. Neurosci.* **2009**, *29*, 12768–12775.
90. Rapicavoli, N.A.; Blackshaw, S. New meaning in the message: Noncoding RNAs and their role in retinal development. *Dev. Dyn.* **2009**, *238*, 2103–2114.

91. Barry, G.; Briggs, J.A.; Vanichkina, D.P.; Poth, E.M.; Beveridge, N.J.; Ratnu, V.S.; Nayler, S.P.; Nones, K.; Hu, J.; Bredy, T.W.; *et al.* The long non-coding RNA Gomafu is acutely regulated in response to neuronal activation and involved in schizophrenia-associated alternative splicing. *Mol. Psychiatr.* **2013**, doi:10.1038/mp.2013.45.
92. Wood, M.; Yin, H.; McClorey, G. Modulating the expression of disease genes with RNA-based therapy. *PLoS Genet.* **2007**, *3*, e109.

PERMISSIONS

All chapters in this book were first published by MDPI; hereby published with permission under the Creative Commons Attribution License or equivalent. Every chapter published in this book has been scrutinized by our experts. Their significance has been extensively debated. The topics covered herein carry significant findings which will fuel the growth of the discipline. They may even be implemented as practical applications or may be referred to as a beginning point for another development.

The contributors of this book come from diverse backgrounds, making this book a truly international effort. This book will bring forth new frontiers with its revolutionizing research information and detailed analysis of the nascent developments around the world.

We would like to thank all the contributing authors for lending their expertise to make the book truly unique. They have played a crucial role in the development of this book. Without their invaluable contributions this book wouldn't have been possible. They have made vital efforts to compile up to date information on the varied aspects of this subject to make this book a valuable addition to the collection of many professionals and students.

This book was conceptualized with the vision of imparting up-to-date information and advanced data in this field. To ensure the same, a matchless editorial board was set up. Every individual on the board went through rigorous rounds of assessment to prove their worth. After which they invested a large part of their time researching and compiling the most relevant data for our readers.

The editorial board has been involved in producing this book since its inception. They have spent rigorous hours researching and exploring the diverse topics which have resulted in the successful publishing of this book. They have passed on their knowledge of decades through this book. To expedite this challenging task, the publisher supported the team at every step. A small team of assistant editors was also appointed to further simplify the editing procedure and attain best results for the readers.

Apart from the editorial board, the designing team has also invested a significant amount of their time in understanding the subject and creating the most relevant covers. They scrutinized every image to scout for the most suitable representation of the subject and create an appropriate cover for the book.

The publishing team has been an ardent support to the editorial, designing and production team. Their endless efforts to recruit the best for this project, has resulted in the accomplishment of this book. They are a veteran in the field of academics and their pool of knowledge is as vast as their experience in printing. Their expertise and guidance has proved useful at every step. Their uncompromising quality standards have made this book an exceptional effort. Their encouragement from time to time has been an inspiration for everyone.

The publisher and the editorial board hope that this book will prove to be a valuable piece of knowledge for researchers, students, practitioners and scholars across the globe.

LIST OF CONTRIBUTORS

Alessandro Rosa
Department of Biology and Biotechnology "Charles Darwin", Sapienza University of Rome, Rome 00185, Italy
Laboratory of Molecular Vertebrate Embryology, The Rockefeller University, New York, NY 10065, USA

Ali H. Brivanlou
Laboratory of Molecular Vertebrate Embryology, The Rockefeller University, New York, NY 10065, USA

Virginie Gébelin, Julie Leclercq and Pascal Montoro
Centre de Coopération Internationale en Recherche Agronomique Pour le Développement, Unité Mixte de Recherche Amélioration Génétique et Adaptation des Plantes Méditerranéennes et Tropicales, Avenue Agropolis, Montpellier F-34398, France

Songnian Hu
Beijing Institute of Genomics, Chinese Academy of Sciences, Beijing 100029, China

Chaorong Tang
Rubber Research Institute, Chinese Academy of Tropical Agricultural Sciences, Danzhou 531737, China

Yoshiro Nagata and Naoki Hibio
Department of Computational Biology, Graduate School of Frontier Sciences, University of Tokyo, 5-1-5 Kashiwanoha, Kashiwa-shi, Chiba-ken 277-8561, Japan

Eigo Shimizu
Department of Biophysics and Biochemistry, Graduate School of Science, University of Tokyo, 7-3-1 Hongo, Bunkyo-ku, Tokyo 113-0033, Japan

Kumiko Ui-Tei
Department of Computational Biology, Graduate School of Frontier Sciences, University of Tokyo, 5-1-5 Kashiwanoha, Kashiwa-shi, Chiba-ken 277-8561, Japan
Department of Biophysics and Biochemistry, Graduate School of Science, University of Tokyo, 7-3-1 Hongo, Bunkyo-ku, Tokyo 113-0033, Japan

Jianghui Hou
Renal Division and Center for Investigation of Membrane Excitability Diseases, Washington University in St. Louis, 660 South Euclid Avenue, St. Louis, MO 63110, USA

Dan Zhao
Division of Pharmacology, PLA 85th Hospital, 1328 Hua Shan Road, Shanghai 20052, China

Giovanni Bussotti and Anton J. Enright
European Bioinformatics Institute, Wellcome Trust Genome Campus, Hinxton, Cambridge CB10 1SD, UK

Cedric Notredame
Bioinformatics and Genomics Program, Centre for Genomic Regulation (CRG), Aiguader, 88, 08003 Barcelona, Spain

Federica Calore, Francesca Lovat and Michela Garofalo
Department of Molecular Virology, Immunology and Medical Genetics, Comprehensive Cancer Center, Ohio State University, Columbus, OH 43210, USA

Ting Shuang, Shuang Chang, Min Wang and Cui Hong Bai
Department of Obstetrics and Gynecology, Shengjing Hospital of China Medical University, Shenyang 110004, China

Chunxue Shi
Department of Obstetrics and Gynecology, Shengjing Hospital of China Medical University, Shenyang 110004, China
Shenyang Maternity and infant Hospital, 90 Qingnian Road, Shenhe District, Shenyang 110014, China

Florian Mayr and Udo Heinemann
Crystallography, Max-Delbrück Center for Molecular Medicine, Robert-Rössle Straße 10, Berlin 13125, Germany
Institute for Chemistry and Biochemistry, Freie Universität Berlin, Takustraße 6, Berlin 14195, Germany

Andoni Garitano-Trojaola and Xabier Agirre
Laboratory of Myeloproliferative Syndromes, Oncology Area, Foundation for Applied Medical Research, University of Navarra, Pamplona 31008, Spain

Felipe Prósper
Laboratory of Myeloproliferative Syndromes, Oncology Area, Foundation for Applied Medical Research, University of Navarra, Pamplona 31008, Spain
Hematology Service and Area of Cell Therapy, University of Navarra Clinic, University of Navarra, Pamplona 31008, Spain

Puri Fortes
Department of Hepatology and Gene Therapy, Foundation for Applied Medical Research, University of Navarra, Pamplona 31008, Spain

Ruxin Zhang, Hong Wang and Jing Wu
Department of Otolaryngology, Huadong Hospital, Fudan University, Shanghai 200040, China

Shaoqing Yu
Department of Otolaryngology, Huadong Hospital, Fudan University, Shanghai 200040, China
Department of Otolaryngology, Tongji Hospital, Tongji University, Shanghai 200065, China

Chunshen Zhu and Jianqiu Cheng
Department of Otolaryngology, Jinan General Hospital of PLA, Jinan 250031, Shandong, China

Chiara Fenoglio, Elisa Ridolfi, Daniela Galimberti and Elio Scarpini
Department of Pathophysiology and Transplantation, University of Milan, "Dino Ferrari" Center, IRCCS Cà Granda Foundation Ospedale Maggiore Policlinico, Via F.Sforza 35, Milan 20122, Italy

Index

www.ingramcontent.com/pod-product-compliance
Lightning Source LLC
LaVergne TN
LVHW080951210926
848485LV00004B/817
* 9 7 8 1 6 4 7 4 0 3 6 2 1 *